P9-CKX-463

ULTRA-WIDEBAND WIRELESS COMMUNICATIONS AND NETWORKS

ULTRA-WIDEBAND WIRELESS COMMUNICATIONS AND NETWORKS

Edited by

Xuemin (Sherman) Shen
University of Waterloo, Canada

Mohsen Guizani
Western Michigan University, USA

Robert Caiming Qiu
Tennessee Technological University, USA

Tho Le-Ngoc
McGill University, Canada

John Wiley & Sons, Ltd

Telephone (+44) 1243 779777

Email (for orders and customer service enquiries): cs-books@wiley.co.uk
Visit our Home Page on www.wiley.com

Other Wiley Editorial Offices

John Wiley & Sons Inc., 111 River Street, Hoboken, NJ 07030, USA

Jossey-Bass, 989 Market Street, San Francisco, CA 94103-1741, USA

Wiley-VCH Verlag GmbH, Boschstr. 12, D-69469 Weinheim, Germany

John Wiley & Sons Australia Ltd, 42 McDougall Street, Milton, Queensland 4064, Australia

John Wiley & Sons (Asia) Pte Ltd, 2 Clementi Loop #02-01, Jin Xing Distripark, Singapore 129809

John Wiley & Sons Canada Ltd, 22 Worcester Road, Etobicoke, Ontario, Canada M9W 1L1

Wiley also publishes its books in a variety of electronic formats. Some content that appears in print may not be available in electronic books.

Library of Congress Cataloging-in-Publication Data

Ultra-wideband wireless communications and networks / edited by Xuemin Shen ... [et al.].
p. cm.
ISBN-13: 978-0-470-01144-7
ISBN-10: 0-470-01144-0
1. Broadband communication systems. 2. Wireless communication systems. 3. Ultra-wideband devices.
TK5103.4.U48 2006
621.384–dc22

2005029359

British Library Cataloguing in Publication Data

A catalogue record for this book is available from the British Library

ISBN-13 978-0-470-01144-7 (HB)
ISBN-10 0-470-01144-0 (HB)

Typeset in 10/12 Times by Laserwords Private Limited, Chennai, India.
Printed and bound in Great Britain by Antony Rowe Ltd, Chippenham, Wiltshire.
This book is printed on acid-free paper responsibly manufactured from sustainable forestry in which at least two trees are planted for each one used for paper production.

Contents

List of Contributors

Jun Cai, PhD
Postdoctoral Fellow
Department of Electrical and Computer Engineering
University of Waterloo
Waterloo, Ontario, Canada

Giuseppe Thadeu Freitas de Abreu, PhD
Research Scientist
Centre for Wireless Communications
University of Oulu
Oulu, Finland

Georgios B. Giannakis, PhD
Professor
Department of Electrical and Computer Engineering
University of Minnesota
Minneapolis, Minnesota, USA

Mohsen Guizani, PhD
Professor
Department of Computer Science
Western Michigan University
Parkview Campus
Kalamazoo, Michigan, USA

Zihua Guo, PhD
Director
Broadband Wireless Technology Lab
Lenovo Corporate R&D
Beijing, P. R. China

Sebastian Hoyos, PhD
Postdoctoral Researcher
Berkeley Wireless Research Center
Department of Electrical Engineering and Computer Sciences
University of California,
Berkeley, California, USA

Hai Jiang
PhD Candidate
Department of Electrical and Computer Engineering
University of Waterloo
Waterloo, Ontario, Canada

Tho Le-Ngoc, PhD
Professor
Department of Electrical and Computer Engineering
McGill University
Montreal, Quebec, Canada

Ye (Geoffrey) Li, PhD
Associate Professor
School of Electrical and Computer Engineering
Georgia Institute of Technology
Atlanta, Georgia, USA

Huaping Liu, PhD
Assistant Professor
School of Electrical Engineering and Computer Science
Oregon State University
Corvallis, Oregon, USA

Jean-Philippe Montillet
Centre for Wireless Communications
University of Oulu
Oulu, Finland

Uzoma A. Onunkwo
PhD Candidate
School of Electrical and Computer Engineering
Georgia Institute of Technology
Atlanta, Georgia, USA

Ian Oppermann, PhD
Director
Product Business
Operations Solutions – Performance
Nokia Networks
Espoo, Finland

Jay E. Padgett, PhD
Senior Research Scientist
Advanced Wireless Signal Processing
Telcordia Technologies Applied Research
Red Bank, New Jersey, USA

Robert C. Qiu, PhD
Associate Professor
Wireless Networking Systems Laboratory
Center for Manufacturing Research/Electrical and Computer Engineering Department
Tennessee Technological University
Cookeville, Tennessee, USA

Alberto Rabbachin
PhD Candidate
Centre for Wireless Communications
University of Oulu
Oulu, Finland

Harri Saarnisaari, PhD
Senior Research Scientist
Centre for Wireless Communications
University of Oulu
Oulu, Finland

Brian M. Sadler, PhD
Army Research Laboratory
AMSRD-ARL-CI-CN
Adelphi, Maryland, USA

Xuemin (Sherman) Shen, PhD, PEng
Professor and Associate Chair for Graduate Study
Department of Electrical and Computer Engineering
University of Waterloo
Waterloo, Ontario, Canada

Zhi Tian, PhD
Associate Professor
Department of Electrical and Computer Engineering
Michigan Technological University
Houghton, Michigan, USA

Zhengyuan Xu, PhD
Associate Professor
Department of Electrical Engineering
University of California
Riverside, California, USA

Liuqing Yang, PhD
Assistant Professor
Department of Electrical and Computer Engineering
University of Florida
Gainesville, Florida, USA

Richard Yao, PhD
Microsoft Research Asia
Beijing, China

Kegen Yu, PhD
Research Scientist
CSIRO ICT Centre
Cnr Vimiera and Pembroke Roads
Marsfield NSW 2122, Australia

Weihua Zhuang, PhD, PEng
Professor
Department of Electrical and Computer Engineering
University of Waterloo
Waterloo, Ontario, Canada

Preface

Ultra-wideband (UWB) transmission has recently received great attention in both academia and industry for applications in wireless communications. It was among the CNN's top 10 technologies to watch in 2004. A UWB system is defined as any radio system that has a 10-dB bandwidth larger than 20 % of its center frequency, or has a 10-dB bandwidth equal to or larger than 500 MHz. The recent approval of UWB technology by Federal Communications Commission (FCC) of the United States reserves the unlicensed frequency band between 3.1 and 10.6 GHz (7.5 GHz) for indoor UWB wireless communication systems. It is expected that many conventional principles and approaches used for short-range wireless communications will be reevaluated and a new industrial sector in short-range (e.g., 10 m) wireless communications with high data rate (e.g., 400 Mbps) will be formed. Further, industrial standards IEEE 802.15.3a (high data rate) and IEEE 802.15.4a (very low data rate) based on UWB technology have been introduced.

UWB technology has many benefits owing to its ultra-wideband nature, which include high data rate, less path loss and better immunity to multipath propagation, availability of low-cost transceivers, low transmit power and low interference, and so on. On the other hand, there exist many technical challenges in UWB deployment, such as the received-waveform distortion from each distinct delayed propagation path, antenna design for and synchronization of very short pulses, performance degradation due to multiple access interference and narrowband jamming, employment of higher order modulation schemes to improve capacity or throughput, and development of link and network layers to take advantage of the UWB transmission benefits at the physical layer. Even though R&D results so far have demonstrated that UWB radio is a promising solution for high-rate short-range wireless communications, further extensive investigation, experiments, and development are necessary for developing effective and efficient UWB communication systems and UWB technology. This book is timely in reporting the results from cutting edge research and state-of-the-art technology in UWB wireless communications.

The first chapter by Qiu, Shen, Guizani, and Le-Ngoc gives an introduction to UWB technology. First, the fundamentals of UWB are overviewed. Then the issues unique to UWB are summarized. The emerging technologies are also identified.

The next four chapters emphasize UWB modulation and signal detection. The chapter by Onunkwo and Li presents single-carrier and orthogonal frequency division multiplexing (OFDM) based modulation and detection for UWB. When a short UWB pulse propagates through a wireless channel, pulse distortion can be caused by frequency dependency of the propagation channel and antennas. The chapter by Qiu addresses the issues related to pulse signal detection of distorted pulses. Accurate timing offset estimation (TOE) also poses major challenges to pulsed UWB systems in realizing their potential bit error rate (BER) performance, capacity, throughput, and network flexibility. The chapter by Tian and Giannakis presents accurate and low-complexity TOE algorithms for UWB impulse radio (IR) receivers, with focus on timing acquisition in dense multipath channels. The chapter by Liu analyzes the error performance of pulsed UWB systems with commonly used data-modulation schemes such as pulse-amplitude modulation (PAM) or pulse-position modulation (PPM).

The succeeding two chapters focus on UWB transceivers. The chapter by Hoyos and Sadler presents the design of mixed-signal communications receivers based on the analog-to-digital converter (ADC) framework obtained via signal expansion. A generalization of the mixed-signal receiver problem is discussed and two frequency-domain receiver design examples based on multicarrier and transmitted-reference signaling are illustrated. An effective UWB transceiver design should consider unique UWB channel characteristics and design constraints, namely severe multipath distortion, low power operation, and low-complexity implementation. The chapter by Xu presents several transceiver design methods, based on two primary categories: digital solutions and mixed analog/digital solutions.

In UWB wireless networks, medium access control (MAC) is essential to coordinate the channel access among competing devices. The MAC has significant effect on the UWB system performance. The chapter by Guo and Yao investigates the performance of the IEEE 802.15.3 MAC. The chapter by Shen, Zhuang, Jiang, and Cai presents a comprehensive overview of UWB radio resource management mechanisms on three important aspects: multiple access, overhead reduction, and power/rate allocation, and identifies some future research issues.

Generally, UWB networks need to coexist with other existing and future narrowband networks. The chapter by Padgett develops models for calculating and simulating the interference at the output of the final IF of a narrowband receiver in response to a UWB input signal. It provides an understanding and the tools necessary to analyze the effect of pulsed UWB interference on any particular type of receiver. The chapter by Yang and Giannakis introduces two UWB multi-access systems that utilize digital single-carrier (SC) or multicarrier (MC) spreading codes. These SC/MC codes lead to baseband operation, and offer flexibility in narrowband interference mitigation by simply avoiding the corresponding digital carriers.

One advantage of UWB technology is its potential in localization. The chapter by Yu, Saarnisaari, Montillet, Rabbachin, Oppermann, and de Abreu provides comprehensive views over UWB localization techniques, including time of arrival (TOA) estimation, positioning approaches, and positioning accuracy analysis.

Finally, as the guest editors, we would like to express our sincere thanks to Mark Hammond and Sarah Hinton from John Wiley & Sons, Ltd., for their support and help in bringing out this special book.

Xuemin (Sherman) Shen, University of Waterloo, Canada
Mohsen Guizani, Western Michigan University, USA
Robert Caiming Qiu, Tennessee Technological University, USA
Tho Le-Ngoc, McGill University, Canada

1

Introduction

Robert Caiming Qiu, Xuemin (Sherman) Shen, Mohsen Guizani
and Tho Le-Ngoc

1.1 Fundamentals

1.1.1 Overview of UWB

Ultra-wideband (UWB) transmission has recently received significant attention in both academia and industry for applications in wireless communications [1, 2]. UWB has many benefits, including high data rate, availability of low-cost transceivers, low transmit power, and low interference. It operates with emission levels that are commensurate with common digital devices such as laptops, palm pilots, and pocket calculators. The approval of UWB technology [3] made by the Federal Communications Commission (FCC) of the United States in 2002 reserves the unlicensed frequency band between 3.1 and 10.6 GHz (7.5 GHz) for indoor UWB wireless communication systems. Industrial standards such as IEEE 802.15.3a (high data rate) and IEEE 802.15.4a (very low data rate with ranging) based on UWB technology have been introduced. On the other hand, the Department of Defense (DoD) UWB systems are different from commercial systems in that jamming is a significant concern. Although R&D efforts in recent years have demonstrated that UWB radio is a promising solution for high-rate short-range and moderate-range wireless communications and ranging, further extensive investigation, experimentation, and development are necessary to produce effective and efficient UWB communication systems. In particular, UWB has found a new application for lower-data-rate moderate-range wireless communications, illustrated by IEEE 802.15.4a and DoD systems with joint communication and ranging capabilities unique to UWB. Unlike the indoor environment in 802.15.3a (WPAN), the new environments for sensors, IEEE 802.15.4a, and DoD systems will be very different, ranging from dense foliage to dense urban obstructions. The application of UWB to low-cost, low-power sensors has a promise. The centimeter accuracy in ranging and communications provides unique solutions to applications, including logistics, security applications, medical applications, control of home

Ultra-wideband Wireless Communications and Networks
Edited by Xuemin (Sherman) Shen, Mohsen Guizani, Robert Caiming Qiu and Tho Le-Ngoc

appliances, search-and-rescue, family communications and supervision of children, and military applications.

1.1.2 History

Although, often considered as a recent breakthrough in wireless communications, UWB has actually experienced well over 40 years of technological developments. The physical cornerstone for understanding UWB pulse propagation was established by Sommerfeld a century ago (1901) when he attacked the diffraction of a time-domain pulse by a perfectly conducting wedge [2]. In fact, one may reasonably argue that UWB actually had its origins in the spark-gap transmission design of Marconi and Hertz in the late 1890s [4–6]. In other words, the first wireless communication system was based on UWB. Owing to the technical limitations, narrowband communication was preferred to UWB. Much like the spread spectrum or the code division multiple access (CDMA), UWB followed a similar path with early systems designed for military covert radar and communication. After the accelerating development since 1994 when some of the research activities were declassified [7, 8], UWB picked up momentum after the FCC notice of inquiry in 1998. The interest in UWB was 'sparked' since the FCC issued a Report and Order allowing its commercial deployment with a given spectral-mask requirement for both indoor and outdoor applications [3].

1.1.3 Regulatory

UWB technology is defined by the FCC as any wireless scheme that occupies a fractional bandwidth $W/f_c \geq 20\,\%$, where W is the transmission bandwidth and f_c is the band center, or more than 500 MHz of absolute bandwidth. The FCC approved [3] the deployment of UWB on an unlicensed basis in the 3.1–10.6 GHz band subject to a modified version of Part 15.209 rules. The essence of the rulings is that power spectral density (PSD) of the modulated UWB signal must satisfy the spectral masks specified by spectrum-regulating agencies. The spectral mask for indoor applications specified by the FCC in the United States is shown in Figure 1.1.

1.1.4 Applications

High data rate (IEEE 802.15.3a): One typical scenario [9] is promising wireless data connectivity between a host (e.g., a desk PC) and associated peripherals such as keyboards, mouse, printer, and so on. A UWB link functions as a 'cable replacement' with transfer data rate requirements that range from 100 Kbps for a wireless mouse to 100 Mbps for rapid file sharing or download of images/graphic files. Additional driver applications relate to streaming of digital media content between consumer electronics appliances (digital TV sets, VCRs, audio CD/DVD, and MP3 players, and so on). In summary, UWB is seen as having potential for applications that to date have not been fulfilled by other wireless short-range technologies currently available, for example, 802.11 LANs and Bluetooth PANs.

Low data rate (IEEE 802.15.4a): Emerging applications of UWB are foreseen for sensor networks that are critical to mobile computing. Such networks combine low- to

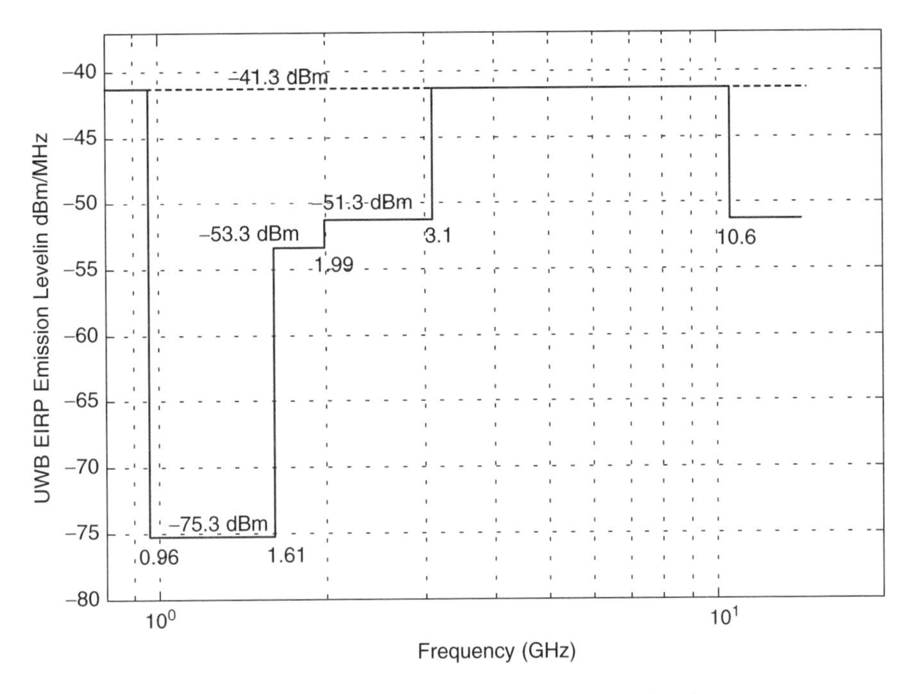

Figure 1.1 FCC spectral mask for indoor applications

medium-rate communications (50 kbps to 1 Mbps) with ranges of 100 m with positioning capabilities. UWB allows centimeter accuracy in ranging as well as in low-power and low-cost implementation of communications systems. In fact, the IEEE 802.15.4a, a standard for low-power, low-date-rate wireless communications, is primarily focused on position location applications. The price point will be in the sub-$1 range for asset tracking and tagging, up to $3 to $4 per node for industrial-control applications. These features allow a new range of applications, including military applications, medical applications (monitoring of patients), family communications/supervision of children, search-and-rescue (communications with fire fighters, or avalanche/earthquake victims), control of home applications, logistics (package tracking), and security applications (localizing authorized persons in high-security areas).

1.1.5 Pulse- or Multicarrier-Based UWB

The main stream papers in the literature deal with pulse-based UWB systems. One reason may be due to the fact that the pulse-based UWB was not sufficiently understood compared to orthogonal frequency-division multiplexing (OFDM) based UWB.

The major reasons for the standard body (IEEE 802.15.3a) to adopt multiband scheme are: (1) it has spectrum flexibility/agility. Regulatory regimes may lack large contiguous spectrum allocation. Spectrum agility may ease coexistence with existing services. (2) Energy collected per RAKE finger scales with longer pulse widths used, which prefers fewer RAKE fingers. (3) Reduced bandwidth after down-conversion mixer reduces power consumption and linearity requirements of receiver. (4) A fully digital solution for signal processing is more feasible than single band solution for the same occupied bandwidth. (5) Transmitter pulse shaping is made easier; longer pulses are easier to synthesize and less distorted by integrated circuits (IC) package and antenna systems. (6) It is capable of utilizing a frequency-division multiple-access (FDMA) mode for severe near–far scenarios.

Multiband pulsed scheme: The main disadvantage of narrow time-domain pulses is that building RF and analog circuits as well as high-speed analog-to-digital converters (ADCs) to process the signal of extremely wide bandwidth is challenging, and usually results in high power consumption [2]. Collection of sufficient energy in dense multipath environments requires a large number of RAKE fingers. The pulsed multiband approach can eliminate the disadvantages associated with a large front-end processing bandwidth, by dividing the spectrum into several subbands. The advantage of this approach is that the information can be processed over a much smaller bandwidth, thereby reducing the complexity of the design and the power consumption, lowering the cost, and improving spectrum flexibility and worldwide compliance. However, it is difficult to collect significant multipath energy using a single RF chain. In addition, very stringent frequency-switching time requirements (<100 ps) are placed at both the transmitter and the receiver.

Multiband OFDM scheme: Multipath energy collection is an important issue as it is a major factor that determines the range of a communication system. The multiband OFDM system transmits information on each of the subbands. This technique has nice properties including the ability to efficiently capture multipath energy with a single RF chain. The drawback is that the transmitter is slightly more complex because it requires an inverse fast Fourier transform (IFFT) and the peak-to-average ratio may be higher than that of the pulse-based multiband approaches. It seems to be very challenging for both the pulse-based and the OFDM-based multiband solutions to meet the target costs imposed by the market.

1.2 Issues Unique to UWB

1.2.1 Antennas

UWB antennas can be modeled as the front-end pulse shaping filters that affect the baseband detection. A narrowband system does not have this problem. In narrowband systems front-end filters are typically designed as matched filters; and since the fractional bandwidth is so small, the antenna frequency response can be designed to lie in the frequency domain. That is not true for a UWB system in which the pulse-waveform distortion is often present [2]. The impulse response of the antenna can change with angle in both elevation and azimuth. Specifically, the transmitted pulse at different elevation angles will be distorted as compared with the pulse observed at bore site.

1.2.2 Propagation and Channel Model

UWB itself represents an unprecedented conceptual revolution. The huge chunk of spectrum (as high as 7.5 GHz) can be legally used for commercial wireless communications to carry information bits up to 480 Mbps or higher that will be propagated via a medium. This medium will attenuate and distort the incident pulse-based signals. The major paradigm shift required for this new concept is to 'think' in the time domain and envision UWB time-domain pulses as basic UWB signals since it is prohibitively inconvenient to envision UWB signals in the frequency domain. The basic building blocks for UWB signals are short 'pulses' rather than the harmonic sine waves. These short-pulse UWB signals of huge transmission bandwidth provide the superresolution of the received signals at the receiver. This fundamental property changes the structure of many basic problems: for example, fading, and time-resolvable multipaths.

Fading is due to the overlapping superposition of unresolved multipaths. Compared with time-resolvable multipaths the major narrowband nuisance of fading is no longer important for UWB: it is only 3–4 dB for UWB in contrast to 30–60 dB for narrowband [10, 11]. Basically, a UWB channel is quasi-static in that the collected total energy is almost constant at each instant [12]. Thus, time-resolvable multipaths will become the dominant mechanism for energy transmission and capture, since a huge number (hundreds) of paths are resolved in contrast to several paths (in a 5 MHz system). The transmitted energy is carried through these time-resolvable multipaths. The UWB pulse signals associated with those individual paths are the basic building blocks of a UWB system and require a careful treatment. Another conceptual difference is in the necessity to consider the per-path pulse distortion [2] that is neglected in Turin's multipath channel model (Turin 1958) widely used in wireless communications [13]. Turin's model is based on empirical experimental data obtained using a typical narrowband measurement system. The generalized multipath model [2] is based on the transient physical mechanisms obtained directly from Maxwell's equations.

1.2.3 Modulations

In choosing modulation schemes for UWB systems [2], one must consider a number of aspects such as data rate, transceiver complexity, spectral characteristics, robustness against narrowband interference, intersymbol interference, error performance, and so on. For pulsed UWB systems, the widely used forms of modulation schemes include pulse amplitude modulation (PAM), on–off keying (OOK), and pulse position modulation (PPM). To satisfy the FCC spectral mask, passband pulses are used to transmit information in pulsed UWB systems. Although these passband pulses could be obtained by modulating a baseband pulse using a sinusoidal carrier signal, the term binary phase-shift keying (BPSK) is still somewhat imprecise in the context of pulsed UWB signaling as the carrier-modulated baseband pulse is usually treated as a single entity – the UWB pulse shape.

The three modulation schemes mentioned above are illustrated in Figure 1.2. For a single-user system with binary PPM signaling, bit '1' is represented by a pulse without any delay and bit '0' is represented by a pulse with a delay τ relative to the time reference. A major factor governing the performance of this system, like any other PPM-based system, is the set of time shifts used to represent different symbols. The most commonly used PPM scheme is the orthogonal signaling scheme for which the UWB pulse shape is orthogonal to its time-shifted version. There also exists an optimal time shift for an M-ary PPM scheme. The time shifts for both the orthogonal and the optimal schemes

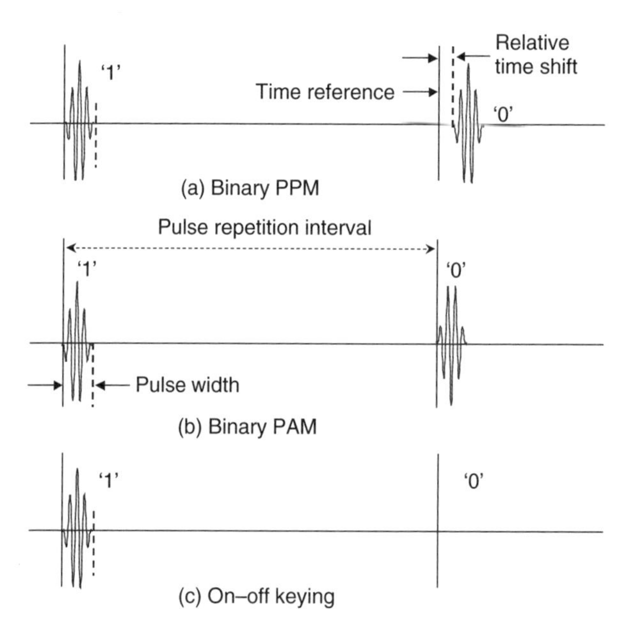

Figure 1.2 UWB modulation options

depend on the choice of the UWB pulse $p(t)$. For binary PAM signaling, information bits modulate the pulse polarity. For OOK signaling, information bit '1' is represented by the presence of a pulse and no pulse is sent for bit '0'.

Binary PAM and PPM schemes have similar performances. The OOK scheme is less attractive than the PAM or PPM scheme because of its inferior error performance in the same environment. However, if receiver complexity is the main design concern, a simple energy-detection scheme can be applied to OOK signals, resulting in a receiver of lowest achievable complexity. PSD of the modulated UWB signal must satisfy the spectral masks specified by spectrum-regulating agencies. In the United States, the spectral mask for indoor applications specified by the FCC is from 3.1 GHz to 10.6 GHz. OOK and PPM signals have discrete spectral lines. These spectral lines could cause severe interference to existing narrowband radios, and various techniques such as random dithering could be applied in PPM to lower these discrete spectral lines and smoothen the spectrum. Because of the random polarities of the information symbols, the PAM scheme inherently offers a smooth PSD when averaged over a number of symbol intervals. In this sense, PAM signaling is attractive.

1.2.4 A/D Sampling

Most of the existing detection schemes require an ADC that operates at a minimum of the Nyquist rate [2]. For high-performance design, the sampling rate is in the multi-GHz range for UWB signaling (minimum bandwidth is 500 MHz). In addition to the extremely high-sampling frequency, the ADC must support a relatively high resolution (e.g., greater than 4–6 bits) to resolve signals from the narrowband interferences. Such a speed mandates the use of an interleaved flash ADC, which tends to be power hungry with the power scaling exponentially increasing with bit precision. Although achievable with today's CMOS technology, such an ADC must be avoided in low-power operations. One technique to avoid using such a high-speed ADC is to implement the correlator in the analog domain. One technical challenge associated with the analog implementation method is the difficulty in obtaining the receiver template waveform and deriving the precise timing of symbols and received paths. In the digital implementation, a bank of correlators that are delayed relative to one another by a fraction of the pulse duration can be applied to correlate with the received digitized signal, and the local peaks corresponding to the possible received pulses are located. With analog implementation, this is difficult to achieve because some sort of analog delay units, and novel multipath tracking methods are needed. Another approach is to represent the samples using fewer bits. One bit (mono-bit) representation of samples, a traditional approach, finds a new application in UWB in reducing ADC speed.

1.2.5 Timing Acquisition

Timing acquisition is the first operating stage in any communication system. It is an especially difficult task in the UWB system mainly owing to limited transmitted power and a high-resolution multipath. Low transmitted power implies a long search time to obtain reliable timing, while a received signal with multipath components can result in a set of 'good' signal phases within the search window, making the decision much more complex. Although timing acquisition is not a new issue in general, in the UWB community it has received attention recently [1, 14, 15]. Basically, the bulk of the research has focused on finding efficient search strategies, reducing search space, and examining sophisticated estimation-based schemes.

1.2.6 Receiver Structures

One of the advantages of pulsed UWB communication is its ability to resolve individual multipath components [2, 9]. However, the large number of resolvable paths in such a system makes it unrealistic to employ the traditional RAKE receiver to capture a significant portion of the energy contained in the received multipath components, and most existing receivers must resort to a partial or a selective RAKE for multipath combining. A partial or a selective RAKE can practically combine up to a few paths, which represent only a small percentage (e.g., less than 10 %) of the total signal energy received in a non-line-of-sight (NLOS) environment. A RAKE with more than 3–5 paths will lead to an exponential increase in complexity because multipath acquisition, multipath tracking, and channel estimation consume too much processing resource. The insufficient multipath energy captured by the receiver results in a poor system range and almost no tolerance to intersymbol interference caused by multipath delay. Suboptimal schemes such as the transmit-reference (TR) scheme [16] or the differential scheme [2] can perform successful multipath energy capture and detection without requiring channel estimation. The primary drawback of these receivers is the significant performance degradation associated with employing noisy received signals as the reference signals for data detection. The net result is that such a receiver could perform much worse than a partial or selective RAKE receiver, which implements only a small number of fingers.

The decision-feedback autocorrelation receiver [17] overcomes the drawbacks of these receivers. In this scheme, the received signal waveform is delayed by using analog delay units. The respective symbol decisions are applied to data-demodulate the outputs of these delay units, which are then summed up, forming the received template waveform. The template waveform quality can be adjusted by changing the number of symbol intervals over which the received waveform is averaged. However, the analog delay units needed in deriving the receiver template could be costly in implementation.

1.2.7 Multiple Access

Multiple-access communication employing pulsed UWB technologies has drawn significant research interest. Various multiple-access schemes and their performances have been reported in the literature [1, 2]. Time hopping (TH) has been found to be a good multiple-access technique for pulsed UWB systems. Direct sequence (DS) spreading is also an attractive method for multiple access in UWB systems. Since pulsed UWB systems are inherently spread-spectrum systems, the use of spreading codes in DS-UWB systems is solely for accommodating multiple users. First, in TH pulsed UWB systems, the pulse duty cycle is very small. Thus, the transmitter is gated off for the bulk of a symbol period. Multiple access can be implemented by employing appropriately chosen hopping sequences for different users to minimize the probability of collisions due to multiple accessing. Second, DS can also be used for multiple access in a PAM-UWB or PPM-UWB system. In such a system, each symbol is represented by a series of pulses that are pulse-amplitude-modulated by a chip sequence. Input symbols are modulated onto either the amplitude or the relative positions of each sequence of pulses.

1.3 Emerging Technologies

Many emerging technologies have been published in the literature. In particular, *IEEE Transactions on Vehicular Technology* has published a special session on UWB [1]. Another special issue appears in [18].

1.3.1 Low-Complexity Noncoherent Receivers

UWB is attractive for lower-data-rate applications, which often require very low power consumption and low-cost implementation due to the densely distributed nature of a sensor network. Some wireless applications require energy scavenging. An average power consumption on the order of $10 - 100\ \mu W$ is required. Noncoherent energy detection-based receivers [19–21] can serve these needs without expensive channel estimation and RAKE filters. Moreover, the required clock timing precision and jitter error can be relaxed greatly. The drawback, however, is noise and interference enhancement. Since the modulation does not experience a correlation in the receiver, signals need to be orthogonal in the time domain. PPM and OOK are popular modulation schemes. A typical receiver includes a bandpass filter, a square-law device, an integrator, and a decision device.

The square-law detector is a diode that outputs DC voltages from very low power (below −20 dBm) RF input signals. All diodes have a square-law region of output but some types will output larger DC voltages than others from a given input. Diodes that are usually considered to be excellent detectors are of the tunnel, Schottky, and germanium variety. These diodes have a sharper I–V slope at zero bias than a silicon diode and can therefore output larger voltages with the same input. When choosing a diode for detection purposes, three things must be taken into consideration: first, the diode must have a sharp I–V curve, second it must be fast, and third it must have small internal capacitance.

For any UWB system with a noncoherent receiver [40], offering antijamming/interference capability is very challenging owing to the large bandwidth.

1.3.2 Location-Based Sensor Networks

Ranging via UWB radios [22] is envisioned to be a major breakthrough in sensor networks owing to its centimeter accuracy and low cost with low power. The ranging network can be viewed as a sensor network [23] where the physical parameter to be sensed is location. In addition, UWB is of strategic value for providing stealth for covert operation by hiding within the noise floor to prevent detection and in situations in which other forms of RF communication find it virtually impossible to operate. UWB's probability of survival increases in a toxic RF battlefield when compared to many other forms of RF. A general purpose UWB test bed for communications and ranging is being built at Tennessee Tech University, in part funded by Army Research Office through a DURIP grant.

One of challenges is to use low-complexity time of arrival (TOA) ranging based on energy detection. This scheme is being standardized within IEEE 802.15.4a. TOA estimation is made on the basis of symbol rate samples that are obtained after a square-law device. Energy detection-based ranging becomes feasible since a simple low-cost diode can be used for a square-law device. Even if it suffers more from noise due to a square-law device, energy detection does not require accurate timing or pulse shapes. Once the energy samples at the output of a square-law device are collected, the TOA estimation can be considered as a problem of leading edge detection in noise.

The merging field of using UWB for sensor networks is a multidisciplinary area that brings together concepts of wireless communications and ad hoc networking, low-power hardware design, signal processing, and embedded software design. The carrier sense multiple-access protocol with collision avoidance (CSMA/CA), a distributed protocol, is the most commonly investigated medium access control (MAC) protocol for mobile ad hoc and sensor networks and is also being used in IEEE 802.15.4. A test bed with CSMA with 12 peer-to-peer wireless sensor devices is reported [23].

1.3.3 Time Reversal

The irreversibility of time is a topic generally associated with fundamental physics [24]. The objective is to exploit time-reversal invariance, a fundamental symmetry that holds everywhere in fundamental particle physics to create useful systems. What one wants is macroscopic time-reversal invariance. Fortunately, this symmetry does hold in both acoustic-wave and electromagnetic-wave phenomena. The essence of time reversal is to exploit multiple scattering between the scatters [25, 26]. After the time-reversal operation, the whole multiple-scattering medium behaves, in effect, as a coherent focusing source whose large angular aperture enhances the resolution at the final focus. Owing to the high resolution of a short UWB pulse, rich resolvable multipath makes the UWB channel act like an underwater acoustical channel [27, 28]. The success of using time reversal for underwater acoustical communications motivates the research UWB radio.

In a time-reversal scheme [29], the channel impulse response (CIR) $h(t)$ is used as a prefilter at the transmitter. After the encoded information bits pass through the channel, the response of the channel is the autocorrelation of the CIR as shown in Figure 1.3. A pilot signal can be used to sound the channel such that the transmitter has the CIR available for precoding. This scheme is used to shift the design complexity from the receiver to the transmitter. In the time-reversal scheme, a signal is precoded such that it focuses both in time and in space at a particular receiver. Owing to temporal focusing, the received power is concentrated within a few taps and the task of equalizer design becomes much simpler than without focusing.

Time reversal can be justified in many ways. Let us use our version of justification. The autocorrelation operation of the matched filter is a process of pulse compression. A UWB channel can be treated as a quasi-static, reciprocal channel [30, 41]. Each multipath can be treated as a 'chip' as in a pseudonoise (PN) sequence. The longer the train of pulses in a rich multipath channel, the sharper is the autocorrelation of the CIR. If we precode the information symbols using the time-reversed version of CIR at the transmitter, the resultant channel response, equal to response of a filter matched to the CIR will output very sharp response equivalent to the autocorrelation of the CIR (similar to PN code) (Figure 1.3(b)). This sharpened matched filter response will reduce ISI greatly if the length of the multipath pulse train is large. This exploits the rich multipath, rather than treating it as a nuisance. This simple precoding scheme yields a concentration of power at only the intended receiver at a particular time. In Figure 1.3(b), the peak accounts for 50 % of the total energy of the CIR. Another important advantage of the time reversal is spatial focusing. Spatial focusing results in very low cochannel interference in a multicell system, resulting in a very efficient use of bandwidth in the overall network. Finally, time reversal reduces the problem of the per-path pulse distortion since the autocorrelation operation (correlating with its time reversal) yields a much smoother symmetric signal waveform at the receiver (see Figure 1.3(b)).

1.3.4 MAC

The major stream of the UWB MAC research includes IEEE 802.15.3 [31] and an alternative MAC specification [32] defined by multiband OFDM alliance (MBOA) [33]. However, to utilize the radio resources more efficiently, the MAC design should take into account the unique UWB characteristics. First, to meet the FCC regulation on PSD and to prolong the battery lifetime, each UWB device needs to transmit with low power. An effective power management mechanism is necessary to achieve this target. On the other hand, low power transmission can also offer benefit to UWB MAC because two nodes with low transmission power can transmit simultaneously as long as they are separated far enough [34]. Second, acquisition time in UWB systems is much longer than that in narrowband systems. Considering the high-rate transmission of data

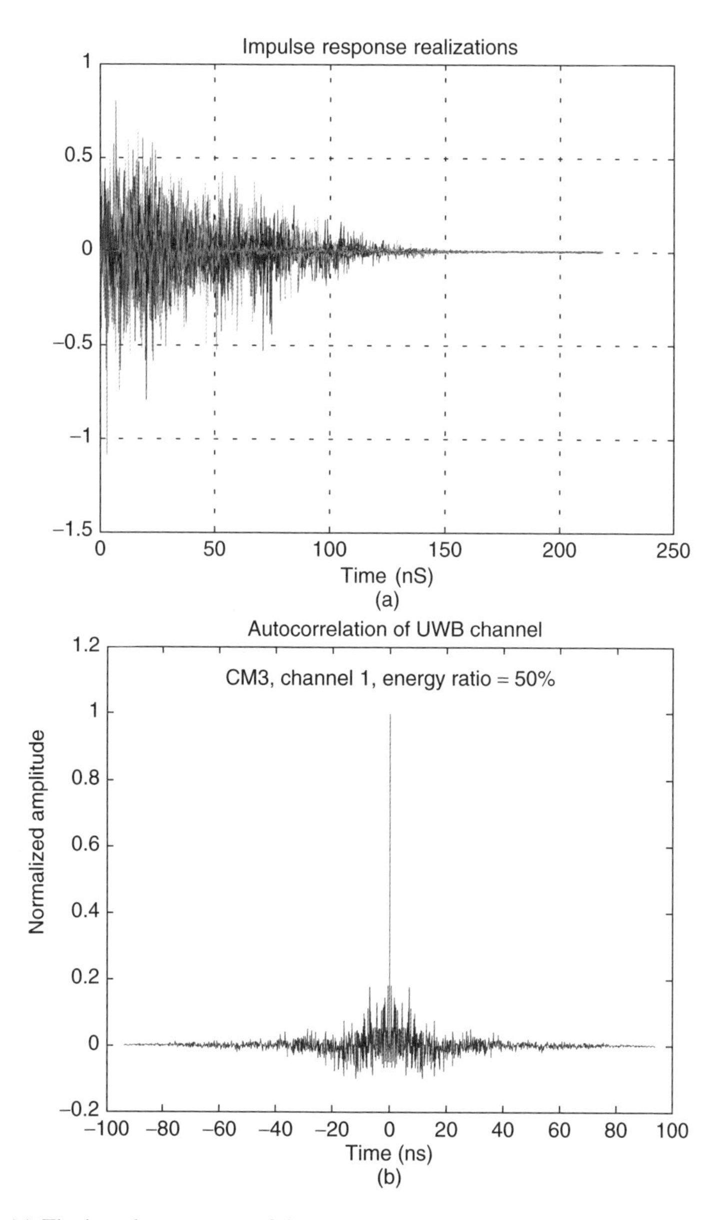

Figure 1.3 (a) The impulse response of the channel model from IEEE 802.15.3a CM3 for indoors. The y-axis is the amplitude of the channel. (b) Autocorrelation of a channel impulse response $R_{hh}(t) = h(t) * h(-t)$ for CM3 channel

payload, the time needed for acquisition may be a large portion of the total time for a packet to be received at the receiver side, resulting in low system efficiency. Actions need to be taken to alleviate the negative effect [35]. Third, UWB technology is capable of accurate positioning. With the help of position information, the UWB MAC performance can be enhanced as routing and power control can be simplified [36].

In traditional MAC protocols such as IEEE 802.11, a single channel is shared by all the nodes. Simultaneous transmissions in a neighbor will lead to collision. However, because of the inherent spread spectrum nature in UWB transmission, simultaneous transmissions can be supported by proper pseudorandom code design. Simultaneous transmissions generate interference to each other. In such an environment, the technologies that allocate the resources (power, rate, and time) and alleviate the induced interference are promising (especially for pulsed UWB) and have been attracting great attention [37].

1.3.5 Future Directions

Theoretically speaking, most of the existing work in wireless communications is still valid [1, 2, 18, 38, 39], since the theory of the linear system that describes signals and noise, and Maxwell's equations that describe short-pulse radiation and propagation are valid. The validity of the linear system theory and Maxwell's equations determines the nature and the framework of UWB communications. Applications of UWB systems are foreseen to be the main stream research efforts in the future. In particular, the fusing of communications with positioning is seen to be a major breakthrough that may play a crucial role in wireless sensor networks. However, the extreme wide bandwidth indeed causes some unique issues. These issues are the following. (1) Pulse distortion caused by antennas and dispersive-propagation environment is critical. The challenge also comes from the fact that it is very difficult to decouple the effect of antennas with propagation environments. A typical pulse-based system usually does not have a front-end filter whereas its narrowband counterpart has. So, pulse distortion serves as a pulse shaping filtering effect that must be included in the system model. (2) The receiver structures need more evaluation. In particular, timing acquisition needs special attention. A RAKE structure is often suggested. Channel estimation and synchronization required for RAKE are too difficult for short UWB pulses, so some suboptimum alternatives are investigated. To avoid these difficulties, the TR paradigm seems to be very promising, especially for DoD's adverse environments and IEEE 802.15.4a. But the noisy template problem and weak capability to counteract intentional interference are two basic problems. (3) UWB antennas are important and often require codesign with other parts of baseband. Communications impose stringent requirements on antennas. The design of low-cost omni antennas are challenging. UWB multiple antennas are rarely understood. (4) Noncoherent energy detection-based receivers for both communications and ranging have in recent times received the most attention for low-power and low-cost applications for sensors networks. (5) The across-layer design between the physical layer and MAC needs consideration. The centimeter accuracy of positioning offered by UWB will change a lot of network design issues such as routing. (6) The enhanced security of UWB should be explored. The CIR of the channel can be regarded as a random PN code. The rich multipath in the UWB channel can be exploited to enhance the security [42]. Time reversal appears to reflect this vision. (7) Test bed is needed to validate the above system and network concepts. The lack of such experimentation is stifling UWB R&D, and can only be remedied with appropriate laboratory infrastructure.

References

[1] R. C. Qiu, R. Scholtz, and X. Shen, "Ultra-Wideband Wireless Communications – A New Horizon," Editorial on Special Session on UWB, *IEEE Trans. Veh. Technol.*, vol. 54, no. 5, Sept. 2005.

[2] R. C. Qiu, H. P. Liu, and X. Shen, "Ultra-Wideband for Multiple Access," *IEEE Commun. Mag.*, vol. 43, no. 2, pp. 80–87, Feb. 2005.

[3] "*Revision of Part 15 of the Commission's Rules Regarding Ultra-Wideband Transmission Systems*," First Note and Order, Federal Communications Commission, ET-Docket 98–153, Adopted February 14, 2002, released April 22, 2002. Available: http// www.fcc.gov/Bureaus/Engineering_Technology/Orders/2002/fcc02048.pdf

[4] T. W. Barrett, "History of UltraWideband Communications and Radar: Part II, UWB Radar and Sensors," *Microwave J.*, Euro-Global Edition, vol. 44, no. 2, pp. 22–52, Feb. 2001.

[5] T. W. Barrett, "History of UltraWideband Communications and Radar: Part I, UWB Communications," *Microwave J.*, Euro-Global Edition, vol. 44, no. 1, pp. 22–56, Jan. 2001.

[6] T. W. Barrett, "History of UltraWideband (UWB) Radar & Communications: Pioneers and Innovators," *Progress in Electromagnetics Symposium 2000 (PIERS2000)*, Cambridge, MA, July, 2000.

[7] R. A. Scholtz, "Multiple Access with Time-hopping Impulse Modulator (invited paper)," *MILCOM '93*, Bedford, MA, Oct. 11–14, 1993.

[8] M. Z. Win and R. Scholtz, "Ultra-Wide Bandwidth Time-Hopping Spread Spectrum Impulse Radio for Wireless Multiple-Access Communications," *IEEE Trans. Commun.*, vol. 48, no. 4, pp. 679–689, Apr. 2000.

[9] S. Roy, J. R. Foerster, V. Somayazulu, and D. Leeper, Ultra-Wideband Radio Design: The Promise of High-Speed, Short Range Wireless Connectivity, *Proc. IEEE*, vol. 92, no. 2, pp. 295–311, Feb. 2004.

[10] F. Ramirez-Mireles, "On the Performance of Ultra-Wideband Signals in Gaussian Noise and Dense Multipath," *IEEE Trans. Veh. Technol.*, vol. 50, no. 1, pp. 244–249, Jan. 2001.

[11] F. Ramírez-Mireles, "Signal Design for Ultra-wideband Communications in Dense Multipath," *IEEE Trans. Veh. Technol.*, vol. 51, no. 6, pp. 1517–1521, Nov. 2002.

[12] D. Porcino and W. Hirt, "Ultra-Wideband Radio Technology: Potential and Challenges Ahead," *IEEE Commun. Mag.*, vol. 41, no. 7, pp. 66–74, July 2003.

[13] R. Rappaport, *Wireless Communications: Theory and Practice*, Prentice Hall, 1995.

[14] E. A. Homier and R. A. Scholtz, "Rapid Acquisition of Ultra-Wideband Signals in the Dense Multipath Channel," *Proc. 2002 IEEE Conference on Ultra Wideband System Technology*, pp. 105–109, 2002.

[15] S. Vijayakumaran and T. Wong, "Best Permutation Search Strategy for Ultra-Wideband Signal Acquisition," *IEEE Trans. Veh. Technol.*, vol. 54, no. 5, Sept. 2005.

[16] J. D. Choi and W. E. Stark, "Performance of Ultra-Wideband Communications with Suboptimal Receiver in Multipath Channels," *IEEE J. Sel. Areas Commun.*, vol. 20, no. 9, pp. 1754–1766, Dec. 2002.

[17] S. Zhao, H. Liu, and Z. Tian, "A Decision Feedback Autocorrelation Receiver for Pulsed Ultra-Wideband Systems," *Proc. IEEE Rawcon '04*, pp. 251–254, 2004.

[18] R. C. Qiu and X. Shen, "Ultra-Wideband (UWB) Wireless Communications for Short Range Communications," Editorial for Special Issue on UWB, Dynamics of Continuous Discrete Impulsive Systems, *Int. J. Theory Appl. Ser. B: Appl. Algorithms*, vol. 12, no. 3, June 2005.

[19] M. Oh, B. Jung, R. Harjani, and D. Park, "A New Noncoherent UWB Impulse Radio Receiver," *IEEE Commun. Lett.*, vol. 9, no. 2, pp. 151–153, Feb. 2005.

[20] I. Oppermann, *et al.* "UWB Wireless Sensor Networks: UWEN – A Practical Example," *IEEE Commun. Mag.*, vol. 42, no. 12, pp. S27–S32, Dec. 2004.

[21] L. Stoica, A. Rabbachin, and I. Oppermann, "An Ultra Wideband System Architecture for Wireless Sensor Networks," *IEEE Trans. Veh. Technol.*, vol. 54, no. 5, Sept. 2005.

[22] S. Gezici, Z. Tian, G. B. Giannakis, H. Kobayahsi, A. F. Molisch, H. V. Poor, and Z. Sahinoglu, Location Via Ultra-Wideband Radios, *IEEE Signal Proc. Mag.*, vol. 22, no. 4, pp. 70–84, July 2005.

[23] N. Patwari, *et al.* "Relative Location Estimation in Wireless Sensor Networks," *IEEE Trans. Signal Proc.*, vol. 51, no. 8, pp. 2137–2148, Aug. 2003.

[24] M. Fink, "Time Reversed Acoustics," *Phys. Today*, pp. 34–40, Mar. 1997.

[25] A. Derode, *et al.* "Taking Advantages of Multiple Scattering to Communicate with Time-Reversal Antennas," *Phys. Rev. Lett.*, vol. 90, no. 1, pp. 014301, Jan. 2003.

[26] G. Lerosey, *et al.* "Time Reversal of Electromagnetic Waves," *Phys. Rev. Lett.*, vol. 92, no. 19, pp. 193904, May 2004.

[27] N. Guo, R. C. Qiu, and B. M. Sadler, "An Ultra-Wideband Autocorrelation Demodulation Scheme with Low-Complexity Time Reversal Enhancement," *Proc. IEEE MILCOM '05*, Atlanta City, NJ, Oct. 17–20, 2005.

[28] A. E. Akogun, R. C. Qiu, and N. Guo, "Demonstrating Time Reversal in Ultra-Wideband Communications Using Time Domain Measurements," *Proc. 51st International Instrumentation Symposium*, Knoxville, TN, May 8–12, 2005.

[29] T. Strohmer, M. Emami, J. Hansen, G. Pananicolaou, and A. J. Paulraj, "Application of Time-Reversal with MMSE Equalizer to UWB Communications," *Proc. IEEE Globecom '04*, pp. 3123–3127, 2004.

[30] M. Weisenhorn and W. Hirt, "Performance of Binary Antipodal Signaling over the Indoor UWB MIMO Channel," *Proc. IEEE ICC '03*, pp. 2872–2878, 2003.

[31] IEEE Std 802.15.3–2003: Wireless Medium Access Control (MAC) and Physical Layer (PHY) Specifications for High Rate Wireless Personal Area Networks (WPANs), Sept. 2003.

[32] *MBOA Wireless Medium Access Control (MAC) Specification for High Rate Wireless Personal Area Networks (WPANs)*, MBOA MAC Specification Draft 0.65, Oct. 15, 2004.

[33] Multiband OFDM Alliance. [online]. Available: http://www.multibandofdm.org/, accessed 2005.

[34] B. Radunovic and J.-Y. Le Boudec, Optimal Power Control, Scheduling, and Routing in UWB Networks, *IEEE J. Sel. Areas Commun.*, vol. 22, no. 7, pp. 1252–1270, Sept. 2004.

[35] K. Lu, D. Wu, Y. Fang, and R. C. Qiu, "On Medium Access Control for High Data Rate Ultra-Wideband Ad Hoc Networks," *Proc. IEEE WCNC '05*, pp. 795–800, 2005.

[36] X. Shen, W. Zhuang, H. Jiang, and J. Cai, "Medium Access Control in Ultra-Wideband Wireless Networks," *IEEE Trans. Veh. Technol.*, vol. 54, no. 5, Sept. 2005.

[37] F. Cuomo, C. Martello, A. Baiocchi, and F. Capriotti, "Radio Resource Sharing for Ad Hoc Networking with UWB," *IEEE J. Sel. Areas Commun.*, vol. 20, no. 9, pp. 1722–1732, Dec. 2002.

[38] X. Shen, M. Guizani, H. H. Chen, R. Qiu, and A. F. Molisch "Ultra-Wideband Wireless Communications," Editorial on Special Issue on UWB, *IEEE J. Sel. Areas Commun.*, vol. 24, 2nd Quarter, 2006.

[39] N. Blefari-Melazzi, M. G. Di Benedetto, M. Gerla, M. Z. Win, and P. Withington, Ultra-Wideband Radio in Multiaccess Wireless Communications, "Editorial on Special Issue on UWB, *IEEE J. Sel. Areas Commun.*, vol. 20, no. 9, Dec. 2002.

[40] M. Weisenhorn and W. Hirt, "Robust Noncoherent Receiver Exploiting UWB Channel Properties," *Proc. IEEE Joint UWBST & IWUWBS*, pp. 156–160, 2004.

[41] R. C. Qiu, C. Zhou, N. Guo, and J. Q. Zhang, "Time Reversal with MISO for Ultra-Wideband Communications: Experimental Results," Invited Paper, IEEE Radio and Wireless Symposium, San Diego, CA, 2006.

[42] R. Scholtz, private communication, MURI Panel Review, Monterey, CA, Nov. 2004.

2

Modulation and Signal Detection in UWB

Uzoma A. Onunkwo and Ye (Geoffrey) Li

In this chapter, we address the concepts of modulation and signal detection in ultra-wide bandwidth (UWB) communications.

After introducing the definition of UWB systems, we present single-carrier and orthogonal frequency division multiplexing (OFDM)–based modulation for UWB. Finally, we briefly summarize the further research topics in this area.

2.1 Overview

In this section, we give an overview of the concepts, history, and applications of UWB communications. After presenting the Federal Communications Commission (FCC)-accepted definition of a UWB system, we briefly introduce its history and the current challenges. We also illustrate the characteristics of the UWB system that differentiate it from narrowband and Code Division Multiple Access (CDMA) systems.

2.1.1 Evolution and Definition

In 2002, FCC formally defined UWB systems as radio systems whose fractional bandwidth is at least 25 % of their center frequency or whose bandwidth is at least 1.5 GHz [1]. In fact, the frequency band that is mostly used for UWB systems ranges from 3.1 to 10.6 GHz [2, 3].

UWB communication is believed to have originated in 1895 from the work of Guglielmo Marconi, known as the *Father of Radio*, on his spark-gap transmitter. At that time, UWB was not really a research area. It is only recently that interests grew in this area when researchers studied the impulse response of microwave circuits [4]. UWB was first called *impulse radio*. However, today, impulse radio only refers to a particular type of modulation

Ultra-wideband Wireless Communications and Networks
Edited by Xuemin (Sherman) Shen, Mohsen Guizani, Robert Caiming Qiu and Tho Le-Ngoc

scheme in UWB communications, which is discussed later. The underlying idea of UWB communication is that it has a relatively large bandwidth.

2.1.2 Major Differences from Narrowband and CDMA Systems

Narrowband communications, as differentiated from UWB communications, have relatively narrow bandwidth compared to its center frequency. The Global System for Mobile (GSM) Communication standard is an example of a narrowband communication system. In GSM, the reverse link and forward link bands are 890–915 MHz and 935–960 MHz, respectively, which refers to a system with a carrier frequency of about 900 MHz and a bandwidth of 25 MHz (approximately 3 % in fractional bandwidth). In narrowband systems, there is a need for an upconverter to convert the baseband signals to a higher frequency range for transmission, which is not necessary in UWB communications.

Moreover, most of the current narrowband systems have bandwidth restriction to avoid interference with other users or communication systems. Therefore, licensing requirements are imposed by appropriate communications regulating authorities (e.g., FCC in the United States). On the other hand, UWB communications use an unlicensed band. This, however, means that a restriction has to be imposed on the power level to avoid drastic impediment, in the form of interference, to existing narrowband communication systems [5].

Similar to CDMA cellular systems, spread spectrum techniques are used in UWB communications to improve their performance. However, UWB systems [6] usually employ much shorter spreading sequences than CDMA systems to achieve higher data rates (the main goal of UWB systems), which causes differences in modulation and signal detection.

2.1.3 Types of UWB Modulation

Essentially, UWB communications come in one of two types – single-band and multiband. Impulse radio, mentioned above, is a single-band UWB system [4]. In impulse radio, the signal that represents a symbol consists of serial pulses with a very low duty cycle. The pulse width is very narrow, typically in nanoseconds. This small pulse width gives rise to a large bandwidth and a better resolution of multipath in UWB channels.

The other type of UWB modulation is multiband-based and is accomplished by using multicarrier or OFDM [3] modulation with Hadamard or other spreading codes [2]. With OFDM-based modulation, systems can effectively deal with delay spread or frequency selectivity of UWB channels.

2.1.4 UWB Applications

UWB is an area that has generated a lot of interest recently. The huge and unlicensed bandwidth (which implies free spectrum) has been the major proponent for research in this field. Therefore, it is an ideal tool for high data rate and short-range communications, which are very desirable for multimedia applications in indoor wireless communications. Other UWB applications are in sensor networks and location estimation. The work by Oppermann *et al.* in [7] shows a practical use of UWB in wireless sensor networks; works by Sakamoto and Sato in [8] show that UWB can be used in target location.

The nanosecond-short pulse widths and low power density used in impulse radio make the UWB signals appear as *noise* to other communication systems/users and hence impulse radio is subtle to intrude. This finds use in high-security communications. Coding and spreading can, in addition, be used to increase the processing gain of the impulse radio, thereby making it immune to interference from other systems. Therefore, UWB has been widely studied in the military field.

As with innovative technologies, it is expected that more applications will arise in the near future.

2.2 Single-Carrier–Based Modulation

In this section, we introduce single-carrier–based modulation, channel estimation, and signal detection for UWB.

2.2.1 Time-Hopping PPM

Pulse-position modulation (PPM) is the scheme employed in impulse radio and is a primitive UWB modulation scheme. It uses a combination of short pulses to represent a symbol. We first present the basic concept of time-hopping (TH) PPM and then address modulation index optimization to improve the performance of TH-PPM systems.

2.2.1.1 Basic TH-PPM

For TH-PPM, the signal for the mth user, $s^{(m)}(t)$, is represented by:

$$s^{(m)}(t) = \sum_{n} w^{(m)}\left(t - nT_s - b_n^{(m)}\delta\right), \tag{1}$$

where T_s is the symbol period, $b_n^{(m)}$ is the nth transmitted data ($b_n^{(m)} \in \{0, 1\}$ for binary data) from the mth user, δ is the modulation index, and $w^{(m)}(t)$, a spreading of basic pulses, is defined as[1]

$$w^{(m)}(t) = \sum_{l=0}^{N-1} p(t - lT_c - c_l^{(m)}\Delta_o), \tag{2}$$

where $c_l^{(m)} \in \{0, 1\}$ is a spreading code unique for every user that enables efficient multiple accessing (similar to CDMA but much shorter); Δ_o is the time-shifting term associated with $c_l^{(m)}$; and N is the number of pulses, $p(t)$, in a symbol duration. Each pulse (called *chip* in CDMA literature) occupies a time interval, T_c; consequently, $T_s = N \cdot T_c$. The modulation index, δ, is an important parameter for TH-PPM that can improve bit error rate (BER) performance, as shown by Onunkwo and Li in [9].

The pulse-shaping function that is mostly adopted for PPM is the Gaussian monocycle or its higher-order derivative, although other functions can be also used. The first-order

[1] If there is only one user, the superscript m will be omitted and the code sequence $\{c_l^{(m)}\}_{l=0}^{N-1}$ will be ignored.

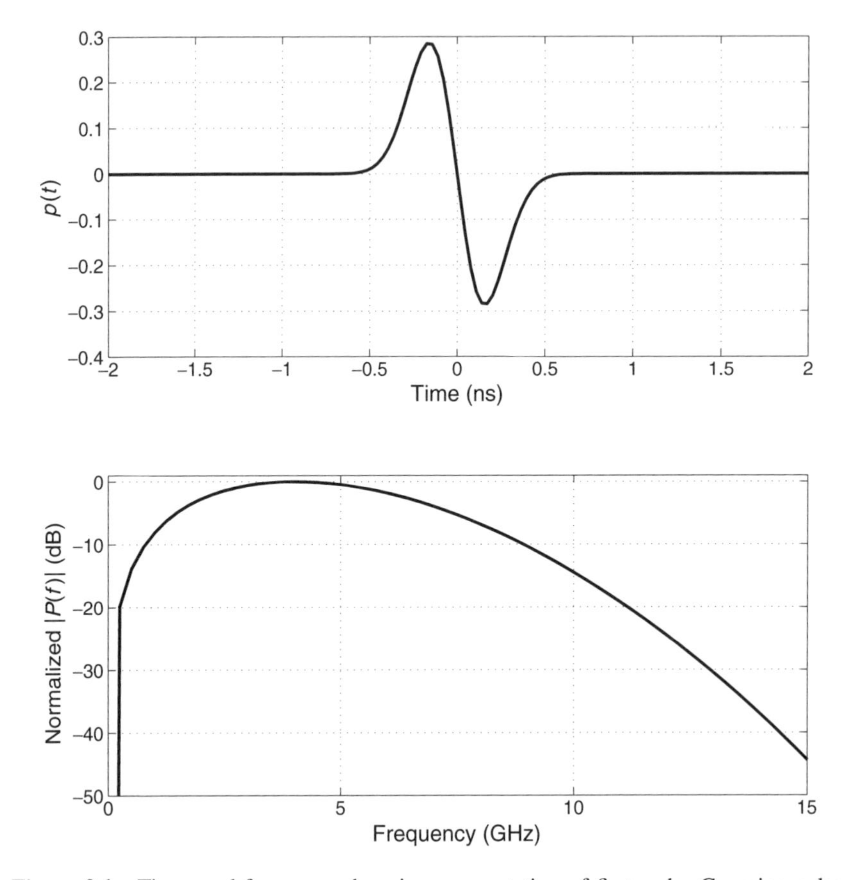

Figure 2.1 Time- and frequency-domain representation of first-order Gaussian pulse

Gaussian monocycle can be expressed as

$$p(t) = \frac{A}{\sigma}\left(c - \frac{t}{\sigma}\right) e^{-0.5\left(\frac{t}{\sigma}-c\right)^2},$$

where the constant A is an amplitude normalization constant, σ is related to the width of the pulse, and c is a time-shifting term. The time- and frequency- domain representations of the first-order Gaussian pulse are shown in Figure 2.1, where $c = 0$ and $2\pi\sigma = 1$ ns. It can be observed from the figure that the Gaussian pulse spreads its power over a wide range of frequency and has virtually no DC and low-frequency components, which makes it suitable for UWB communication [10].

2.2.1.2 Modulation Index Optimization

It has been previously mentioned that the modulation index, δ, has an important role in determining the BER performance of UWB. It was Onunkwo and Li in [9] who demonstrated its importance.

Without loss of generality, consider the basic Gaussian pulse function given in this section with $c = 0$,

$$p(t) = -\frac{At}{\sigma^2} e^{-\frac{t^2}{2\sigma^2}}.$$

If we use $p(t)$ to represent '0' and $p(t - \delta)$ to represent '1' for binary data, the normalized correlation function of the basic pulse over a frame duration becomes

$$\begin{aligned} R_1(\delta) &= \frac{1}{E_p} \int_{-\frac{T_f}{2}}^{\frac{T_f}{2}} p(t)p(t-\delta)\,dt, \\ &\approx \frac{1}{E_p} \int_{-\infty}^{\infty} p(t)p(t-\delta)\,dt \qquad \left(p(t) \approx 0 \text{ for } |t| \in \left(\frac{T_f}{2}, \infty\right)\right) \\ &= \frac{A^2\sqrt{\pi}}{\sigma E_p}\left(0.5 - \pi^2\alpha^2\right) e^{-\pi^2\alpha^2}, \end{aligned} \tag{3}$$

where $\delta = 2\pi\sigma \cdot \alpha$ has been substituted and E_p is the energy per pulse, defined as:

$$E_p = \int_{-\frac{T_f}{2}}^{\frac{T_f}{2}} |p(t)|^2\,dt.$$

Assuming additive white Gaussian noise (AWGN) and single-user environment, the problem of reducing the BER (same as increasing the minimum distance between symbols) becomes the problem of finding the δ that reduces the correlation in (3) above. The modulation index that yields the least possible correlation is called the *optimum* modulation index, δ_{opt}. To obtain δ_{opt}, (3) is differentiated and set to zero:

$$\begin{aligned} R_1'(\delta) &= \left.\frac{dR_1(\delta)}{d\delta}\right|_{\delta=2\pi\sigma\alpha} \\ &= \frac{A^2\sqrt{\pi}}{\sigma E_p}\left(-2\pi^2\alpha\, e^{-\pi^2\alpha^2}\right) \cdot \left(\frac{3}{2} - \pi^2\alpha^2\right) = 0\,. \end{aligned}$$

Further simplification shows that, for the first-order Gaussian monocycle, the solution, $\alpha_{\text{opt}} = \frac{\sqrt{6}}{2\pi}$; therefore, $\delta_{\text{opt}}, = \sqrt{6}\sigma \approx 0.38985(2\pi\sigma)$.

Figure 2.2 from [9] shows the significance of using the optimum modulation index where the pulse width, $T_p, = 2\pi\sigma = 1.0$ ns, $N = 1$, and $T_s = T_c = 100$ ns. As can be seen from the figure, using the optimum modulation index for the first-order Gaussian monocycle yields a 2-dB improvement compared with the typically used modulation index, $\delta = 0.2T_p$ [9].

The derivation above is for the first-order Gaussian monocycle. The same technique can be used on a higher-order Gaussian monocycle by finding the modulation index that minimizes the normalized correlation of the kth-order Gaussian pulse, $p_k(t) = \frac{d^k}{dt^k} p(t)$, which is given by

$$R_k(\delta) \approx \frac{1}{E_{p_k}} \int_{-\infty}^{\infty} p_k(t)p_k(t-\delta)\,dt \tag{4}$$

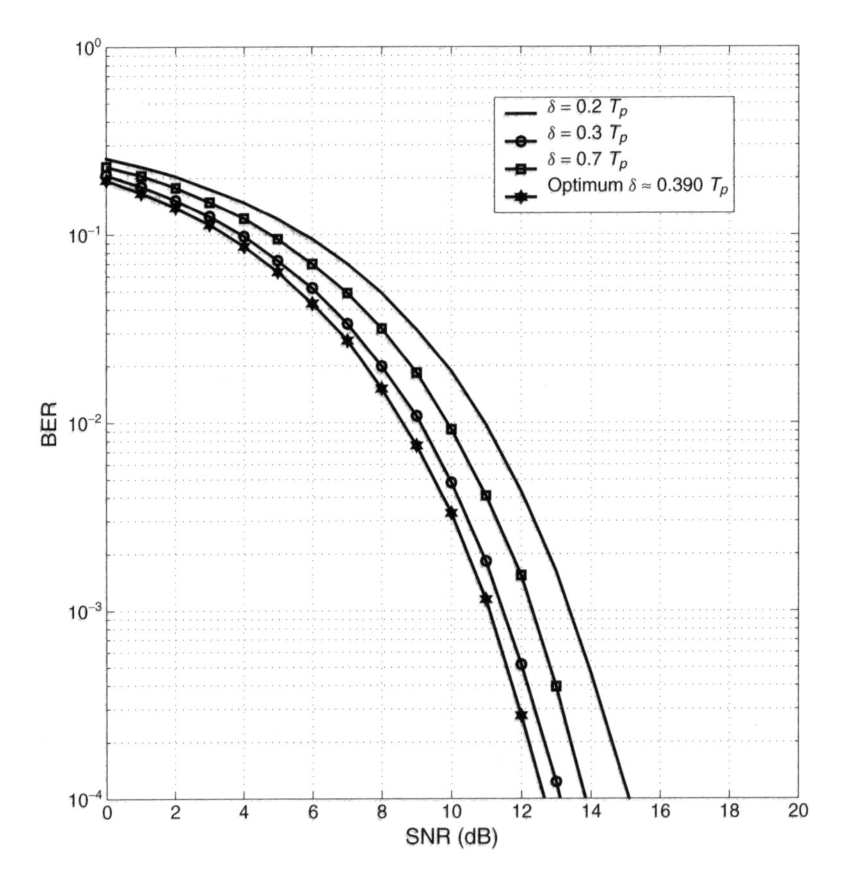

Figure 2.2 Performance of PPM with different modulation indexes, δ's

Table 2.1 Optimum modulation indices $\delta_{\text{opt}} = 2\pi\sigma \cdot \alpha_{\text{opt}}$ and corresponding normalized correlation for various orders of Gaussian monocycle

Pulse-order	α_{opt}	Normalized correlation ($R_{\min}$)
First	$\frac{\sqrt{6}}{2\pi} \approx 0.3899$	−0.4463
Second	$\frac{\sqrt{10-2\sqrt{10}}}{2\pi} \approx 0.3051$	−0.6183
Third	0.2599	−0.7086
Fourth	0.2304	−0.7644

A recursive formula can be used instead of (4) to obtain the correlation formula [9]

$$R_k(\delta) = -R''_{k-1}(\delta) = (-1)^{k-1} R_1^{(2k-2)}(\delta).$$

Solving these recursive equations for the second-, third-, and fourth-order Gaussian monocycles gives the result summarized in Table 2.1.

As seen from the table, the higher-order Gaussian monocycles at their optimum modulation indexes achieve lower normalized correlation. In other words, the higher the Gaussian monocycle order, the lower the achievable BER. However, this improvement in BER performance comes at the expense of a higher time sensitivity to the receiver synchronization and receiver complexity. This is because the higher-order Gaussian pulses exhibit correlation properties that change considerably more with little timing offset in the received signal pulse. This gradient becomes higher as the order of the Gaussian pulse increases.

2.2.2 Other Types of Modulations

In this section, we present other types of single-carrier–based modulation used in UWB communications, including amplitude modulation (AM), orthogonal pulse modulation (OPM), and pseudochaotic time-hopping (PCTH) modulation.

2.2.2.1 Amplitude Modulation (AM)

Amplitude modulation (AM) is sometimes also used for UWB modulation besides TH-PPM. For AM, the modulated signal for the mth user in Equation (1) is replaced by:

$$s^{(m)}(t) = \sum_{n} d_n^{(m)} \cdot w^{(m)}(t - nT_s), \tag{5}$$

where

$$w^{(m)}(t) = \sum_{l=0}^{N-1} (2c_l^{(m)} - 1)p(t - lT_c),$$

and the amplitude modulating gain, $d_n^{(m)}$, can be chosen from the set $\{\pm 1, \pm 3, \ldots, \pm(2M-1)\}$ for an M-ary system, which is determined by the binary sequence, $b_n^{(m)}$, to be transmitted by the user m. For example, if binary AM is used, then $d_n^{(m)} = 2b_n^{(m)} - 1$. The primary difference between PPM and AM is that the PPM uses the position of pulse to carry information, while AM uses the pulse sign and amplitude to carry information.

2.2.2.2 Orthogonal Pulse Modulation (OPM)

One of the questions that may be asked is why use only Gaussian monocycles for single-carrier–based modulation. Is it possible to use other pulse functions to implement modulation in UWB with considerably good performance? Michael, Ghavami, and Kohno in [11] have used orthogonal pulses generated from Hermite polynomials as modulation pulses. In their work, a set of orthogonal pulses was used to represent the symbols for M-ary modulation; therefore, it is called orthogonal pulse modulation (OPM).

The Hermite polynomial is defined as

$$f_0(t) = 1$$

$$f_n(t) = (-1)^n e^{\frac{t^2}{2}} \frac{d^n}{dt^n}\left(e^{-\frac{t^2}{2}}\right),$$

for $n = 1, 2, \ldots$. Since these polynomials are not orthogonal to each other, Gram–Schmidt procedure is used to obtain a new set of orthogonal polynomials, called modified Hermite polynomial (MHP) [11], which can be expressed as

$$\begin{aligned} g_n(t) &= e^{-\frac{t^2}{4}} f_n(t) \\ &= (-1)^n e^{\frac{t^2}{4}} \frac{d^n}{dt^n}\left(e^{-\frac{t^2}{2}}\right). \end{aligned} \tag{6}$$

Using these MHPs, a multiple-access scheme can be designed. Different users in the multiple-access environment are assigned different pairs of polynomials for binary data transmission. Each pair, in itself, contains two orthogonal polynomials. In other words, each user uses an orthogonal pair of pulse polynomials, which is in turn orthogonal to other users' signals.

The BER performance in a two-user environment using the MHP-based OPM has been studied in [11], where the signal from one of the users acts as an interference. Figure 2.3, from [11], shows that the MHP-based OPM has a significant performance improvement

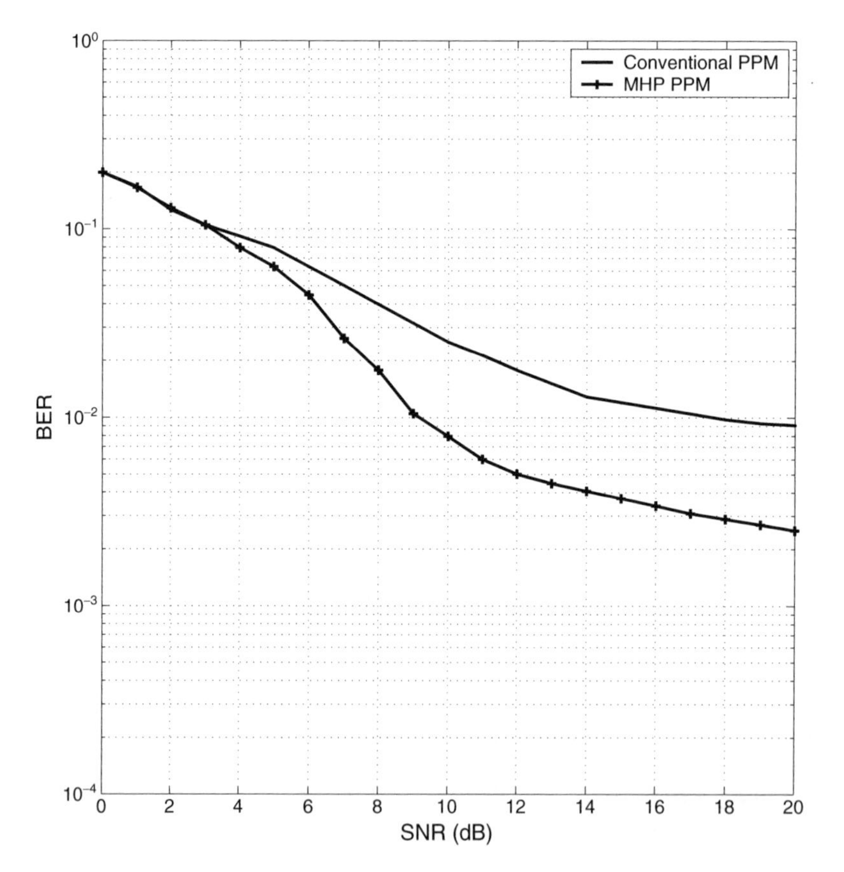

Figure 2.3 Performance comparison between traditional PPM and OPM in a two-user environment

compared with the traditional PPM, since the MHPs in OPM have better immunity to multiple-access interference than the traditional TH-PPM.

However, it is worthy to note that MHP pulses suffer more from timing jitters. In a broadcast environment with a controlling base station, the timing could be accurate so that the effect of the timing jitter is negligible. However, in an ad hoc setup, it is very difficult to control the timing of each user, thereby making it difficult to preserve the orthogonality of user signals.

2.2.2.3 Pseudochaotic Time-Hopping PPM

Pseudochaotic time-hopping (PCTH) PPM, recently proposed in [12, 13], is another interesting candidate for UWB modulation. In the PCTH-PPM, the nth binary data to be transmitted by user m, $b_n^{(m)}$, is first transformed into new symbols $m_n^{(m)}$ by

$$m_n^{(m)} = \sum_{l=0}^{L-1} 2^{L-1-l} b_{n-l}^{(m)}.$$

From the above definition, it is obvious that $m_n^{(m)}$ could take any integer value between 0 and $2^L - 1$. In particular, if $b_n^{(m)} = 0$, then $0 \leq m_n^{(m)} \leq 2^{L-1} - 1$. Otherwise, $2^{L-1} \leq m_n^{(m)} \leq 2^L - 1$. From the $m_n^{(m)}$, the transmitted signal for the mth user can be generated by

$$s^{(m)}(t) = \sum_n w^{(m)}(t - nT_s - m_n^{(m)}\delta),$$

where $w^{(m)}(t)$ could be just a Gaussian pulse, or its spreading version given by

$$w^{(m)}(t) = \sum_{l=0}^{N-1} c_l^{(m)} p(t - lT_c).$$

As indicated in [12] and [13], the PCTH-PPM increases the immunity to interception. This is particularly useful in a high-security network, as in the military. This is still an ongoing research work, and details on this topic can be found in [12] and [13] and the references therein.

2.2.3 Channel Estimation

To detect a UWB signal, the parameters of UWB channel have to be estimated at the receiver. Therefore, we discuss channel estimation for single-carrier–based modulation here.

UWB channel response can modeled by

$$h(t) = \sum_{l=1}^{L} \gamma_l \delta(t - \tau_l), \tag{7}$$

where γ_l and τ_l are the gain and the delay of the lth path, respectively, L is the total number of resolvable propagation paths, and $\delta(.)$ is the Dirac delta function. When a signal, $s(t)$, passes through a UWB channel, the channel output can be expressed as

$$r(t) = h(t) * s(t) + n(t) = \sum_{l=1}^{L} \gamma_l s(t - \tau_l) + n(t), \tag{8}$$

where $n(t)$ is AWGN. A training sequence is used to estimate the channel parameters. In this case, the transmitted signal, $s(t)$, is known to the receiver.

A number of channel estimation techniques have been proposed. We describe two channel estimation approaches – one by Li, Molisch, and Zhang in [14] and the other by Lottici, D'Andrea, and Mengali in [15].

2.2.3.1 Model-Based ML Channel Estimation

From (7), once the amplitude gains, γ_ls, and delays, τ_ls, are known, the channel's impulse response can be determined. The algorithm in [15] makes the maximum likelihood estimation to γ_ls and τ_ls.

During the training period, the transmitted signal, $s(t)$, is known to the receiver. By comparing it with the received signal, $r(t)$, a log-likelihood function is obtained, which is determined by

$$\log[\Lambda(\vec{\gamma}, \vec{\tau})] = 2\int_0^{KT_s} r(t)\left(\sum_{l=1}^{L} \tilde{\gamma}_l s(t - \tilde{\tau}_l)\right) dt - \int_0^{KT_s} \left(\sum_{l=1}^{L} \tilde{\gamma}_l s(t - \tilde{\tau}_l)\right)^2 dt, \tag{9}$$

where K is the number of training symbols, and $\{\tilde{\gamma}_l\}_{l=1}^{L}$ and $\{\tilde{\tau}_l\}_{l=1}^{L}$ are the realization of the gains and delays,

$$\vec{\gamma} = (\tilde{\gamma}_1, \tilde{\gamma}_2, \ldots, \tilde{\gamma}_L),$$

and

$$\vec{\tau} = (\tilde{\tau}_1, \tilde{\tau}_2, \ldots, \tilde{\tau}_L).$$

The values of the vectors $\vec{\gamma}$ and $\vec{\tau}$ can be obtained by maximizing the log-likelihood function (9).

Note that the likelihood function is a complicated function of $\vec{\gamma}$ and $\vec{\tau}$ and that there is no closed form or simple solution to the problem. Following is a practical approach to deal with the problem. First, for a given $\vec{\tau}$, find $\vec{\gamma}_o(\vec{\tau})$ that maximizes the likelihood function, and then, by exhaustive searching, find $\vec{\tau}_o$ that maximizes the log-likelihood function $\log[\Lambda(\vec{\gamma}_o(\vec{\tau}), \vec{\tau})]$. Once the optimum delay vectors, $\vec{\tau}_o$, are found, the optimum gain vector can be found by $\vec{\gamma}_o = \vec{\gamma}_o(\vec{\tau}_o)$. Details on this approach can be found in [15].

2.2.3.2 Low-Complexity Channel Estimation

The ML channel estimation introduced above has a very good performance. However, it is too complicated to be implemented in UWB systems, which usually require low-complexity receivers. Here, we present a low-complexity approach developed in [14]. The receiver structure used for both the channel estimation and signal detection is shown in Figure 2.4.

For simplicity of discussion, the AM is assumed to be used. However, with only minor modifications, the approach can be used in other types of modulation. Assume that $d_k \in \{+1, -1\}$ for $k = 0, \ldots, K-1$, is a pseudorandom training sequence. Then, the transmitted signal is

$$s(t) = \sum_{k=0}^{K-1} d_k w(t - kT_s),$$

where $w(t)$ is the spreading of the basic pulses. The expression of the received signal is similar to (8).

As shown in Figure 2.4, to reduce the complexity at the receiver, an analog matched filter, matching $w(t)$, is used. Consequently, the output of the matched filter is

$$\begin{aligned} y(t) &= r(t) * w(-t) \\ &= \sum_{k=0}^{K-1} d_k \left(\sum_{l=1}^{L} \gamma_l \int_{-\infty}^{\infty} w(t - kT_s - \tau_l + \tau) w(\tau)\, d\tau \right) + \tilde{n}(t) \\ &= \sum_{k=0}^{K-1} d_k h(t - kT_s) + \tilde{n}(t) \end{aligned} \tag{10}$$

where

$$h(t) = \sum_{l=1}^{L} \gamma_l \int_{-\infty}^{\infty} w(t - \tau_l + \tau) w(\tau)\, d\tau,$$

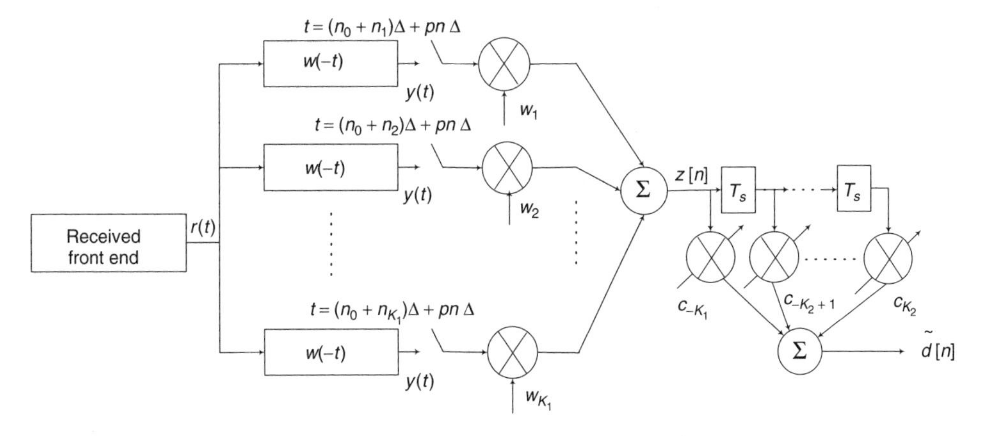

Figure 2.4 Equivalent receiver structure (rake + equalizer) for channel estimation and signal detection

and

$$\tilde{n}(t) = n(t) * w(-t).$$

The baud-rate sampled version of the analog matched-filter output at each finger can be observed and used in channel estimation and signal detection, which can be expressed as

$$y(t_i + nT_s) = \sum_{k=0}^{K-1} d_k h(t_i + nT_s - kT_s) + \tilde{n}(t_i + nT_s),$$

where t_i is the timing of the finger. However, for signal detection, the UWB channel used should be much denser than the baud-rate sampled version of the same. Therefore, we have to estimate $h(l\Delta)$, where $\Delta = \frac{T_s}{p}$ for some integer p.

To estimate, $h(l\Delta)$, we need to obtain a denser sampled version of the matched filter $y(l\Delta)$ first. To obtain a denser sampling, we use the channel sounding method proposed in [14, 16], which repeatedly sends the same training sequence and adjusts the timing of each finger, t_i, to obtain $y(l\Delta)$. Once $y(l\Delta)$ is obtained, the channel parameter can be estimated by

$$\tilde{h}(l\Delta) = \frac{1}{K} \sum_{k=0}^{K-1} d_k y(l\Delta + kT_s).$$

For pseudorandom training sequence, d_k,

$$\frac{1}{K} \sum_{k=0}^{K-1} d_k d_{k-k'} \approx \delta[k'],$$

then

$$\begin{aligned}
\tilde{h}(l\Delta) &= \frac{1}{K} \sum_{k=0}^{K-1} d_k \left(\sum_{m=0}^{K-1} d_m h(l\Delta + kT_s - mT_s) + \tilde{n}(l\Delta + kT_s) \right) \\
&= \frac{1}{K} \sum_{k=0}^{K-1} \sum_{m=0}^{K-1} d_k d_m h(l\Delta + (k-m)T_s) + \frac{1}{K} \sum_{k=0}^{K-1} d_k \tilde{n}(l\Delta + kT_s) \\
&= \sum_{k'} \underbrace{\frac{1}{K} \sum_{k=0}^{K-1} d_k d_{k-k'}}_{\approx \delta[k']} h(l\Delta + k'T_s) + \widehat{n}(l\Delta) \\
&\approx h(l\Delta) + \widehat{n}(l\Delta), \qquad (11)
\end{aligned}$$

where

$$\widehat{n}(l\Delta) = \frac{1}{K} \sum_{k=0}^{K-1} d_k \tilde{n}(l\Delta + kT_s)$$

is the effect of channel noise.

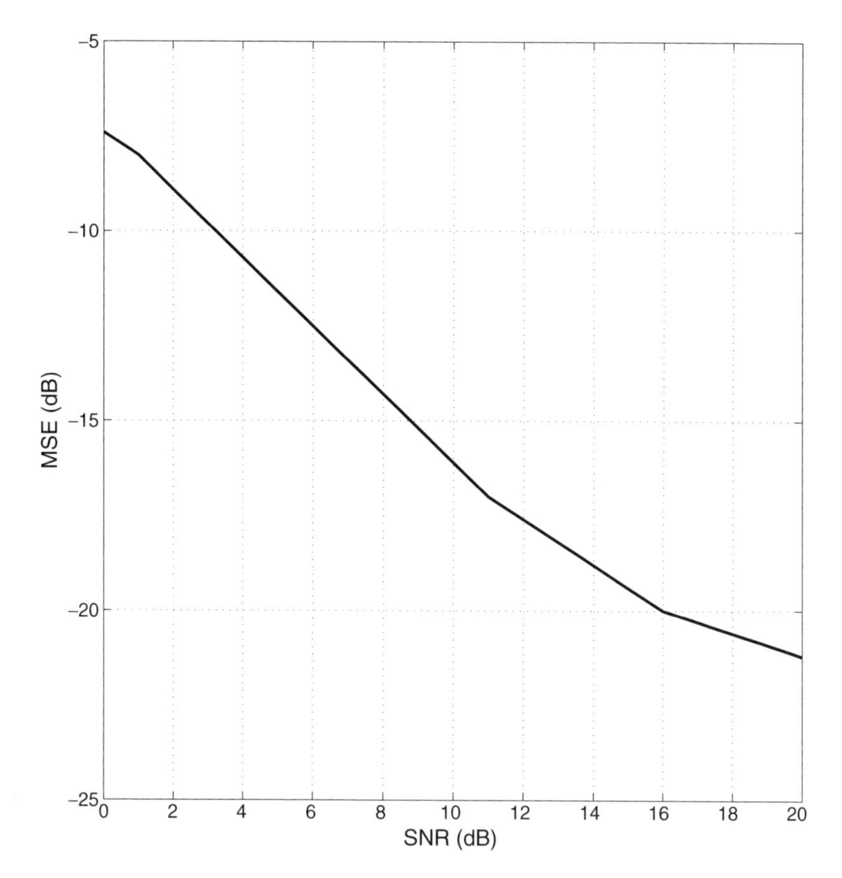

Figure 2.5 Performance of the low-complexity channel estimation technique in [14]

Figure 2.5, from [14], shows the performance of the low-complexity channel estimator. In the simulated system, the fifth derivative of the Gaussian function, with $\sigma = 5.28 \times 10^{-11}$ s, is used. The chip duration is $T_c = 0.625$ ns. The length of the spreading sequence is $N = 8$; therefore, the symbol duration is $T_s = 8T_c = 5$ ns. The impulse response of UWB channel is oversampled 32 times ($p = 32$). The analog rake receiver has $K_1 = 10$ fingers. The training sequence, consisting of 511 symbols, is repeated four times to obtain oversamples 32 times. From the figure, we can see that the mean square error (MSE) of the channel estimator is much smaller than that of channel noise. Therefore, the effect of channel estimation error on the signal detection will be negligible.

2.2.4 Signal Detection

Figure 2.4 shows a signal detector with K_1 fingers for the rake receiver and $2K_2 + 1$ taps for the equalizer of a UWB system with AM, which is a combination of rake receiver and channel equalizer.

Let $h(n_k\Delta)$, for $k = 1, \ldots, K_1$, be the K_1 taps with the largest $|\widehat{h}(n_k\Delta)|$. If the weights for the rake fingers are $\{w_k\}_{k=1}^{K_1}$, then the rake output can be expressed as

$$z[n, n_o] = \sum_{k=1}^{K_1} w_k y((n_o + n_k)\Delta + nT_s),$$

where $n_o\Delta$ is the timing of the whole rake receiver.

For the traditional maximal ratio (MR) rake receiver, $n_o = 0$ and $w_k = h^*(n_k\Delta)$. To improve its performance, minimum mean-square-error (MMSE) rake combining can be used. In this case, the weights and timing are determined by minimizing

$$MSE(\mathbf{w}, n_o) = \frac{1}{K}\sum_{k=0}^{K-1} |z[k, n_o] - d_k|^2, \tag{12}$$

where $\mathbf{w} = (w_1, \ldots, w_{K_1})$ is the rake weights vector. Therefore, for a fixed timing, n_o, the optimum weight vector, can be determined by

$$\mathbf{w}_o[n_o] = \left(\mathbf{Y}[n_o]\mathbf{Y}^H[n_o]\right)^{-1}\left(\mathbf{Y}[n_o]\mathbf{d}^H\right),$$

where $\mathbf{d} = (d_0, \ldots, d_{K-1})$ is the training vector and

$$\mathbf{Y}[n_o] = \begin{pmatrix} y((n_1 + n_o)\Delta) & \ldots & y((K-1)T_s + (n_1 + n_o)\Delta) \\ y((n_2 + n_o)\Delta) & \ldots & y((K-1)T_s + (n_2 + n_o)\Delta) \\ \vdots & \ldots & \vdots \\ y((n_{K_1} + n_o)\Delta) & \ldots & y((K-1)T_s + (n_{K_1} + n_o)\Delta) \end{pmatrix}.$$

The optimum timing is then found by minimizing $MSE(\mathbf{w}_o[n_o], n_o)$.

To further mitigate the residual intersymbol interference of the rake receiver output, a short MMSE equalizer is used after the receiver. The coefficients of the MMSE equalizer are found by minimizing

$$MSE(\mathbf{c}, n_o) = \frac{1}{K}\sum_{n=0}^{K-1}\left|\sum_{k=-K_2}^{K_2} c_k z[n-k, n_o] - d_n\right|^2,$$

where $\mathbf{c} = (c_{-K_2}, \ldots, c_0, \ldots, c_{K_2})$ is the equalizer coefficient vector. Therefore,

$$\mathbf{c} = \left(\sum_{k=0}^{K-1} \mathbf{z}_k {\mathbf{z}_k}^T\right)^{-1}\left(\sum_{k=0}^{K-1} \mathbf{z}_k d_k\right), \tag{13}$$

where the output vector

$$\mathbf{z}_n = \begin{pmatrix} z[n + K_2, n_o] \\ \vdots \\ z[n - K_2, n_o] \end{pmatrix}.$$

More details on this signal detector can be found in [14]. Figure 2.6 compares the BER performance of MMSE and MR rake combining for CM-3 and CM-4 channel models [17], respectively. The UWB system is the same as that for channel estimation (at the end of

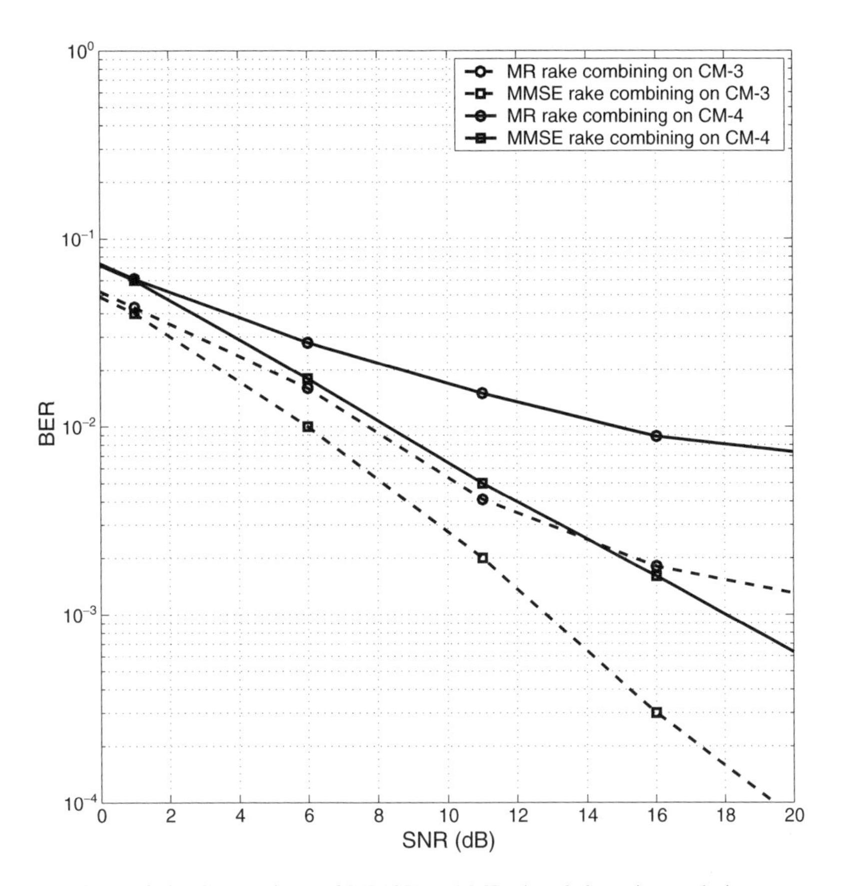

Figure 2.6 Comparison of MMSE and MR signal detection techniques

Section 2.2.3). The rake receiver for signal detector uses $K_1 = 10$ fingers implemented by analog device, and the equalizer has five taps ($K_2 = 2$). From the figure, we can see that the MMSE rake receiver has much better performance than the MR rake receiver. In particular, the required SNR for a 1 % BER has about 6 dB improvement for the CM-4 channel model.

2.3 OFDM-Based Modulation

In the previous section, we have described single-carrier–based modulation. To deal with delay spread, OFDM can be used in UWB communications. In this section, we address OFDM-based modulation, signal detection, and interference suppression.

2.3.1 Basic OFDM for UWB

The OFDM technique used in UWB is fundamentally the same as that in normal OFDM, which was first discussed in [18]. A baseband OFDM signal can be presented by

$$s(t) = \sum_{k=0}^{N-1} d_k e^{j2\pi f_k t}, \tag{14}$$

for $0 \leq t \leq NT_s$, where N is the number of subcarriers per OFDM block, which also equals the number of transmitted symbols per OFDM block; d_k represents the complex symbol to be transmitted; $f_k = f_0 + k\Delta f$ is the subcarrier frequency; and Δf is the frequency spacing between two adjacent subcarriers. To demodulate OFDM signals, it is required that $NT_s\Delta f = 1$, which is called the *orthogonal condition*.

When $f_o = 0$, it can be proved that the sampled version of the OFDM signal can be expressed as,

$$s(nT_s) = \sum_{k=0}^{N-1} d_k e^{j2\pi \frac{nk}{N}}.$$

Therefore, inverse discrete Fourier transform (DFT) can be used to convert the transmitted symbols into OFDM signal, which has an efficient implementation – fast Fourier transform (FFT).

With this condition, the transmitted symbol can be detected by

$$\begin{aligned}\frac{1}{NT_s}\int_0^{NT_s} s(t)e^{-j2\pi f_k t}dt &= \frac{1}{NT_s}\int_0^{NT_s}\left(\sum_{l=0}^{N-1} d_l e^{j2\pi f_l t}\right)e^{-j2\pi f_k t}dt \\ &= \sum_{l=0}^{N-1} d_l\left(\frac{1}{NT_s}\int_0^{NT_s} e^{j2\pi f_l t}e^{-j2\pi f_k t}dt\right) \\ &= \sum_{l=0}^{N-1} d_l\delta[l-k] = d_k. \end{aligned} \tag{15}$$

It can be also proved that the transmitted symbols can be obtained by considering DFT as $\{s(nT_s)\}_{n=0}^{N-1}$, the T_s-sampled version of OFDM signal, $s(t)$.

In the above discussion, we have ignored the channel's delay spread and noise. For UWB channels with delay spread, cyclic extension needs to be used. With proper cyclic extension, the demodulated signal can be expressed as [18]

$$\hat{r}_k = H_k d_k + n_k, \tag{16}$$

where H_k is the UWB channel's frequency response at the kth subcarrier, and n_k represents the effect of channel noise.

Once H_k in (16) is known, the transmitted symbol can be estimated. In the following sections, we will discuss channel estimation, and interference suppression for OFDM-based UWB.

2.3.2 Channel Estimation

Channel estimation for OFDM-based wireless communications has been investigated in [19] and the references therein. However, for OFDM-based UWB communications, a low-complexity channel estimate is desired. Here, we present the estimator developed in [20].

In UWB communications, channels can be regarded as being stationary or quasi-stationary. Therefore, channel tracking is not necessary. Channel estimation is performed while using the training sequence. During the training period, the transmitted symbols are known to the receiver, which can be used to get a temporal estimation of channel parameters by

$$\tilde{H}_k = \frac{r_k}{d_k} = H_k + \tilde{n}_k,$$

where

$$\tilde{n}_k = \frac{n_k}{s_k}.$$

Then, taking an inverse DFT (I-DFT), we have $\{\tilde{h}_n\}_{n=0}^{N-1} = \text{I-DFT}\{\tilde{H}_k\}_{k=0}^{N-1}$.

Let the delay profile of UWB channel be p_n, which is determined by the environment and can be estimated in practical situations. With the delay profile, it is possible to further reduce the noise level by defining a new estimate

$$\widehat{h}_n = \frac{p_n}{p_n + \Delta}\tilde{h}_n,$$

where Δ is noise (including interference) power and can be estimated by

$$\Delta = \frac{1}{N_2 - N_1 - 1}\sum_{n=N_1}^{N_2}\left|\tilde{h}_n\right|^2.$$

Figure 2.7 shows the MSE of the above channel estimator. The parameters of the OFDM-based UWB are described in [20]. It can be observed from the figure that the effect of channel estimation error on the system performance is negligible because the MSE is much smaller for this effect than that of the channel noise.

2.3.3 Interference Suppression

For an OFDM-based UWB system with sub-band repetition diversity [20] or receiver antenna arrays, the system also has the ability to suppress interference besides diversity gain. In this case, the received signal at the kth subcarrier can be expressed as

$$r_k^{(l)} = H_k^{(l)} d_k + n_k^{(l)}, \tag{17}$$

for $l = 1, \ldots, L$, where l is the antenna or sub-band index; $n_k^{(l)}$ includes AWGN and cochannel or multiuser interference, which is therefore not *spatially* white any more. The spatial correlation of $n_k^{(l)}$ can be used to improve the system performance by finding combining weights, $w_k^{(l)}$'s, to minimize

$$MSE(\mathbf{w}_k) = E\left|\mathbf{w}_k^H \mathbf{r}_k - d_k\right|^2, \tag{18}$$

where

$$\mathbf{r}_k = \begin{pmatrix} r_k^{(1)} \\ \vdots \\ r_k^{(L)} \end{pmatrix} \quad \text{and} \quad \mathbf{w}_k = \begin{pmatrix} w_k^{(1)} \\ \vdots \\ w_k^{(L)} \end{pmatrix}.$$

In reality, time average is used instead of ensemble average in (18) to find the combining weights. Approaches have been developed for single-carrier [21] and for OFDM [22].

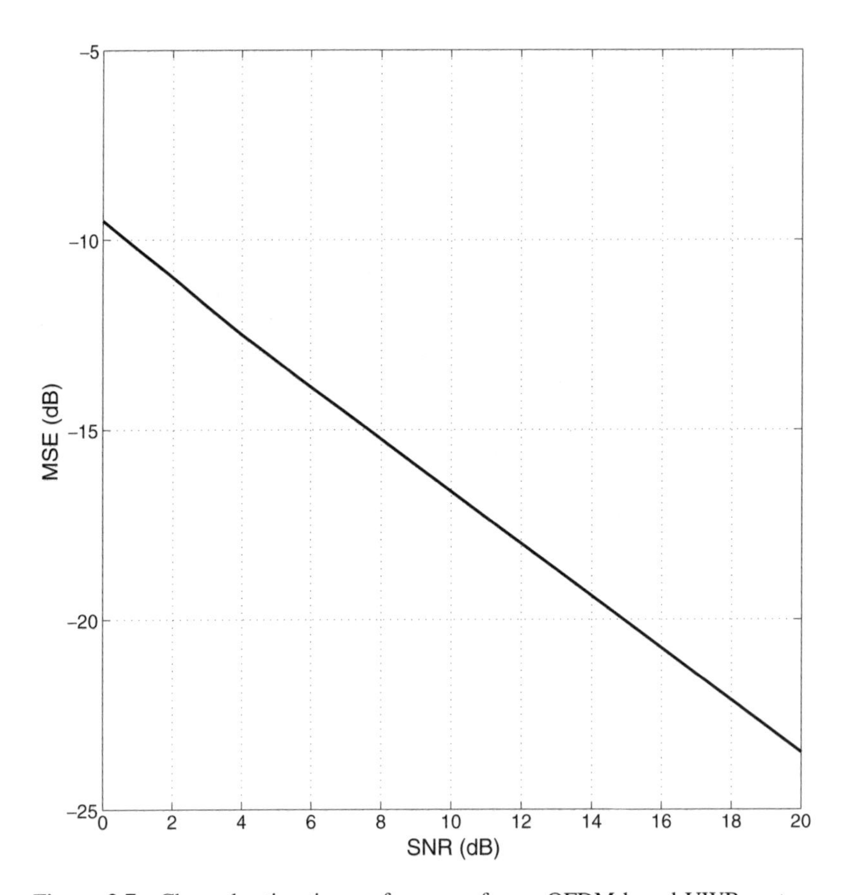

Figure 2.7 Channel estimation performance for an OFDM-based UWB system

However, as indicated in [20], these approaches cannot be directly used in OFDM-based UWB systems. Therefore, we introduce the approach in [20] here.

To exploit the correlation of UWB channel's frequency responses at different subchannels, we estimate combining weights to minimize the following cost function

$$C(\mathbf{w}_k) = \frac{1}{\sum_n \lambda^{|n-k|}} \sum_n \lambda^{|n-k|} \left| \mathbf{w}_k^H \mathbf{r}_n - d_n \right|^2, \tag{19}$$

where λ is a forgetting factor between 0 and 1, depending on the channel's frequency selectivity. Its value is smaller for channels with larger frequency selectivity. Direct calculation of (19) yields

$$\mathbf{w}_k = \mathbf{R}_k^{-1} \mathbf{a}_k,$$

where the correlation matrix and vector are defined respectively as,

$$\mathbf{R}_k = \frac{1}{\sum_n \lambda^{|n-k|}} \sum_n \lambda^{|n-k|} \mathbf{r}_n \mathbf{r}_n^H$$

and

$$\mathbf{a}_k = \frac{1}{\sum_n \lambda^{|n-k|}} \sum_n \lambda^{|n-k|} \mathbf{r}_n s_n^*.$$

Once the combining coefficients are found, the signals received from different antennas or sub-bands can be combined by

$$x_k = \mathbf{w}_k^H \mathbf{r}_k.$$

Figure 2.8 shows significant improvement of the above developed scheme over maximal ratio (MR) combining [20] for an OFDM-based UWB system similar to the one submitted for IEEE proposal [24]. In particular, it uses packet transmission, in which each packet contains 8096 (=2^{13}) bits. A rate-1/3 convolutional code with generator sequences 133, 145, and 171 is used. Consequently, the length of each code word is about 24 300 bits. The coded sequence is then converted into 12 150 (complex) QPSK symbols. Each OFDM block transmits 50 symbols; therefore, there are 243 OFDM data blocks in each slot. Another three OFDM blocks, each for one of the three sub-bands, are used for training.

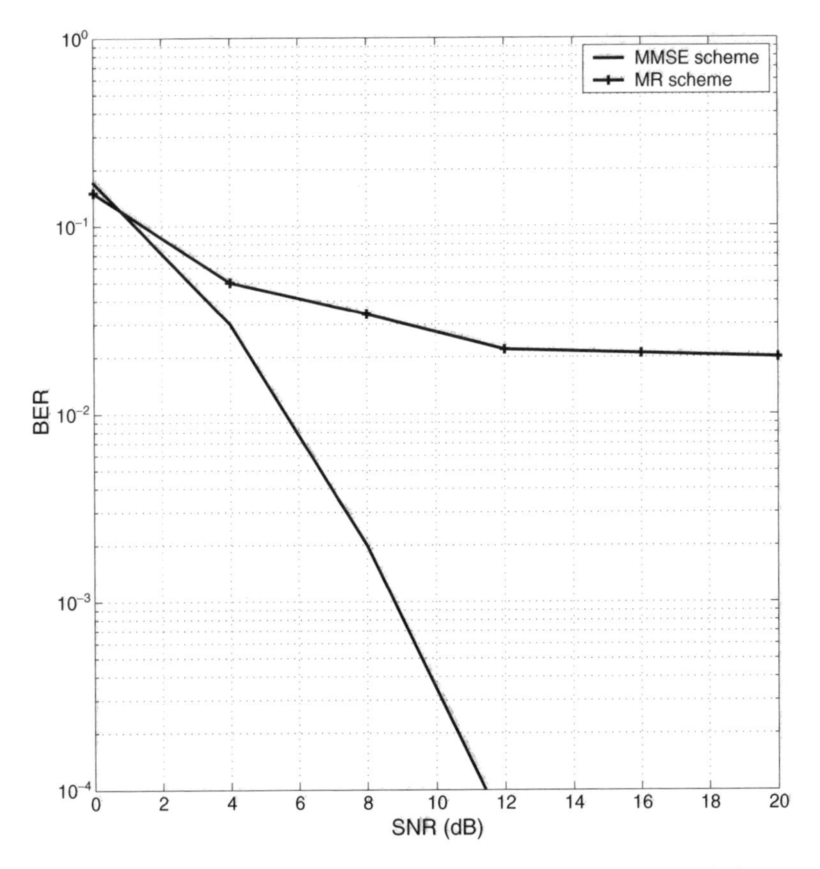

Figure 2.8 Performance of interference suppression for an OFDM-based UWB system

Hence, there are totally 246 OFDM blocks. Each consists of 128 tones, 100 of them being information tones, 12 of them being pilots, and the rest being null tones (transmitting no signal) for peak-to-average power ratio (PAPR) reduction. From the figure it can be observed that the system with MMSE combining can successfully suppress multiuser interference, while the system with MR combining cannot.

Besides the above OFDM-based UWB modulation, a modified version of OFDM has been also proposed in [3, 25] for multiple access of UWB systems.

2.4 Conclusion and Further Reading

This chapter has discussed the modulation and signal detection techniques in UWB communication. We have discussed the history of UWB communication, from the traditional impulse radio (for which PPM was the dominating modulation scheme) to the multiband-based UWB. Multiband UWB, more specifically UWB-OFDM, is gaining grounds as the favorable UWB scheme because of its ability to mitigate the effect of delay spread. However, PPM is still an ongoing research topic, and there is interest in improving its spectral efficiency.

The current drive for UWB technology is due to its applications. UWB promises high data rate at the unlicensed band, which will make more multimedia applications possible. UWB also provides an inherent security because of its the low power.

Many research works are active in UWB technology. The area of pulse shaping to improve PPM modulation for UWB has been studied in [9], [23], [26], and [27]. In [28] and [29], the capacity of UWB-OFDM systems has been studied. Among these topics are the search for better interference cancellation schemes for UWB systems with narrowband interferers and better pulse shapes to meet FCC spectral mask. As with innovative ideas, UWB is a very promising technology for wireless personal area networks.

References

[1] FCC Notice of Proposed Rule Making "Revision of part 15 of the commission's rules regarding ultrawideband transmission systems," ET-Docket 98–153, 1998.

[2] X. Gao, R. Yao, and Z. Feng, "Multi-band UWB System with Hadamard Coding," *IEEE Conference on Vehicular Technology*, Orlando, FL, vol. 2, pp. 1288–1292, Oct. 2003.

[3] E. Saberinia and A. H. Tewfik, "Multi-user UWB-OFDM Communications," *IEEE Pacific Rim Conference on Communications, Computers and Signal Processing*, Victoria, Canada, vol. 1, pp. 127–130, Aug. 2003.

[4] R. Harjani, J. Harvey, and R. Sainati, "Analog/RF Physical Layer Issues for UWB Systems," *Proc. 17th International Conference VLSI Design*, Mumbai, India, pp. 941–948, January 2004.

[5] J. Tang and K. K. Parhi, "On the Power Spectrum Density and Parameter Choice of Multi-Carrier UWB Communications," *37th Asilomar Conference on Signals, Systems and Computers*, Pacific Grove, CA, vol. 2, pp. 1230–1234, Nov. 2003.

[6] A. F. Molisch, Y. (G.), Li, Y. P. Nakache, P. Orlik, M. Miyake, Y. Wu, S. Gezici, H. Sheng, S. Y. Kung, H. Kobayashi, H. V. Poor, A. Haimovich, and J. Zhang, "A Low-Cost Time-Hopping Impulse Radio System for High Data Rate Transmission," *EURASIP J. Appl. Signal Process.*, vol. 1, pp. 397–412, Mar. 2005.

[7] I. Oppermann, L. Stoica, A. Rabbachin, Z. Shelby, and J. Haapola, "UWB Wireless Sensor Networks: UWEN – A Practical Example," *IEEE Commun. Mag.*, vol. 42, no. 12, pp. S27-S32, Dec. 2004.

[8] T. Sakamoto and T. Sato, "An Estimation Method of Target Location and Scattered Waveforms for UWB Pulse Radar Systems," *Proc. IEEE International Geoscience and Remote Sensing Symposium*, Toulouse, France, vol. 6, pp. 4013–4015, July 2003.

[9] U. Onunkwo and Y. G., Li, "On the Optimum Pulse-Position Modulation Index for Ultra-Wideband Communication," *Int. J. Dyn. Contin. Ser. B: Appl. Algorithms Spec. Issue UWB Commun.*, vol. 12, no. 3, pp. 353–362, June 2005, also in *Proc. IEEE Circuits and Systems, Symposium on Emerging Technologies: Frontiers of Mobile and Wireless Communications*, Shanghai, China, vol. 1, pp. 77–80, May-June 2004.

[10] X. Chen and S. Kiaei, "Monocycle Shapes for Ultra Wideband System," *IEEE International Symposium Circuits and Systems*, Scottsdale, AZ, vol. 1, pp. 597–600, May 2002.

[11] L. Michael, M. Ghavami, and R. Kohno, "Multiple Pulse Generator for Ultra-Wideband Communication Using Hermite Polynomial Based Orthogonal Pulses," *IEEE Conference on UWB Systems and Technologies*, Baltimore, MD, pp. 47–51, May 2002.

[12] G. M. Maggio, N. Rulkov, and L., Reggiani, "Pseudo-Chaotic Time Hopping for UWB Impulse Radio," *IEEE Trans. Circuits Syst.-I.*, vol. 48, no. 12, pp. 1424–1435, Dec. 2001.

[13] G. M. Maggio, D. Laney, F. Lehmann, and L. Larson, "A Multiple-Access Scheme for UWB Radio Using Pseudo-Chaotic Time Hopping," *IEEE Conference on UWB Systems and Technologies*, Baltimore, MD, pp. 225–230, May 2002.

[14] Y. (G.), Li, A. F. Molisch, and J. Zhang, "Channel Estimation and Signal Detection for UWB," *International Symposium on Wireless Personal Multimedia Communications*, Kanagawa, Japan, October 2003.

[15] V. Lottici, A. D'Andrea, and U. Mengali, "Channel Estimation for Ultra-Wideband Communications," *IEEE J. Sel. Areas Commun.*, vol. 20, no. 9, pp. 1638–1645, Dec. 2002.

[16] D. C. Cox, "Delay Doppler Characteristics of Multipath Propagation at 910 MHz in Suburban Mobile Radio Environment" *IEEE Trans. Antennas Propag.*, vol. 20, pp. 625–635, Sept. 1972.

[17] IEEE P802.15 Working Group for WPAN, "Channel Modeling Sub-committee Report Final," *IEEE P802.15-02/490rl-SG3a*, 7 February, 2003.

[18] S. B. Weinstein and P. M. Ebert, "Data Transmission by Frequency-Division Multiplexing Using the Discrete Fourier Transform," *IEEE Trans. Commun. Technol.*, vol. 19, no 5, pp. 628–634, Oct. 1971.

[19] Y. G. Li, L. J. Cimini Jr., and N. R. Sollenberger, "Robust Channel Estimation for OFDM Systems with Rapid Dispersive Fading Channels," *IEEE Trans. Commun.*, vol. 46, no. 7, pp. 902–915, July 1998.

[20] Y. G., Li, A. F. Molisch, and J. Zhang, "Practical approaches to channel estimation and interference suppression for OFDM based UWB communications," *Proc. IEEE Circuits and Sys. Symposium Emerging Tech.: Frontiers of Mobile and Wireless Commun.*, Shanghai, China, May–June 2004, vol. 1, pp. 21–24.

[21] J. H., Winters, "Signal Acquisition and Tracking with Adaptive Arrays in the Digital Mobile Radio System IS-136 with Flat Fading," *IEEE Trans. Veh. Technol.*, vol. 42, pp. 377–384, Nov. 1993.

[22] Y. G., Li and N., Sollenberger, "Adaptive Antenna Arrays for OFDM Systems with Cochannel Interference," *IEEE Trans. Commun.*, vol. 47, pp. 217–229, Feb. 1999.

[23] L. Bin, E. Gunawan, and L. C. Look, "On the BER Performance of TH-PPM UWB Using Parr's Monocycle in the AWGN Channel," *IEEE Conference on UWB Systems and Technologies*, Reston, VA, pp. 403–407, Nov. 2003.

[24] A., Batra, J. Balakrishnan, and A. Dabak, "Multiband-OFDM Physical Layer Proposal", IEEE P802.15-03/268r2, Nov. 2003.

[25] A. Batra, J. Balakrishnan, G. R. Aiello, J. R. Foerster, and A. Dabak, "Design of a Multiband OFDM System for Realistic UWB Channel Environments," *IEEE Trans. Microw. Theory Tech.*, vol. 52, no. 9, pp. 2123–2138, Sept. 2004.

[26] R. S. Dilmaghani, M. Ghavami, B. Allen, and H. Aghvami, "Novel UWB Pulse Shaping Using Prolate Spheroidal Wave Functions," *IEEE International Symposium on Personal, Indoor and Mobile Radio Communications*, Beijing, China, vol. 1, pp. 602–606, Sept. 2003.

[27] F. Ramirez-Mireles, "Signal Design for Ultra Wideband PPM Communications," *Proc. Military Communications Conference (MILCOM)*, Anaheim, CA, vol. 2, pp. 1085–1088, Oct. 2002.

[28] R. Gupta and A. H. Tewfik, "Capacity of Ultra-Wideband OFDM," *57th IEEE Vehicular Technology Conference*, Jeju, Korea, vol. 2, pp. 1420–1424, Apr. 2003.

[29] E. Saberinia and A. H. Tewfik, "Outage Capacity of Pulsed-OFDM Ultra Wideband Communications," *IEEE Conference on UWB Systems and Technologies '2004*, Kyoto, Japan, pp. 323–327, May 2004.

3

UWB Pulse Propagation and Detection

Robert Caiming Qiu

3.1 Introduction

Radio propagation is essential to wireless communications [1]. When a short pulse is used for a Ultra-wideband (UWB) system [2], some unique features occur [3]. When a short UWB pulse propagates through a channel, multiple pulses are received via multipath. This is true for both narrowband and UWB. However, unlike narrowband systems, in the case of UWB, these pulses in general have pulse shapes different from the incident short pulse. This phenomenon is called *pulse distortion*. Pulse distortion can be caused by frequency dependence of the propagation channel and antennas. The per-path impulse response is introduced to describe pulse distortion for each individual path. The impact of pulse distortion on the baseband transmission has been investigated in the past [3–11]. This chapter is intended to address these issues related to pulse signal detection of distorted pulses.

3.2 UWB Pulse Propagation

3.2.1 Generalized Multipath Model

On the basis of the Geometric Theory of Diffraction (GTD) framework and principle of locality, we establish the fact that the total response from a complex multipath channel is modeled by the sum of the impulse responses of local scattering centers [7, 9, 10]:

$$h(\tau, \theta, \varphi) = \sum_{n=1}^{N_{\mathrm{GO}}} a_n(\theta, \varphi)\delta(\tau - \tau_n) + \sum_{n=1}^{N_{\mathrm{D}}} h_n(\tau, \theta, \varphi) * \delta(\tau - \tau_n), \tag{1}$$

where N_{GO} and N_{D} represent the number of geometric optics rays and diffracted rays, respectively. The operation '*' denotes convolution in (1). The transient responses of the

Ultra-wideband Wireless Communications and Networks
Edited by Xuemin (Sherman) Shen, Mohsen Guizani, Robert Caiming Qiu and Tho Le-Ngoc

diffracted rays (the second term) can be obtained through exact, experimental, numerical, and asymptotic methods. A geometric optics ray can be treated as a generalized diffracted ray with $h_n(\tau, \theta, \varphi) = \delta(\tau, \theta, \varphi)$. In general, the per-path impulse response $h_n(\tau, \theta, \varphi)$ is obtained from the solutions of the time-domain Maxwell equations. It is analytically impossible to give the solution for an arbitrary generalized diffracted ray that is valid for arbitrary configurations. Rather, we are interested in the expressions valid for a large class of such rays. These expressions are sufficient for most configurations in our engineering applications. Fortunately, we can define the field in the neighborhood of the singularity $\tau = \tau_\alpha$ only. The field component of a generalized ray has the behavior

$$h_n(\tau) = \begin{cases} \xi(\tau_\alpha - \tau)\sum_{n=0}^{\infty} \frac{C_n}{n!}(\tau_\alpha - \tau)^n, \tau < \tau_\alpha \\ \eta(\tau_\alpha - \tau)\sum_{n=0}^{\infty} \frac{D_n}{n!}(\tau_\alpha - \tau)^n, \tau > \tau_\alpha \end{cases},$$

then the corresponding transfer function has the form [7, 9, 10]

$$H_n(\omega) = \sum_{n=0}^{\infty} \left\{ \frac{D_n}{n!}\frac{1}{(j\omega)^n} \int_0^{\infty} \eta\left(\frac{t}{j\omega}\right) t^n e^{-t}\, dt - \frac{C_n}{n!}\frac{1}{(-j\omega)^n} \int_0^{\infty} \xi\left(\frac{t}{-j\omega}\right) t^n e^{-t}\, dt \right\} \tag{2}$$

where $\xi(\tau)$ and $\eta(\tau)$ are rather general and need definition for a specific configuration. Note that these two functions are only dependent on the local properties of a ray such as the edge of a building. For the important special case in which both $\xi(\tau)$ and $\eta(\tau)$ are $\tau^{1/2}$, the asymptotic series contains fractional powers of $(1/\omega)$, for example, $1/\omega^{n+\frac{1}{2}}$ where n is an integer. For $\omega \to \infty$, each term of the series vanishes. This is as expected, for there is no classical diffracted geometric optics field. The leading term for large ω may well serve as the geometric optics diffracted field. For a special case with an integer n, via inverse Laplace transform we obtain

$$h_n(\tau) = \begin{cases} \sum_{n=0}^{\infty} \frac{D_n}{n!}(\tau - \tau_0)^n, \tau > \tau_0 \\ 0, \tau < \tau_0 \end{cases} \qquad H_n(\omega) = \sum_{n=0}^{\infty} \frac{D_n}{(j\omega)^{n+\frac{1}{2}}}\Gamma\left(n + \frac{3}{2}\right). \tag{3}$$

Equation (3) is valid for a special class of rays. All the rays diffracted by a single wedge as well as multiply diffracted by several wedges can be modeled by (3). This model is sufficiently general for commonly encountered shapes predicted by GTD or UTD. Heuristically, we can postulate that a more generalized class of rays has the per-path frequency response of $(j\omega)^{\alpha_n}$, where α_n is a real number. This generalization was first postulated a decade ago [12, 13]. If a slow-varying frequency response $H_n(\omega)$ can be fitted approximately by the curve of $(j\omega)^{\alpha_n}$ for the nth path, this postulation can be assumed valid. The justification of this postulation is its mathematical simplicity.

On the basis of the above postulation, the generalized model of a form can be expressed as

$$H(\omega) = \sum_{n=1}^{N} A_n (j\omega)^{\alpha_n} e^{j\omega\tau_n} \qquad h(\tau) = \sum_{n=1}^{N} \frac{A_n}{\Gamma(-\alpha_n)} \tau^{-(1+\alpha_n)} U(\tau) * \delta(\tau - \tau_n), \tag{4}$$

where α is a real value. This mode is asymptotically valid for incident, reflected, singly diffracted, and multiply reflected/diffracted ray path field. For practical problems of pulses propagating through walls and buildings, a more generic function like the one given in (4) is needed. Note that the parameter α_n can be a random variable, as first suggested in [13]. We are measuring this function in the lab. The results will be reported elsewhere. The motivation of using the function in (4) is its simplicity and feasibility for a large category of problems. This form can allow us to use the powerful mathematical tool of fractional calculus [14] to conveniently describe our problems.

3.2.2 IEEE 802.15.4a Channel Model

A special form of (4) is adopted in the IEEE 802.15.4a channel model [15, 16]. The frequency dependency of each path is assumed to be

$$H_n(j\omega) = (j\omega)^{\alpha}, \tag{5}$$

where α lies between 0.8 and 1.4 (including antennas effects), while excluding antennas effects, α is found to be between -1.4 (in industrial environments) and $+1.5$ (in residential environments). The first suggestion of using (4) was made in [13]. Alternative modeling of the frequency dependency of the transfer function for each path includes a frequency-dependent pathloss exponent $n(f)$ and an exponential dependency $\exp(-\chi f)$, with χ between 1.0 (LOS) and 1.4 (NLOS) [16].

3.3 UWB Pulse Signal Detection

3.3.1 Optimum Receiver

This section investigates the performance of the mismatched filter in the receiver [4]. For convenience, the generalized channel model (1) is *postulated* as

$$h(\tau) = \sum_{n=1}^{L} A_n h_n(\tau) * \delta(\tau - \tau_n), \tag{6}$$

where L generalized paths are associated with amplitude A_n, delay τ_n, and per-path impulse response $h_n(\tau)$. The $h_n(\tau)$ represents an arbitrary function that has finite energy. Equation (6) together with (4) is sufficient for most practical use. When a pulse $p(t)$ passes through the path given by (4), the distorted pulse is given by

$$y_n(t) = p(t) * h_n(t) = \left(\frac{d}{dt}\right)^{\alpha_n} p(t), \tag{7}$$

where $\left(\frac{d}{dt}\right)^{\alpha_n}$ is the fractional differential of the $p(t)$. Fractional calculus is a powerful tool in calculation and manipulation [14].

One goal of this chapter is to investigate how pulse distortion would affect the system performance if no compensation for this kind of distortion is included. To achieve this, we first present a general system model and give its performance expression. In Figure 3.1, we present a general binary baseband system model for the physics-based signals. The

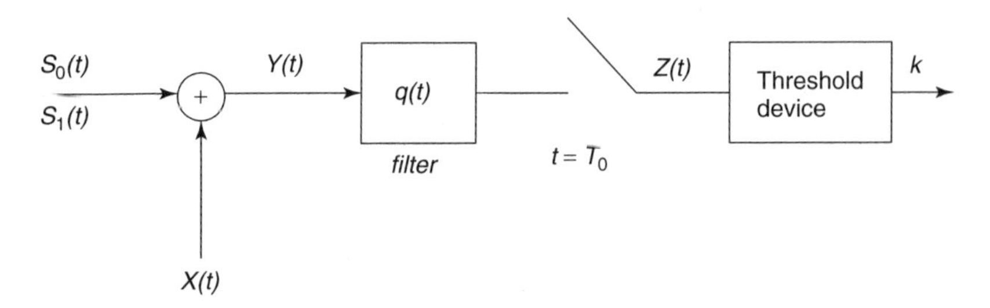

Figure 3.1 General model for binary baseband data transmission

signals considered are given by

$$s_0(t) = A\psi_0(t), \quad s_1(t) = A\psi_1(t) \tag{8}$$

where $\{\psi_0, \psi_1\}$ is a binary signal set. Here, ψ_0 and ψ_1 are finite energy[1], time-limited signals of duration T. (When a pulse $p_i(t)$, $i = 0, 1$, is transmitted at the transmitter, the received signal $s_i(t) = p_i(t) * h(t)$, $i = 0, 1$, where $h(t)$ is the impulse response of the channel, as given in the next section.)

The received signal $Y(t)$ is the sum of the noise $X(t)$ and the signal $s_i(t)$, where $i = 0$ if the binary digit '0' is sent and $i = 1$ if the binary digit '1' is sent. The filter shown in Figure 3.1 is a time-invariant linear filter with impulse response $q(t)$. The output of this filter, which is denoted as $Z(t)$, is sampled as time T_0. The output $Z(T_0)$ of the sampler is then compared with an arbitrary threshold γ in order to make a decision between two alternatives 0 and 1. In Figure 3.1, the noise $X(t)$ is an additive Gaussian noise (AGN) channel, not necessarily white. The channel noise is stationary with zero mean and is independent of the input to the receiver. The filter $q(t)$ is not necessarily matched to the signal $s_i(t)$. It is the optimization of the $q(t)$ that motivates this research.

Following the steps of [4, 17], we can obtain the probability of error. The probability of error when 0 is sent and when 1 is sent is respectively given by

$$P_{e,0} = Q\left(\left[\mu_0(T_0) - \gamma\right]/\sigma\right) \tag{9}$$

and

$$P_{e,1} = Q\left(\left[\gamma - \mu_1(T_0)\right]/\sigma\right) \tag{10}$$

where $\mu_i(t) = s_i(t) * q(t)$, $i = 0, 1$, and $\sigma^2 = (q(t) * q(-t) * R_X(t))\,|_{t=0}$. $Q(x)$ is defined as $Q(x) = \int_x^\infty \frac{1}{\sqrt{2\pi}} \exp(-y^2/2)\,dy$. $R_X(t)$ is the autocorrelation of the AGN $X(t)$. For the AWGN noise, we have $\sigma^2 = \frac{1}{2}N_0 \int_{-\infty}^{\infty} q^2(t)\,dt$, where $\frac{1}{2}N_0$ is the spectral density for the white noise process $X(t)$.

In (9 10), the γ is arbitrary. When the Bayes decision criterion is used [17], the threshold is given by

$$\bar{\gamma} = \frac{\mu_0(T_0) + \mu_1(T_0)}{2} + \frac{\sigma^2 \ln(\pi_1/\pi_0)}{\mu_0(T_0) - \mu_1(T_0)}, \tag{11}$$

where π_0 and π_1 are the probabilities that 0 and 1 are sent, respectively.

[1] The finite-energy condition is important. We will find that the received physics-based signals satisfy this condition. Although the signals are singular at some points, their energy is finite.

The average probability is thus given by

$$\begin{aligned}\bar{P}_e &= \pi_0 P_{e,0}(\bar{\gamma}) + \pi_1 P_{e,1}(\bar{\gamma}) \\ &= \pi_0 Q\left[SNR - (2SNR)^{-1}\ln(\pi_1/\pi_0)\right] + \pi_1 Q\left[SNR + (2SNR)^{-1}\ln(\pi_1/\pi_0)\right], \quad (12)\end{aligned}$$

where the SNR is the signal-to-noise ratio at the input to the threshold device, given by

$$SNR = \frac{\mu_0(T_0) - \mu_1(T_0)}{2}. \quad (13)$$

When 0 and 1 are sent with equal probability, then $\pi_0 = \pi_1 = 1/2$. As a result, the average probability is reduced to

$$\bar{P}_e = Q(SNR). \quad (14)$$

Now we can optimize $q(t)$ in terms of SNR defined by

$$SNR = \frac{(s_0(t) * q(t))\,|_{t=T_0} - (s_1(t) * q(t))\,|_{t=T_0}}{\sqrt{2N_0}||q||}, \quad (15)$$

where $||q|| = \left(\int_{-\infty}^{\infty} q^2(t)\,dt\right)^{1/2}$ is the norm of $q(t)$. $s_0(t) = p_0(t) * h(t)$ and $s_1(t) = p_1(t) * h(t)$ can be singular but their energies are limited, as will be observed in (35).

Equations (12) and (14) are valid when the filter impulse response, $q(t)$, is arbitrary. The receiver optimum filter is known to be the matched filter that is matched to the signal $s_0(t)$ and $s_1(t)$, the pulse waveforms received at the receiver antennas. However, sometimes our receivers are designed to be matched to $p_0(t)$ and $p_1(t)$, the pulse waveforms at the transmitter. This mismatch practice will result in performance degradation in terms of SNR, which will be illustrated by numerical results.

UWB antennas usually distort the transmitted pulse [3]. The transmitter and receiver antennas as a whole can be modeled as a linear filter with impulse response, $h_a(t)$. Their impact can be absorbed in the new transmitted pulse waveforms, $\tilde{p}_0(t) = p_0(t) * h_a(t)$, $\tilde{p}_1(t) = p_1(t) * h_a(t)$. So, the received signals are $\tilde{s}_0(t) = \tilde{p}_0(t) * h(t)$ and $\tilde{s}_1(t) = \tilde{p}_1(t) * h(t)$. To simplify the analysis, we assume $h_a(t) = \delta(t)$. In other words, we ignore the antennas' impact at this point and will report elsewhere.

3.3.2 Generalized RAKE Receiver

All the signal processing algorithms require knowledge of the channel parameters in order to detect the signal. The channel must be estimated prior to the actual detection. We use a data-aided (DA) approach [18, 19] where the data frame begins with a sequence of known data, the so-called *pilot signal*. The generalized RAKE receiver can be used to replace the matched filter in the presence of intersymbol interference [8] and multiuser detection [5]. To avoid the channel estimation of the generalized RAKE, we can use the autocorrelation receiver based on Transmitted Reference [20].

3.3.2.1 Two-Dimensional Tap-Delayed Line Model

The key of our generalized RAKE is to use an FIR filter to represent the per-path impulse response $h_n(\tau)$ in (6):

$$h_n(\tau) = \sum_{m=1}^{M} \beta_{mn}\delta(\tau - \tau_{mn}). \tag{16}$$

This FIR filter has M taps with tap spacing T_s. The received signal is sampled every T_s seconds. Note that an FIR representation of pulse distortion was used in channel modeling [21]. As a result, we obtain many (say M) discrete taps for each generalized path. Equation (6) is rewritten as

$$h(\tau) = \sum_{n=1}^{P}\sum_{m=1}^{M} \tilde{a}_{mn}\delta(\tau - \tilde{\tau}_{mn}) \tag{17}$$

where $\tilde{a}_{mn} = A_n\beta_{mn}$ is the real amplitude of each tap corresponding to τ_{mn}. With mapping, we can reduce the two-dimensional model to a one-dimensional discrete model

$$h(\tau) = \sum_{l=1}^{L} a_l\delta(\tau - \tau_l), \tag{18}$$

where

$$\begin{aligned}
&L = MP \\
&\tau_l = \tau_{[m+(n-1)M]} = \tilde{\tau}_{mn}, \\
&a_l = \tilde{a}_{mn} = A_n\beta_{mn}, \\
&l = m + (n-1)M, m = 0, 1, \ldots, M, n = 0, 1, \ldots, P.
\end{aligned}$$

The one-dimensional discrete model can be handled using conventional channel estimation algorithm that is used for a narrowband system. Thus, the FIR representation reduces our channel estimation for the generalized RAKE to that of the conventional RAKE.

3.3.2.2 Optimal ML Channel Estimation

Following steps of [5, 18, 19], the received signal can be expressed as

$$r = Da + \eta, \tag{19}$$

where

$$[D]_{jl} = \sum_{i=0}^{N_P-1} b_P(i)g_k^0(jT_s - iT - \tau_l)$$

η is AWGN with two-sided spectral density $N_0/2$ and a is the vector of the channel amplitude a_l defined in (18). g_k^0 is the spreading waveform with duration T. For multiuser detection, g_k^0 is given by (38). We assume that a frame consists of N_p known pilot symbols b_P. The received signal $\boldsymbol{r}$ is Gaussian with mean $\boldsymbol{Da}$ and covariance matrix $\boldsymbol{C}$

that has terms of noise variance on its diagonal and zeros elsewhere. The optimal channel estimation is to maximize a function $\Lambda(a, \tau)$ given by

$$\Lambda(a, \tau) = 2r'C^{-1}Da - a'D'C^{-1}Da \tag{20}$$

where τ is a vector of channel path delays τ_l corresponding to amplitudes a_l. The search for the optimum is complex and we will use a suboptimum algorithm in the following.

3.3.2.3 Successive Channel (SC) Estimation

We can use the above optimal channel estimation for a one-tap channel. The estimated delay and amplitude are

$$\hat{\tau} = \arg\max \left\{ \frac{\left|\xi'(\tau)C^{-1}r\right|^2}{\xi'(\tau)C^{-1}\xi(\tau)} \right\} \tag{21}$$

$$\hat{a} = \frac{\xi'(\hat{\tau})C^{-1}r}{\xi'(\hat{\tau})C^{-1}\xi(\hat{\tau})} \tag{22}$$

where

$$[\xi(\tau)]_m = \sum_{i=0}^{N_P-1} b_P(i)g_k^0(mT_s - iT - \tau), \qquad 1 \le m \le M$$

where $g_k^0(t)$ has a duration T. The above scheme can be performed iteratively for the multipath channel defined in (18). The algorithm is summarized by the following four steps in [5, 18, 19].

1. Initialization: Set *threshold* and $c(\tau) = 0$ for $\tau_{\min} \le \tau \le \tau_{\max}$.
2. Perform the search for the strongest tap $\hat{\tau}$ and calculate $\hat{\alpha}$ by using the above equations,

$$c(\hat{\tau}) \Longleftarrow c(\hat{\tau}) + \hat{\alpha},$$

$$r \Longleftarrow r - \hat{\alpha}\xi\left(\hat{\tau}\right).$$

3. If $\hat{\alpha} \ge$ *threshold*, go to step 2; otherwise set $\hat{h}(\tau) = c(\tau)$ and stop.

Using the above SC estimation algorithm, the channel impulse response $\hat{h}(\tau)$ is obtained. With (16) and (17), the FIR representation of the per-path impulse response is estimated as $\hat{h}_n(\tau)$. For the nth path, the pulse waveform is $q_n(\tau) = p(\tau) * \hat{h}_n(\tau)$. Let us consider two cases:

(1) If one tap is used in (16), $h_n(\tau) = \hat{\beta}_{1n}\delta(\tau - \hat{\tau}_1)$, and thus $q_n(\tau) = \hat{\beta}_{1n}p(\tau - \hat{\tau}_1)$. So the matched filter can be implemented with an impulse response of $p(\tau)$, the transmitted pulse waveform. This special case is just the conventional RAKE receiver used in narrowband and UWB scenario.

(2) If several taps (say three) are used in (16), then $\hat{h}_n(\tau) = \sum_{m=1}^{M} \hat{\beta}_{mn}\delta(\tau - \hat{\tau}_{mn})$ and thus the received pulse waveform for the nth path is estimated as

$$\hat{q}_n(\tau) = \sum_{m=1}^{M} \hat{\beta}_{mn} p(\tau - \hat{\tau}_{mn}) \tag{23}$$

The matched filter for each user should be designed to match $\hat{q}_n(\tau)$, instead of $p(\tau)$. In other words, the front-end filter impulse response is equal to $\hat{q}_n(\tau)$, not $p(\tau)$. The generalized RAKE structure is obtained here.

3.3.3 Optimum Receiver with Intersymbol Interference

We will incorporate the per-path impulse response into the optimal receiver when ISI is present [3, 8]. Without pulse distortion, the work in presence of ISI is done in [22, 23]. In the following framework, we pay special attention to the $h_n(\tau)$. For Pulse Amplitude Modulation (PAM) signals in a single-user system, the transmitted signal is expressed as

$$s(t) = \sum_{n=-\infty}^{\infty} b_n x(t - nT_s), \tag{24}$$

where b_n represents the nth discrete information symbol with duration of T_s. The received signals are represented as

$$r(t) = \sum_{k=0}^{\infty} b_n y(t - nT_s) + n(t), \tag{25}$$

where $n(t)$ is AWGN. It has been shown for a narrowband system [24] that the optimum receiver structure is illustrated in Figure 3.2.

The received signal $r(t)$ first passes the matched filter followed by a sampler with a sampling rate of $1/T_s$. The sampled sequence is further processed by a maximum-likelihood sequential estimator (MLSE) detector that was first studied by Forney (1972). This receiver structure is optimal in terms of minimizing the probability of transmission error in detecting the information sequence. If in the UWB receiver design we assume

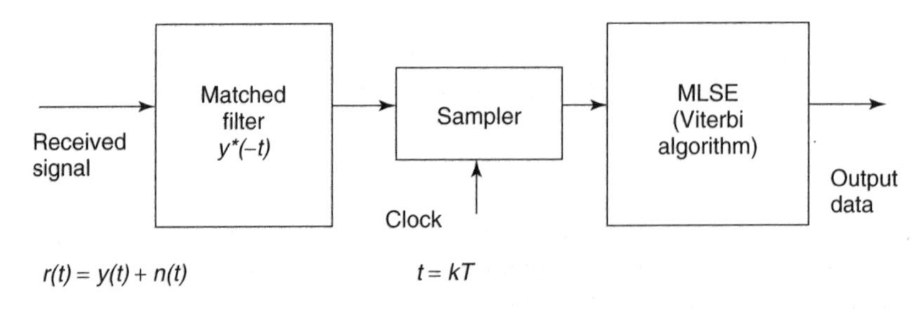

Figure 3.2 Optimal receiver structure in presence of ISI in an AWGN channel

that $y(t)$ is of finite energy, the structure of Figure 3.2 can immediately be applied to our UWB problem [3, 8]. The output of the matched filter is expressed as

$$q(t) = \sum_{n=0}^{\infty} b_n R_{yy}(t - nT_s) + v(t), \tag{26}$$

where

$$R_{yy}(t) = y(t) \otimes y^*(-t) = R_{xx}(t) \otimes h(t) \otimes h^*(-t) \tag{27}$$

is the autocorrelation of $y(t)$, and $v(t)$ is the response of the matched filter to AWGN noise $n(t)$. $R_{xx}(t)$ is the autocorrelation of $x(t)$. Denote

$$R_{k-n} = R_{yy}(t - nT_s)|_{t=kT_s}. \tag{28}$$

When the matched filter $y^*(-t)$ is replaced by a generalized RAKE receiver described in the previous text, the performance will be degraded. When this is done, the rest of detection remains the same.

After the sampler in Figure 3.2, the discrete signals at times $t = kT_s$ of (26) are given by

$$q_k = \sum_{n=0}^{\infty} b_n R_{k-n} + v_k, k = 0, 1, 2, \ldots. \tag{29}$$

R_0 (the energy of the nth symbol) can be regarded as an arbitrary scale factor and conveniently set to 1; from (29) we obtain

$$q_k = b_k + \sum_{n=0, n \neq k}^{\infty} b_n R_{k-n} + v_k, \tag{30}$$

where b_k term represents the expected information symbol of the kth sampling period, the 2nd term is the ISI, and v_k is the additive Gaussian variable at the kth sampling point. The sequence of q_k will be further processed by MLSE (Viterbi algorithm) before being sent to decision circuits.

If the duration of each symbol is smaller than the multipath spread T_D, that is, $T_s > T_D$, these is no ISI; thus, $R_{k-n} = 0$ for $n \neq k$. Then, (30) reduces to

$$q_k = b_k + v_k. \tag{31}$$

It is, therefore, sufficient to detect the symbols independently one by one for each given instant $t = kT_s$. This is the famous matched filter for isolated symbol detection. As expected, once matched, the detection depends on the energy $R_0 = R_{yy}(0)$ of the symbol, not the received composite pulse waveform $y(t)$. Thus, per-path pulse distortion has no impact on the single symbol matched filter detector. However, according to (30), we will find that this is not the case when ISI is present. We will evaluate this effect using some closed-form expressions for system performance.

If the number of the overlapping symbols is less than 10, the MLSE (Figure 3.2) may be even feasible for the state-of-the-art signal processing capability.

3.3.3.1 Suboptimum Detection of Physics-Based Signals

Let us simplify optimum receiver structures in two directions: simplifying matched filter and suboptimum decisions. First, the matched filter can be simplified in a suboptimum manner. Since the per-path pulse waveform has been included in the channel impulse, the resultant matched filter is sometimes called '*Generalized RAKE*' receiver structure [5, 6]. When per-path pulse distortion is included, the suboptimum implementation of the matched filter is the RAKE receiver structure. In the absence of per-path pulse distortion, the matched filter is identical to the RAKE receiver of Price & Green (1958). Different suboptimum structures can be used to approximate the front end or the mathematical operation of $R_{yy}(t)$ in (29) [20, 22, 23].

Let us follow Kailath and Poor [25] and restrict ourselves to a linear detector with a matched filter front end. The detected bits are

$$\hat{b}_n = sign(z_n), \tag{32}$$

where $z = My$, matrix M is arbitrary, and vector y has elements of $y(t)|_{t=nT_s}$. Further, we assume b_n to take on the values of ± 1 with equal probabilities. For zero-forcing, the probability that b_n is in error is simply

$$P_e(n) = Q\left(\frac{a_n}{\sigma\sqrt{(R^{-1})_{n,n}}}\right), \tag{33}$$

where σ is the noise density, matrix $R = [R_{k-n}]$ defined in (29), and a_n^2 is the energy of $y(t - nT_s)$.

Under some general conditions, the error probability of the linear MMSE detector

$$P_e(n) \approx Q\left(\frac{a_n M_{n,n}}{\sigma\sqrt{(MRM)_{n,n}}}\right), \tag{34}$$

where $M = (R + \sigma^2 D_a^{-2})^{-1}$ and $D_a = \text{diag}(a_n)$.

It is observed from (33) and (34) that per-path pulse distortion affects the system performance through the output of the matched filter. The time-domain overlapping of distorted pulses make the matrix R nondiagonal. When R is diagonal, the per-path pulse distortion affects each simple energy separately. Interpulse overlapping combined with per-path pulse distortion makes receiver design very challenging.

To gain some insights, let us consider a two-symbol case. Denote

$$R^{-1} = \begin{bmatrix} 1 & \rho \\ \rho & 1 \end{bmatrix}^{-1} = \frac{1}{1-\rho^2}\begin{bmatrix} 1 & -\rho \\ -\rho & 1 \end{bmatrix}, \tag{35}$$

where $\rho = R_1 = R_{-1} = R_{yy}(\tau)|_{\tau=\pm T_s} \leq 1$.

In (33), the performance only depends on the diagonal elements of R^{-1}, which are inversely proportional to $1 - \rho^2$. Compared with the isolated single symbol detector, the performance loss in SNR for zero-forcing detector can be defined as

$$\beta_n \triangleq 1/\left(R^{-1}\right)_{n,n} \tag{36}$$

and $\beta_n = 1 - \rho^2$ for two-symbol case. Thus, β_n depends on per-path pulse distortion $h_n(\tau)$ through $R_{yy}(\tau)$ in (35) and (29). The β_n depends on the whole physics-based channel model and symbol rate.

3.3.4 Receiver with Time-Reversal Channel Impulse Response

The optimum receiver structure of Figure 3.2 is used as a heuristic approach to derive some new structures. In Figure 3.3, we show the autocorrelation of the channel impulse response, $R_{hh}(t) = h(t) \otimes h(-t)$, for a typical IEEE 802.15.3a model. The $R_{hh}(t)$ is normalized by $R_{hh}(0)$. The plot shows $R_{hh}(t)/R_{hh}(0)$. It is interesting to notice that the first peak has an energy that is about 50 %. We can look at each pulse as a chip for a spread-spectrum system, and the channel impulse response can be regarded as a pseudonoise (PN) code where the width of multipath delay spread corresponds to the length of the PN code. This observation first made in [26] intuitively suggests that it is simpler to base the detection on $R_{hh}(t)$, rather than the noisy $h(t)$.

As given in (29), the output of the matched filter is

$$R_{yy}(t) = R_{xx}(t) \otimes R_{hh}(t). \tag{37}$$

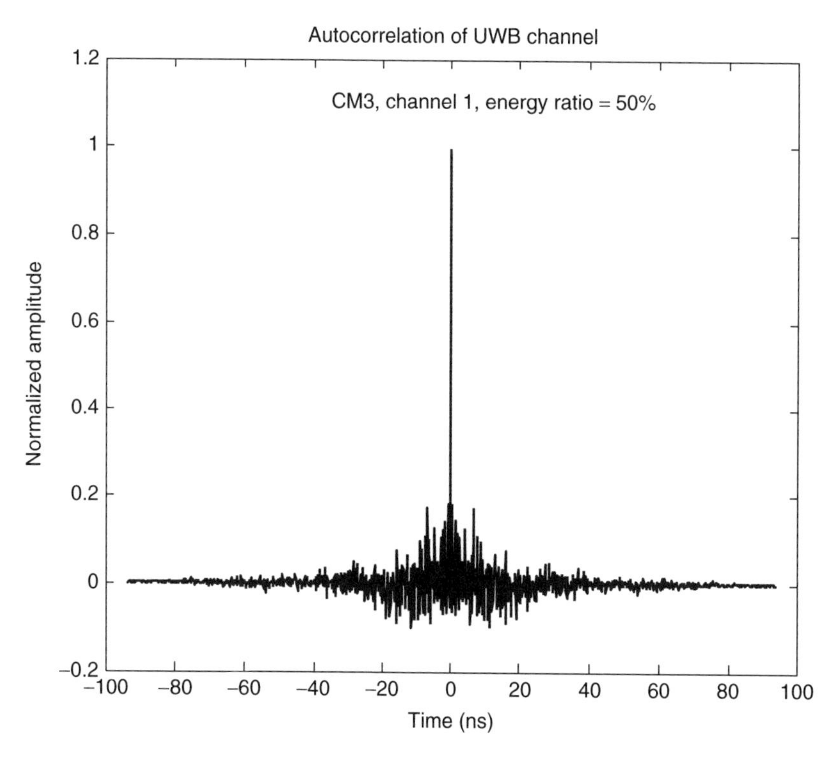

Figure 3.3 Autocorrelation of a channel impulse response $R_{hh}(t) = h(t) * h(-t)$ for IEEE 802.15.3a CM3 channel

$R_{xx}(t)$ is typically of a smooth shape. Finally, a time-reversal scheme is used to shift the design complexity from the receiver to the transmitter [26]. In time reversal, a signal is precoded in such a way that it focuses both in time and in space at a particular receiver. Because of temporal focusing, the received power is concentrated within a few taps and the task of equalizer design becomes much simpler than without focusing. If we use the prefilter (at the transmitter), which has an impulse response of $h(t)$, the output of the matched filter (at the receiver) is $R_{yy}(t) = R_{xx}(t) \otimes R_{hh}(t)$. If the first tap is used (shown in Figure 3.3), the collected energy is about 50 % of the total energy. Implementation of the prefilter can be simplified using the mono-bit A/D technology [23].

3.3.5 Optimum Receiver with Multiuser Detection

Let us consider a DS-CDMA UWB channel that is shared by K simultaneous users. Each user is assigned a signature waveform $g_k^0(t)$ of duration T, where T is the symbol interval. A transmitted signature waveform for the kth user may be expressed as

$$g_k^0(t) = p(t) * \sum_{n=0}^{N_c-1} c_k(n)\delta(t - nT_c), \quad 0 \le t \le T, \tag{38}$$

where $\{c_k(n), 0 \le n \le N_c - 1\}$ is a PN code sequence consisting of N_c chips that take values $\{\pm 1\}$, $p(t)$ is a pulse of duration T_c, the chip interval. Without loss of generality, we assume that K signature waveforms have unit energy.

For simplicity, we assume that binary antipodal signals are used to transmit the information from each user. Consider a block of N consecutive bits for each user in an observation window. Let the information sequence of the kth user be denoted by $b_k(m)$, where the value of each information bit may be ± 1. It is convenient to consider the transmission of a block of some arbitrary length, say N. The data block from the kth user is

$$b_k = \sqrt{E_k}[b_k(1) \ldots b_k(N)]^T, \tag{39}$$

where E_k is the transmitted energy of the kth user for each bit. The transmitted waveform is

$$x_k(t) = \sqrt{E_k} \sum_{i=1}^{N} b_k(i) g_k^0(t - iT). \tag{40}$$

The composite transmitted signal for the K users is

$$x(t) = \sum_{k=1}^{K} x_k(t - T_k) = \sum_{k=1}^{K} \sqrt{E_k} \sum_{i=1}^{N} b_k(i) g_k^0(t - iT - T_k), \tag{41}$$

where T_k are the transmission delays, which satisfy the condition $0 \le T_k \le T$ for $1 \le k \le K$. Without loss of generality, we assume that $0 \le T_1 \le T_2 \le \cdots \le T_k < T$. This is the model in an asynchronous mode. For synchronous mode, $T_k = 0$ for $1 \le k \le K$. We assume that the receiver knows T_k.

At the receiver end, the corresponding equivalent low-pass, received waveform may be expressed as

$$r(t) = y(t) + n(t), \tag{42}$$

where $n(t)$ is AWGN, with power spectral density of $\frac{1}{2}N_0$. The received signal is

$$y(t) = \sqrt{E_k} \sum_{k=1}^{K} \sum_{i=1}^{N} b_k(i) g_k(t - iT - T_k), \tag{43}$$

where $g_k(t)$ is the received signature waveform given by

$$g_k(t) = g_k^0(t) * h^{(k)}(t) = \left[p(t) * h^{(k)}(t)\right] * \sum_{n=0}^{N_c-1} c_k(n)\delta(t - nT_c), \tag{44}$$

where $h^{(k)}(t)$ is the impulse response of the kth user given in (6). The received signature waveform is given by $g_k(t)$, in contrast to the transmitted signature waveform $g_k^0(t)$. The pulse response of the front-end filter (used in forming $g_k(t)$) is denoted by $y^{(k)}(t) = p(t) * h^{(k)}(t)$.

When $h^{(k)}(t) = \delta(t), \forall k$, (44) reduces to the conventional case [24, 27–29] and the conventional RAKE is thus reached. In simulations, the estimated channel impulse response, $\hat{h}^{(k)}(t)$, will be used to replace the $h^{(k)}(t)$. With pulse distortion included in $\hat{h}^{(k)}(t)$, we call our new receiver structure (in Figure 3.4) the generalized RAKE structure.

The optimum receiver is defined as the receiver that selects the most probable sequence of bits $\{b_k(n), 0 \leq n \leq N, 1 \leq k \leq K\}$ given the received signal $r(t)$ observed over the time interval $0 \leq t \leq NT + 2T$.

The cross-correlation between pairs of signature waveforms play an important role in the metrics for the signal detector and on the performance. The pulse distortion affects

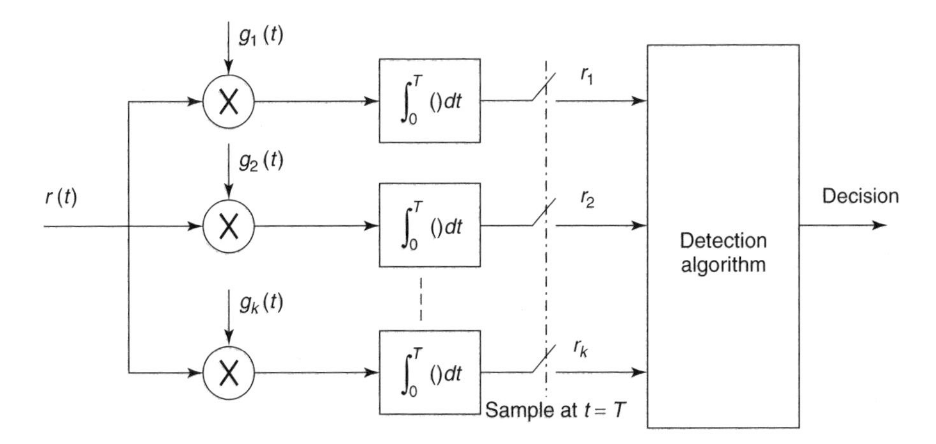

Figure 3.4 Generalized RAKE for multiuser detection. Estimated channel impulse response is used in forming the signature waveform $g_k(t)$ for the kth user

the system through the cross-correlation. We define, where $0 \leq \tau \leq T$ and $i < j$,

$$\rho_{ij}(\tau) = \int_{\tau}^{T} g_i(t) g_j(t-\tau)\, dt \qquad \rho_{ji}(\tau) = \int_{0}^{\tau} g_i(t) g_j(t+T-\tau)\, dt. \tag{45}$$

Similarly, we may define

$$\rho_{ij}^{0}(\tau) = \int_{\tau}^{T} g_i^0(t) g_j^0(t-\tau)\, dt \qquad \rho_{ji}^{0}(\tau) = \int_{0}^{\tau} g_i^0(t) g_j^0(t+T-\tau)\, dt. \tag{46}$$

It is important to connect these cross-correlation functions through the channel impulse response via (44). As a result, we obtain

$$\rho_{ij}(\tau) = \rho_{ij}^{0}(\tau) * [h^{(i)}(t) * h^{(j)}(-t)]. \tag{47}$$

When $h^{(i)}(t) = h^{(j)}(t)$, (47) reduces to the familiar autocorrelation form. Further, when $h^{(i)}(t)$ and $h^{(j)}(t)$ can be modeled by the Turin's model, (42) reduces to $r(t) = x(t) + n(t)$, which is the conventional form in [24].

Let us consider the synchronous transmission. In AWGN, it is sufficient to consider the signal received in one signal interval, say $0 \leq t \leq T$, and determine the optimum receiver. The optimum maximum-likelihood receiver is based on the log-likelihood function

$$\Lambda(b) = \int_{0}^{T} \left[r(t) - \sum_{k=1}^{K} \sqrt{E_k} b_k(1) g_k(t) \right]^2 dt, \tag{48}$$

and selects the information sequence $\{b_k(1), 1 \leq k \leq K\}$ that minimizes $\Lambda(b)$. Expanding (48) leads to

$$\begin{aligned}\Lambda(b) = {} & \int_{0}^{T} r^2(t)\, dt - 2 \sum_{k=1}^{K} \sqrt{E_k} b_k(1) \int_{0}^{T} r(t) g_k(t)\, dt \\ & + \sum_{j=1}^{K} \sum_{k=1}^{K} \sqrt{E_j E_k} b_j(1) b_k(1) \int_{0}^{T} g_j(t) g_k(t)\, dt.\end{aligned} \tag{49}$$

The term $r_k = \int_0^T r(t) g_k(t)\, dt$ represents the cross-correlation of the received signal with the kth signature. In practice, through (44), $\hat{g}_k(t) = g_k^0(t) * \hat{h}^{(k)}(t)$ is used where $\hat{h}^{(k)}(t)$ is the estimated channel impulse response. The equation (49) may be expressed compactly as

$$\Omega(r_K, b_K) = 2 b_K' r_K - b_K' R_s b_K \tag{50}$$

where

$$r_K = [r_1 r_2 \ldots r_K]'$$

$$b_K = \left[\sqrt{E_1} b_1(1) \ldots \sqrt{E_K} b_K(1) \right]'$$

and the prime denotes matrix transpose, R_s is the correlation matrix with elements $\rho_{jk}(0)$. Pulse distortion enters the statistic through $\rho_{jk}(0) = \int_0^T g_j(t) g_k(t)\, dt$.

The received signal vector r_K represents the output of K correlators or matched filters – one for each of the K signatures. In the generalized RAKE structure, these matched filters matches the shapes of the received distorted multipath pulses, instead of the shapes of the transmitted pulses. The r_K is given by

$$r_K = R_s b_K + n_K, \tag{51}$$

where

$$b_K(i) = [\sqrt{E_1} b_1(i) \sqrt{E_2} b_2(i) \ldots \sqrt{E_K} b_K(i)]'$$
$$n_K = [n^T(1) n^T(2) \ldots n^T(N)]'.$$

The noise vector n_K has a covariance

$$E\left[n_K n_K'\right] = \tfrac{1}{2} N_0 R_s \tag{52}$$

where the prime denotes matrix transpose.

The rest of detection follows the standard steps in [5, 24, 30]. The channel estimation is addressed in the section of generalized RAKE receiver.

References

[1] H. Bertoni, *Radio Propagation for Modern Wireless Systems*, Prentice Hall, 2000.

[2] M. Z. Win and R. Scholtz, "Ultra-Wide Bandwidth Timing-Hopping Spread Spectrum Impulse Radio for Wireless Multiple-Access Communications," *IEEE Trans. Commun.*, vol. 48, pp. 36–38, Apr. 2000.

[3] R. C. Qiu, H. P. Liu, and X. Shen, "Ultra-Wideband for Multiple Access," *IEEE Communications Mag.*, pp. 2–9, Feb. 2005.

[4] R. C. Qiu and C. Zhou, "Physics-Based Pulse Distortion for Ultra-Wideband Signals," *IEEE Trans. Veh. Technol.*, vol. 54, No. 5, invited paper, Sept. 2005.

[5] R. C. Qiu, J. Q. Zhang, and N. Guo, "Detection of physics-based ultra-wideband signals using generalized RAKE and multi-user detection (MUD)," *IEEE J. Select. Areas Commun.*, pp. 1546–1555, 2nd Quarter 2006.

[6] R. C. Qiu, "UWB Wireless Communications," *Design and Analysis of Wireless Networks*, Editor: Y. Pan and Y. Xiao, Nova Science Publishers, 2004.

[7] R. C. Qiu, Chapter "UWB Pulse Propagation Processes," *UWB Wireless Communications - A Comprehensive Overview*, Editor: T. Kaiser (Germany), Eurasip, 2005.

[8] R. C. Qiu, "Optimum and Sub-Optimum Detection of Physics-Based Ultra-Wideband Signals – A Tutorial Review," *Dynamics of Continuous, Discrete and Impulsive Systems (DCDIS)–An International Journal for Theory and Applications (Series B): Special Issue on UWB Wireless Communications*, vol. 12, No. 3, June 2005.

[9] R. C. Qiu, "A Generalized Time Domain Multipath Channel and its Application in Ultra-Wideband (UWB) Wireless Optimal Receiver Design: Part III System Performance Analysis," *IEEE Trans. Wireless Commun.*, to appear.

[10] R. C. Qiu, "A Generalized Time Domain Multipath Channel and its Application in Ultra-Wideband (UWB) Wireless Optimal Receiver Design: Part II Wave-Based System Analysis," *IEEE Trans. Wireless Commun.*, vol. 3, No. 11, pp. 2312–2324, Nov. 2004.

[11] R. C. Qiu, "A Study of the Ultra-Wideband Wireless Propagation Channel and Optimum UWB Receiver Design, Part I" *IEEE J. Select. Areas Commun. (JSAC)*, the First Special Issue on UWB Radio in Multiple Access Wireless Comm., vol. 20, No. 9, pp. 1628–1637, Dec. 2002.

[12] R.C. Qiu and I. T. Lu, "Multipath Resolving with Frequency Dependence for Broadband Wireless Channel Modeling," *IEEE Trans. Veh. Technol.*, vol. 48, No. 1, pp. 273–285, Jan. 1999.

[13] R. C. Qiu and I-T. Lu, "Wideband Wireless Multipath Channel Modeling with Path Frequency Dependence," *IEEE International Conference on Communications (ICC'96)*, Dallas, TX, June 23-27, 1996.

[14] K. B. Oldham and J. Spanier, *The Fractional Calculus*, Academic Press, New York, 1974.

[15] Channel Model Subcommittee, "Status of models for UWB propagation channel," IEEE 802.15.4a Channel Model (Final Report), http://www.ieee802.org/15/pub/TG4a.html, Aug. 2004.

[16] R. C. Qiu, C. Zhou, Q. Liu, "Physics-Based Pulse Distortion for Ultra-Wideband Signals," *IEEE Trans. Veh. Tech.*, invited paper, Vol. 54, No. 5, pp. 1–10, Sept. 2005.

[17] M. B. Pursley, *Introduction to Digital Communications*, Prentice-Hall, 2004.

[18] A. A. D'Amico, U. Mengali, and M. Morelli, "Multipath Channel Estimation for the Uplink of a DS-CDMA System, " *IEEE ICC 2002*, pp. 16–20, 2002.

[19] M. Wessman (Editor), "Delivery D4.2 transceiver design and link level simulation results," Report W-04-03-0025-R07, IST Ultrawaves Project (IST-2001-35189), Dec. 2003.

[20] N. Guo and R. Qiu, "Improved Autocorrelation Demodulation Receivers Based on Multiple-Symbol Detection for UWB Communications," *IEEE Trans. Wireless Commun.*, to appear.

[21] R. M. Buehrer, A. Safaai-Jazi, and W. Davis, D. Sweeney, "Ultra-Wideband Propagation Measurements and Modeling," Final Report, DAPRA NETEX Program, Virginia Tech, Jan. 2004.

[22] S. Zhao, H. Liu, and Z. Tian, "A Decision Feedback Autocorrelation Receiver for Pulsed Ultra-Wideband Systems," *Proc. IEEE Rawcon'04*, Sept. 2004.

[23] S. Hoyos, B. M. Sadler, and G. R. Arce, "Mono-Bit Digital Receivers for Ultra-Wideband Communications," *Wireless Communications, IEEE Transactions*, vol. 4, No. 4, pp. 1337–1344, July 2005.

[24] J. Proakis, *Digital Communications*, 4th edition, McGraw-Hill, 2000.

[25] T. Kailath and H. V. Poor, "Detection of Stochastic Processes," *IEEE Trans. Inf. Theory*, vol. 44, No. 6, pp. 2230–2259, Oct. 1998.

[26] T. Strohmer, M. Emami, J. Hansen, G. Pananicolaou, and A. J. Paulraj, "Application of Time-Reversal with MMSE Equalizer to UWB Complications," *Proc. IEEEGlobecom'04*, Dallas, TX, Dec. 2004.

[27] Q. Li and L. A. Rusch, "Multiuser Detection for DS-CDMA UWB in the Home Environment," *IEEE J. Select. Areas Commun.*, vol. 20, pp. 1701–1711, Dec. 2002.

[28] Y. C. Yoon and R. Kohno, "Optimum Multi-User Detection In Ultra-Wideband (UWB) Multiple Access Communications Systems," *Proc. ICC 2002*, pp. 812–816, May 2002.

[29] J. Choi, "Multiuser Detection with a Modified Time-Hopping Pulse Position Modulation for UWB," *IEEE Trans. Wireless Commun.*, to appear.

[30] S. Verdu, *Multiuser Detection*, Cambridge University Press, New York, 1998.

4

Timing Synchronization for UWB Impulse Radios

Zhi Tian and Georgios B. Giannakis

4.1 Introduction

Ultra-wideband (UWB) impulse radios (IRs) convey information over a stream of impulse-like, carrier-less pulses of very low power density and ultrashort width – typically, a few tens of picoseconds to a few nanoseconds [1, 2]. The huge (several GHz) bandwidth endows signals with fine time resolution and offers the potential for ample multipath diversity. These properties make UWB a favorable candidate for short-range, high-speed wireless communication and ad hoc networking, with simple baseband circuitry and the capability to overlay legacy wireless systems [3].

However, the unique advantages of UWB signaling are somewhat encumbered by the stringent timing requirements because the transmitted pulses are narrow and have low power density under the noise floor. Accurate timing-offset estimation (TOE) imposes major challenges to pulsed UWB systems in realizing their potential bit error rate (BER) performance, capacity, throughput, and network flexibility [4–6]. In a UWB IR, every symbol is repeatedly transmitted with a low duty cycle over a large number of frames with one pulse per frame to gather adequate symbol energy while maintaining low power density [1]. Such a transmission structure entails a twofold TOE task: one is a coarse frame-level acquisition to identify when the first frame in each symbol starts and the other is a fine pulse-level tracking to find the location of a pulse within a frame. UWB radios operating in diverse propagation environments encounter different timing tolerances to acquisition and tracking errors [6].

In traditional radar and ranging applications, UWB signals typically propagate through direct-path or sparse-multipath channels. Tracking accuracy is critical under such scenarios to capture those ultrashort pulses arriving at the receiver with a low duty cycle. It has been shown that even a slight timing jitter (tracking error) of a tenth of the pulse duration can decrease the system throughput to zero [5]. Fine-scale tracking of sparse-multipath

Ultra-wideband Wireless Communications and Networks
Edited by Xuemin (Sherman) Shen, Mohsen Guizani, Robert Caiming Qiu and Tho Le-Ngoc

channels faces major challenges in complexity because exhaustive search over thousands of fine-scale bins is required. Recent efforts in the quest for fast tracking include a coarse bin search for direct-path AWGN channels [7], a bit reversal search for the noiseless case [8], a special beacon code that increases the search intervals in conjunction with a bank of correlators that operate in the absence of multipath [9], and a frequency-domain treatment of UWB timing in sparse-multipath channels [10–12]. Among them, a notable approach for channel and timing estimation is based on the notion of finite innovation rates [10, 11], which aims at recovering the delays and gains of (a few) dominant channel taps by using a subspace-based spectral harmonic retrieval approach. As such, the sampling rates of innovation are lower, yet still comparable to the Nyquist rate (GHz).

Emerging interest in the commercial applications of UWB radios focuses on dense multipath indoor wireless communications [13]. In a dense multipath channel, the received signal no longer has a low duty cycle, as each transmitted pulse is spread out by a large number of closely spaced channel taps to occupy almost the entire frame duration. As a result, a judiciously designed correlation receiver is relatively robust to fine-scale tracking errors [6], and frame-level acquisition becomes the more critical timing task that affects the system's performance, especially when frame-dependent time-hopping (TH) codes are employed for smoothing the transmit spectrum and for enabling multiple access [2]. Dense multipath entails large diversity and can enhance the energy capture, but this is challenging during the synchronization phase because channel and timing information is not available [14–17]. Effective timing acquisition thus faces two primary challenges: one is to reduce the acquisition time and complexity, which calls for synchronizers that operate at a low sampling rate, preferably at the frame rate and the other is to provide the desired acquisition accuracy at a reasonable transmission power, which can be achieved only when a 'timing-blind' synchronizer is able to capture a sufficient part of the energy provided by the dense multipath channel. Furthermore, timing acquisition has to be performed successfully in multiple-access environments, which generally have to rely on the timing information for user separation.

This chapter presents accurate and low-complexity TOE algorithms for UWB IR receivers, with focus on timing acquisition in dense multipath channels. In Section 4.2, we describe the signal model and state the synchronization problem for UWB radios. Detection of UWB signals from ambient noise is discussed in Section 4.3. The challenges in UWB synchronization necessitate a fresh look at correlation receivers, for which we discuss a mistimed aggregate template (MAT) and a synchronized aggregate template (SAT) in Section 4.4. We show that the MAT is an asymptotically optimal correlation template that enables sufficient multipath energy capture at low complexity in the absence of timing or channel knowledge.

Sections 4.5–4.7 present various TOE algorithms based on the MAT concept. The idea is to start with a low-complexity coarse algorithm to obtain the initial timing-offset estimate, followed by a high-resolution local search algorithm. In Section 4.5, we present a conditional maximum-likelihood (CML) estimator that yields coarse timing-offset estimates derived in closed form from low symbol-rate correlator output samples. In Section 4.6, we rely on sample mean-square (SMS) values of the correlator outputs to develop a fast TOE algorithm that permits flexible timing resolution, even fine-scale tracking, by adjusting the search stepsize. This algorithm can also be imposed on cross-correlation output samples, leading to timing with dirty templates (TDT). The focus of Section 4.7 is on timing synchronization in multiuser setups, for which we highlight a TOE algorithm based on simultaneous synchronization and SAT recovery, which is resilient to both intersymbol interference (ISI) and multiple-user interference (MUI) and is particularly suitable for low-complexity rapid timing in ad hoc multiple-access networks.

The accuracy of a timing synchronization algorithm has to be put in perspective with respect to its purpose. For a UWB communication system, in Section 4.8, we describe the optimal detector under residual mistiming and evaluate analytically the system-level BER performance, which is the most relevant figure of merit. The mean-square error (MSE) of the TOE itself is more relevant for clock synchronization and target localization applications. The analytical MSEs for various TOE algorithms are presented in the respective sections.

4.2 Signal Model

A UWB impulse radio transmits a stream of ultrashort pulses $p(t)$ of duration T_p in the nanosecond scale, occupying a GHz-wide bandwidth. Every symbol is conveyed by repeating over N_f frames, one pulse per frame of duration T_f ($\gg Tp$) [2]. Every frame contains N_c chips, each of chip duration T_c. Once the timing is acquired at the receiver, user separation can be accomplished with user-specific pseudorandom TH codes $\{c_j\} \in [0, N_c - 1]$, which time-shift the pulse position in each frame by multiples of T_c [2]. The transmit symbol-waveform comprising N_f frames is given by

$$p_T(t) := \sum_{j=0}^{N_f-1} p(t - jT_f - c_j T_c) \tag{1}$$

and has a duration of $T_s = N_f T_f$. Both N_f and N_c can be quite large, on the order of tens to hundreds, leading to low duty-cycle transmissions in which the symbol rate $1/T_s$ can be considerably smaller than the chip rate $1/T_c$ or the pulse rate $1/T_p$. Let $s[k] \in \{\pm 1\}$ be independent and identically distributed (*i.i.d.*) binary symbols, each with energy $\mathcal{E}_s$ spread over N_f frames. By focusing on pulse amplitude modulation (PAM) [18], we can express the transmitted UWB waveform as [cf Figure 4.1(a)]:

$$u(t) = \sqrt{\mathcal{E}_s} \sum_{k=0}^{\infty} s[k] p_T(t - kT_s). \tag{2}$$

The signal $u(t)$ propagates through a multipath fading channel, whose impulse response $h(t) := \sum_{l=0}^{L-1} \alpha_l \delta(t - \tau_l)$ has coefficients α_l and delays τ_l, obeying $\tau_0 < \tau_1 < \cdots < \tau_{L-1}$ [14, 27, 37]. The timing information of interest refers to the first arrival time τ_0, which has to be estimated prior to symbol detection. To isolate τ_0, we define $\tau_{l,0} := \tau_l - \tau_0$ as the relative time-delay of each channel tap, where $\tau_{L-1,0}$ is the channel delay spread. The aggregate receive-pulse within each frame is then given by $p_r(t) := \sum_{l=0}^{L-1} \alpha_l p(t - \tau_{l,0})$; the pulses with dashed lines are shown in Figure 4.1(b). The waveform at the output of the receiver antenna is:

$$r(t) = \sqrt{\mathcal{E}_s} \sum_{k=0}^{\infty} s[k] p_R(t - kT_s - \tau_0) + w(t), \tag{3}$$

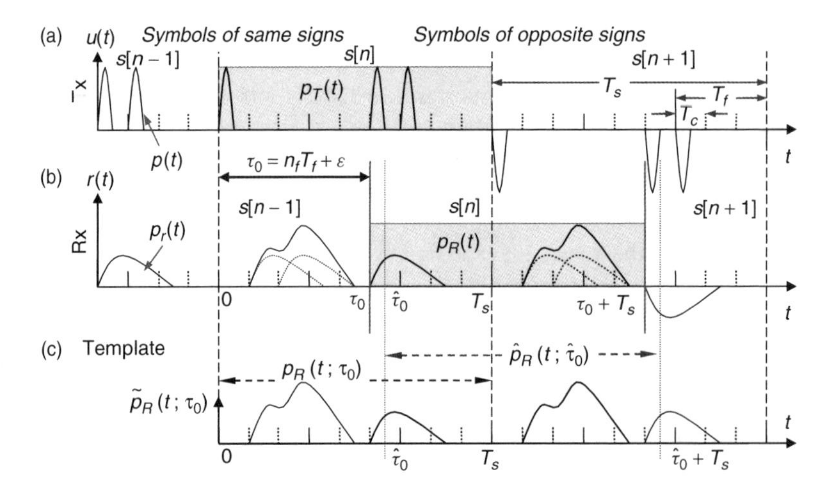

Figure 4.1 Transmit (Tx) and Receive (Rx) UWB signal model $N_f = N_c = 3, T_f = 3T_c$, $\{c_j\}_{j=0}^{N_f-1} = [0, 2, 0]$, $n_s = 0$, $n_f = 1$, $\epsilon = (1/3)T_f$, and $\mathcal{E}_s = 1$. The dashed pulses in each frame in (b) is the received $p_r(t)$ after channel distortion and the solid pulse in each symbol is the composite waveform $p_R(t)$ after mixing TH-shifted $p_r(t)$s. The template $\tilde{p}_R(t; \hat{\tau}_0)$ in (c) is the periodic extension of $p_R(t)$ that always matches the received waveform in $r(t)$. Solid vertical lines in (b) are actual symbol boundaries unknown prior to timing, dashed vertical lines are T_s-apart asynchronous boundaries according to the receiver's clock, and dotted vertical lines are synchronized symbol boundaries after recovering $\hat{\tau}_0$

where

$$p_R(t) := \sum_{j=0}^{N_f-1} p_r(t - jT_f - c_jT_c) = \sum_{l=0}^{L-1} \alpha_l p_T(t - \tau_{l,0}) \tag{4}$$

denotes the aggregate receive-waveform of each symbol with duration T_s; the composite symbol-waveform with solid lines is shown in Figure 4.1(b). The term $w(t)$ accounts for both ambient noise and MUI, whose composite effect is approximated as a zero-mean white Gaussian process, with the two-sided power spectral density (PSD), σ^2, and bandwidth, B, dictated by the ideal low-pass, front-end filter's cutoff frequency. To avoid ISI, we select T_f to satisfy the following condition:

$$\text{ISI-free condition :} \quad T_f \geq (c_{N_f-1} - c_0)T_c + \tau_{L-1,0} + T_p, \tag{5}$$

which can be conveniently realized by setting $T_f > \tau_{L-1,0} + T_p$ and $c_{N_f-1} \leq c_0$. The ISI-free condition only concerns the first and last chips of each symbol's TH code. As such, the nonzero support of $p_R(t)$ is no greater than that of T_s. Notice that we allow interframe interference (IFI), which may arise because of TH and/or multipath.

The timing offset τ_0 can be expressed in different time scales as $\tau_0 = n_sT_s + n_fT_f + \epsilon$, where $n_s = \lfloor \tau_0/T_s \rfloor \geq 0$ denotes the symbol-level timing-offset parameter, $n_f = \lfloor (\tau_0 - n_sT_s)/T_f \rfloor \in [0, N_f - 1]$ the frame-level offset parameter, and $\epsilon = (\tau_0 \bmod T_f) \in [0, T_f)$ the pulse-level offset; here, $\lfloor \cdot \rfloor$ represents the floor operation and (mod) denotes the modulo operation. These different levels of timing offset are illustrated in Figure 4.1

via an example with $\tau_0 = (n_s = 0) \cdot T_s + (n_f = 1) \cdot T_f + (\epsilon = T_f/3)$. The task of estimating τ_0 can thus be decomposed into two subtasks: timing acquisition, which amounts to estimating n_s and n_f, and tracking, which refers to recovering the residual ϵ. It is worth noting that a synchronizer operating in a blind mode cannot distinguish time delays that are separated by multiple symbol intervals. Therefore, UWB timing acquisition can only recover n_f in a non–data-aided (NDA) mode, but both n_s and n_f can be obtained in a data-aided (DA) mode.

4.3 Signal Detection and Symbol-Level Acquisition

The first step in receiver processing is to detect the received UWB signal $r(t)$ and identify the symbol-level timing-offset parameter n_s. As we mentioned earlier, the latter requires training. Suppose that $r(t)$ contains N_s training symbols, which arrive at the receiver within the time interval $[n_s T_s, n_s + (N_s + 1)T_s]$. For the preceding $t \in [0, n_s T_s)$, only noise is received, since $n_s T_s \leq \tau_0$. Various forms of energy change detection can be utilized during this phase.

4.3.1 Analog Energy Detectors

A conventional square-law (SL) energy detector detects $r(t)$ and decides $\hat{n}_s$ by

$$\hat{n}_s = \arg \max_{n_s} J_{SL}(n_s), \quad J_{SL}(n_s) := \int_{n_s T_s}^{(n_s+N_s)T_s} r^2(t)\, dt. \tag{6}$$

The presence of $r(t)$ is declared when $J_{SL}(\hat{n}_s) \geq \eta_{SL}$, where η_{SL} is a threshold determined by the desired probability of false alarms [19].

The objective function $J_{SL}(n_s)$ in (6) involves a noise-square term that may raise the noise floor. To alleviate this problem, we may adopt a cross-correlation (CC) detector that correlates $r(t)$ with its own shifted copy. With N_s all-one pilot symbols transmitted, the detection rule is

$$\hat{n}_s = \arg \max_{n_s} J_{CC}(n_s), \quad J_{CC}(n_s) := \int_{n_s T_s}^{(n_s+N_s-1)T_s} r(t) r(t + T_s)\, dt. \tag{7}$$

The presence of $r(t)$ is declared when $J_{CC}(\hat{n}_s) \geq \eta_{CC}$, where η_{CC} is a threshold determined by the desired probability of false alarms. Notice that $J_{CC}(n_s)$ has a noise–noise term, $n(t)n(t + T_s)$, inside the integral, which causes little increase in the noise floor because of the independence between shifted noise terms. For a given false alarm rate, the cross-correlation based signal detection requires a lower threshold $\eta_{CC} < \eta_{SL}$, thus resulting in a better probability of detection [20].

4.3.2 Discrete-Time Energy Detectors

Consider digital receivers whose samples are formed at the output of a symbol-rate sliding correlator. For a general correlation template $f(t)$ with nonzero time-support no greater than T_s, the discrete-time correlator output is

$$y[n] = \int_0^{T_s} r(t + nT_s) f(t)\, dt. \tag{8}$$

Recognizing that any T_s-long segment of $r(t)$ can only entail up to two consecutive symbols in the presence of timing offset, we can express each symbol-rate sample $y[n]$ as a linear combination of two transmitted symbols in the form [6, 21]:

$$y[n] = A_\epsilon (s[n - n_s](1 - \nu_\tau) + s[n - n_s - 1]\nu_\tau) + w[n], \tag{9}$$

where the indices of contributing symbols are determined by the symbol-level offset n_s. The sample amplitude A_ϵ captures the effects of the transmit-filter, the multipath channel, and the receive-template, while the weighting scalar $\nu_\tau \in [0, 1]$ is related to the timing-offset parameters n_f and ϵ. Being independent of n_s, the values of A_ϵ and ν are not needed for signal detection, even though A_ϵ determines the detection SNR.

Depending on whether successive symbols have identical or opposite signs, we can partition the training symbols into the two nonoverlapping subsets: one is $\mathcal{G}_+ := \{s[n] : s[n] = s[n-1]\}$ and the other is $\mathcal{G}_- := \{s[n] : s[n] = -s[n-1]\}$. The corresponding cardinalities of these two data sets are denoted by $N_+ := |\mathcal{G}_+|$ and $N_- := |\mathcal{G}_-|$ respectively, both being determined solely by the training data pattern.

For $s[n] \in \mathcal{G}_+$, (9) reduces to $y[n + n_s] = A_\epsilon s[n] + w[n + n_s]$. On performing a binary hypothesis test on this model, it is immediately revealed that the input–output cross-correlation $R_{ys}^+(\hat{n}_s) := N_+^{-1} \sum_{s[n]\in\mathcal{G}_+} y[n + \hat{n}_s]s[n]$ peaks at $\hat{n}_s = n_s$ [22]. Similarly, for the subset $\mathcal{G}_-$, $R_{ys}^-(\hat{n}_s) := N_-^{-1} \sum_{s[n]\in\mathcal{G}_-} y[n + \hat{n}_s]s[n]$ also peaks at $\hat{n}_s = n_s$. When both training subsets are present, a discrete-time energy (DE) detector following a generalized likelihood ratio test (GLRT) is given by [21, 22]

$$\hat{n}_s = \arg\max_{n_s} J_{DE}(n_s), \quad J_{DE}(n_s) := N_+ \left[R_{ys}^+(n_s)\right]^2 + N_- \left[R_{ys}^-(n_s)\right]^2. \tag{10}$$

As before, the presence of $r(t)$ is declared when $J_{DE}(\hat{n}_s)$ exceeds a threshold.

Despite the simplicity of discrete-time processing after analog–digital conversion, the major challenge for a digital receiver is to construct a proper template $f(t)$ needed to form $y[n]$ in (8). This challenge is appreciated if we recognize that a proper $f(t)$ should match the aggregate receive-waveform resulting from convolving $p_T(t)$ with $h(t - \tau_0)$, where the channel $h(t)$ and the timing τ_0 are unknown at this stage. To bypass this problem, a cross-correlation receiver can be used as the counterpart of (6) in discrete time to generate symbol-rate samples as follows:

$$\begin{aligned} y[n] &= \int_0^{T_s} r(t + nT_s) r(t + (n+1)T_s)\, dt \\ &= \check{A}_\epsilon \big(s[n - n_s]s[n - n_s + 1](1 - \check{\nu}_\tau) + s[n - n_s - 1]s[n - n_s]\check{\nu}_\tau\big) + \check{w}[n], \end{aligned} \tag{11}$$

where, similar to (9), $\check{A}_\epsilon$ and $\check{\nu}_\tau \in [0, 1]$ are independent of n_s. The training symbols are partitioned in two nonoverlapping subsets of different patterns: one is $\check{\mathcal{G}}_+ := \{s[n] : s[n-1] = s[n+1]\}$ of size $\check{N}_+ := |\check{\mathcal{G}}_+|$ and the other is $\check{\mathcal{G}}_- := \{s[n] : s[n-1] = -s[n+1]\}$ of size $\check{N}_- := |\check{\mathcal{G}}_-|$. Mimicking (10), a discrete-time cross-correlation (DC) energy detector yields

$$\hat{n}_s = \arg\max_{n_s} J_{DC}(n_s), \quad J_{DC}(n_s) := \check{N}_+ \left[\check{R}_{ys}^+(n_s)\right]^2 + \check{N}_- \left[\check{R}_{ys}^-(n_s)\right]^2, \tag{12}$$

where $\check{R}_{ys}^+ := \check{N}_+^{-1} \sum_{n \in \check{\mathcal{G}}_+} y[n+n_s]s[n]s[n-1]$ and $\check{R}_{ys}^+ := \check{N}_-^{-1} \sum_{n \in \check{\mathcal{G}}_-} y[n+n_s] \times s[n]s[n-1]$. It turns out that $\check{A}_\epsilon = \mathcal{E}_s \int_0^{T_s} p_R^2(t)dt$, which is the maximum energy of the (unknown) receive symbol-waveform $\sqrt{\mathcal{E}_s} p_R(t)$. As a result, the detector in (12) has a higher detection SNR than that in (10).

In the rest of this chapter, we suppose that signal detection has been carried out and that n_s has been identified successfully. This allows us to henceforth confine τ_0 within one symbol duration; that is, from now on, we have $\tau_0 \in [0, T_s)$.

4.4 SAT and MAT: Templates with and without Timing

For correlation-based receivers, employing an appropriate template $f(t)$ in (8) is critical not only for data demodulation but also during timing synchronization. Under perfect timing and when the ISI-free condition in (5) is satisfied, the ideal template for demodulation is the received symbol-waveform $p_R(t)$. Meanwhile, the estimation of τ_0 can be performed optimally by a sliding correlator, with its template matched to $p_R(t)$ [23]. However, the challenge here is that no clean template is available for matching, since the multipath channel (and thus $p_R(t)$) is unknown. One could resort to joint estimation of channel taps and delays and use the estimates $\{\hat{\alpha}_l, \hat{\tau}_l\}$ to form an L_r-finger RAKE receiver template $f_{\text{RAKE}}(t) = \sum_{l=0}^{L_r-1} \hat{\alpha}_l p_T(t - \hat{\tau}_l)$. To this end, an ML estimator would search over the following objective function formed over a synchronization window of length T_0 in the absence of data modulation [16]:

$$\{\hat{\alpha}_l, \hat{\tau}_l\} = \arg \max_{\{\alpha_l, \tau_l\}_{l=0}^{L_r-1}} \int_0^{T_0} \left| r(t) - \sum_{l=0}^{L_r-1} \alpha_l p_T(t - kT_s - \tau_l) \right|^2 dt. \tag{13}$$

Unfortunately, this ML estimator is quite costly to realize for a practical UWB receiver, as it requires sampling at or above the Nyquist rate, resulting in formidably high sampling rates (over 10 GHz) [16]. For a dense multipath channel, there is a very large number of closely spaced channel taps that must be estimated accurately, which incurs high computational complexity and long synchronization time [14, 16, 17]. The estimation performance is further degraded by per-pulse distortion in UWB propagation, which leads to nonideal receive-templates.

We are thus motivated to look for low-complexity means of recovering ideal templates for the correlator in (8). To this end, it is instrumental to transmit a train of N_s all-one pilot symbols, which is also the preferred training pattern for signal detection, as we saw in Section 4.3. Since the transmission is free of data modulation, the expectation of the received $r(t)$ is simply its noise-free waveform that we denote as $\tilde{p}_R(t; \tau_0)$:

$$\tilde{p}_R(t; \tau_0) := (1/\sqrt{\mathcal{E}_s}) \mathrm{E}\{r(t)\}|_{s[n]=1} = \sum_k p_R(t - kT_s - \tau_0). \tag{14}$$

Here, $\tilde{p}_R(t; \tau_0)$ is the T_s-periodic extension of $p_R(t)$ after being time-shifted by τ_0. Because $\tilde{p}_R(t; \tau_0)$ matches $r(t)$ perfectly, the ideal template $f_{\text{opt}}(t) \propto p_R(t; \tau_0)$ that maximizes multipath energy capture even under mistiming is a T_s-long period of $\tilde{p}_R(t; \tau_0)$ according to the receiver's clock [27]; the T_s-apart mistimed symbol boundaries are shown in Figure 4.1(c) (dashed vertical lines) at $t = nT_s, \forall n$. Mathematically, this period $p_R(t; \tau_0)$ can be expressed by

$$p_R(t; \tau_0) := \tilde{p}_R(t; \tau_0)\big|_{t\in[0,T_s]} = \begin{cases} p_R(t - \tau_0), & t \in [\tau_0, T_s]; \\ p_R(t + T_s - \tau_0), & t \in [0, \tau_0). \end{cases} \tag{15}$$

In essence, $p_R(t; \tau_0)$ is the aggregate receive-waveform $p_R(t)$ after being circularly shifted by τ_0 to confine it within $[0, T_s)$, that is, $p_R(t; \tau_0) = p_R\big((t - \tau_0) \bmod T_s\big)$. Because of the periodicity of $\tilde{p}_R(t; \tau_0)$, $p_R(t; \tau_0)$ can be estimated using the mean-square sense (mss) consistent sample average across N_s segments of $r(t)$, each of size T_s:

$$\hat{p}_R(t; \tau_0) = \frac{1}{N_s} \sum_{n=0}^{N_s-1} r(t + nT_s) = \sqrt{\mathcal{E}_s}\, p_R(t; \tau_0) + \bar{w}(t), \qquad t \in [0, T_s] \tag{16}$$

where the averaged noise $\bar{w}(t) = (1/N_s) \sum_{n=0}^{N_s-1} w(t + nT_s)$ has zero mean and reduced variance $\bar{\sigma}^2 := \sigma^2/N_s$. Obviously, $\hat{p}_R(t; \tau_0)$ is an asymptotically optimal template under mistiming, since it approaches the ideal template, $\sqrt{\mathcal{E}_s}\, p_R(t; \tau_0)$, as $N_s \to \infty$. We term $\hat{p}_R(t; \tau_0)$ as the MAT.

After $\hat{\tau}_0$ is recovered, a SAT $\hat{p}_R(t; \hat{\tau}_0)$ can be obtained from the MAT by simple circular shifting to yield

$$\hat{p}_R(t; \hat{\tau}_0) = \hat{p}_R\big((t + \hat{\tau}_0) \bmod T_s; \tau_0\big) = \sqrt{\mathcal{E}_s}\, p_R(t; \tau_\Delta) + \bar{w}(t), \tag{17}$$

where $\tau_\Delta := [(\tau_0 - \hat{\tau}_0) \bmod T_s]$ indicates the closeness between τ_0 and $\hat{\tau}_0$. In the noise-free case, $\hat{p}_R(t; \hat{\tau}_0)$ is a period of $\sqrt{\mathcal{E}_s}\, \tilde{p}_R(t; \tau_0)$ according to $\hat{\tau}_0$; these T_s-apart estimated symbol boundaries are shown in Figure 4.1(c) (dotted vertical lines) at $t = \hat{\tau}_0 + nT_s, \forall n$. Therefore, given $\hat{\tau}_0$, the SAT is an asymptotically optimal template for data demodulation. As $N_s \to \infty$, the SAT approaches $\sqrt{\mathcal{E}_s}\, p_R(t; \tau_\Delta)$, which is proportional to $p_R(t; 0) = p_R(t)$ when $\hat{\tau}_0 = \tau_0$ [cf (15)].

In the ensuing Sections 4.5–4.7, we present various timing synchronizers that employ the MAT in the absence of timing knowledge. The SAT is used for coherent symbol detection, which is discussed in Sections 4.7 and 4.8. Without complex tap-by-tap channel estimation, a symbol-rate correlation receiver adopting the MAT or SAT as its template matches $r(t)$ perfectly to yield maximum sample energy in the noise-free case, thus effecting sufficient energy capture even in the presence of TH and dense multipath.

4.5 Coarse Synchronization Using Symbol-Rate Samples

Coarse TOE which is useful at the initial stage of synchronization, can be derived in closed form from symbol-rate correlator output samples. We present a DA CML estimator for frame-level timing acquisition, along with the associated estimation performance bounds. In Section 4.3.2, we saw two distinct training subsets $\mathcal{G}_+$ and $\mathcal{G}_-$, depending on whether two adjacent symbols have identical or opposite signs. During synchronization, we first transmit N_s all-one symbols belonging to $\mathcal{G}_+$, which are used for signal detection and MAT estimation. Afterwards, we transmit N symbols belonging to $\mathcal{G}_-$, which are used to estimate τ_0, or coarsely, $n_f T_f$.

4.5.1 Discrete-Time Correlator Output Model under Mistiming

The CML synchronizer adopts the sliding correlator in (8) to generate samples at the symbol rate of $y[n] = \int_0^{T_s} r(t + nT_s) f(t)\, dt$, where a practical realization of the ideal mistimed template $p_R(t;\, \tau_0)$ is the MAT $\hat{p}_R(t;\, \tau_0)$ estimated via (16). Each correlator output $y[n]$ obeys the general expression in (9) with $n_s = 0$, that is,

$$y[n] = A_\epsilon\big(s[n](1 - \nu_\tau) + s[n-1]\nu_\tau\big) + w[n], \tag{18}$$

where the sample amplitude A_ϵ, the weighting scalar ν_τ, and the noise term $w[n]$ can be deduced for $f(t) = \hat{p}_R(t;\, \tau_0) = \sqrt{\mathcal{E}_s}\, p_R(t;\, \tau_0) + \bar{w}(t)$ as

$$A_\epsilon = \int_0^{T_s} \left(\sqrt{\mathcal{E}_s}\, p_R(t;\, \tau_0)\right)^2 dt = \mathcal{E}_s \int_0^{T_s} p_R^2(t)\, dt := \mathcal{E}_s \mathcal{E}_{\max},$$

$$\nu_\tau = \frac{1}{A_\epsilon} \int_0^{\tau_0} \left(\sqrt{\mathcal{E}_s}\, p_R(t;\, \tau_0)\right)^2 dt = \frac{1}{\mathcal{E}_{\max}} \int_{T_s - \tau_0}^{T_s} p_R^2(t)\, dt \in [0, 1],$$

$$w[n] = \int_0^{T_s} \sqrt{\mathcal{E}_s}\, p_R(t;\, \tau_0) w(t + nT_s)\, dt + \int_0^{T_s} \sqrt{\mathcal{E}_s}\, p_R(t;\, \tau_0) \bar{w}(t)\, dt + \int_0^{T_s} w(t + nT_s) \bar{w}(t)\, dt.$$

The correlator output captures full multipath diversity, since A_ϵ collects the maximum energy of the receive-waveform $p_R(t)$ via $\mathcal{E}_{\max} := \int_0^{T_s} p_R^2(t)\, dt$. The weighting coefficient ν_τ indicates the fraction of total energy $\mathcal{E}_{\max}$ captured by the interfering symbol $s[n-1]$ that is present in $y[n]$ only because of mistiming. The noise sample $w[n]$ contains two noise terms and one noise–noise term contributed from both the noisy signal $r(t)$ and the noisy MAT. This composite noise is similar to that in the transmit-reference (TR) [24] and pilot-assisted waveform modulation (PWAM) schemes [25] and has been to shown to be well approximated as zero-mean Gaussian with correlation function:

$$\mathrm{E}\{w[n;\, \tau] w[m;\, \tau]\} = \left[\mathcal{E}_s \mathcal{E}_{\max}(1 + 1/N_s)\sigma^2 + BT_s\sigma^4/N_s\right]\delta_{n,m} := \sigma_w^2 \delta_{n,m}, \tag{19}$$

where σ^2 is defined as the double-sided PSD of the noise process $w(t)$.

To delve into ν_τ that is related to the timing-offset parameters n_f and ϵ via τ_0, it should be noticed that the integrands in A_ϵ and ν_τ are the same, but the integration range in ν_τ is $[0, \tau_0]$, which is a fraction of the $[0, T_s]$ range used in A_ϵ. Because the fraction $\tau_0/T_s = (n_f + \epsilon/T_f)/N_f$ belongs to $[n_f, n_f + 1]/N_f$, ν_τ approximates the timing-offset ratio τ_0/T_s and determines the frame-level timing-offset parameter n_f by [21]

$$\nu_\tau = (n_f + \lambda_\epsilon)/N_f \approx \tau_0/T_s, \qquad n_f = \lfloor \nu_\tau N_f \rfloor. \tag{20}$$

Bounded in $[0, 1]$, the scalar λ_ϵ is a small adjustment term to n_f that is needed to account for the channel delay spread, the TH code, and the unknown tracking offset.

4.5.2 CML Timing Synchronization

With (20), timing synchronization can be acquired by estimating ν_τ on the basis of $y[n]$ samples obeying (18). Specifically, we seek the (conditional) maximum-likelihood estimates (MLEs) of A_ϵ and ν_τ, which are treated as unknown but deterministic in the presence of Gaussian distributed noise $w[n]$. For the two nonoverlapping training symbol subsets $\mathcal{G}_+$ and $\mathcal{G}_-$ described in Section 4.3.2, (18) can be simplified to

$$y[n] = \begin{cases} A_\epsilon s[n] + w[n], & s[n] \in \mathcal{G}_+; \\ A_\epsilon (1 - 2\nu_\tau) s[n] + w[n], & s[n] \in \mathcal{G}_-. \end{cases} \tag{21}$$

Amplitude estimation has been suggested by (21) to estimate the channel-dependent amplitude A_ϵ and the timing-offset ratio ν_τ from the training subsets $\mathcal{G}_+$ and $\mathcal{G}_-$, respectively. In fact, the MLEs of A_ϵ and ν_τ can be derived from these two disjoint data subsets separately. For $s[n] \in \mathcal{G}_+$, $y[n]$ in (21) does not contain ν_τ. On the other hand, for any $y[n]$ resulting from $s[n] \in \mathcal{G}_-$, A_ϵ and ν_τ are always coupled in a nonidentifiable manner. Estimating ν_τ separately from A_ϵ maintains the optimality of both $\hat{\nu}_\tau$ and $\hat{A}_\epsilon$, since no useful data are discarded [6].

It is well known that CML-based amplitude estimation relies on the cross-correlation of input and output samples [22]. On the basis of the separability property, we define $R_{ys}^+ := N_+^{-1} \sum_{s[n]\in\mathcal{G}_+} y[n]s[n]$ as the average cross-correlation for the subset $\mathcal{G}_+$ and $R_{ys}^- := N_-^{-1} \sum_{s[n]\in\mathcal{G}_-} y[n]s[n]$ as that for the other subset $\mathcal{G}_-$. The solution for CML-based TOE is summarized as follows [6]:

CML Timing Acquisition *Data-aided timing acquisition can be performed in the absence of ISI by a symbol-rate sliding correlator that employs the MAT. By relying on the input–output cross-correlation over two nonoverlapping training subsets $\mathcal{G}_+$ and $\mathcal{G}_-$, the optimum parameters can be estimated successively – $\hat{A}_\epsilon$ first, followed by $\hat{n}_f = \hat{\nu}_\tau N_f$ in closed form – as follows:*

$$\hat{A}_\epsilon = R_{ys}^+; \tag{22}$$

$$\hat{n}_f = \frac{N_f}{2} - \frac{N_f R_{ys}^-}{2\hat{A}_\epsilon}. \tag{23}$$

The CML timing algorithm has low implementation complexity, as it performs simple cross-correlation operations over symbol-rate $y[n]$ samples in closed form.

4.5.3 Analytic and Simulated Performance

The estimators in (22) and (23) can asymptotically reach their Cramer–Rao bounds (CRBs) because they possess ML optimality. It can be straightforwardly shown that the CRBs of $\hat{A}_\epsilon$ and $\hat{n}_f$ conditioned on A_ϵ are:

$$\mathrm{CRB}(\hat{A}_\epsilon) = \frac{\sigma_w^2}{N_+}, \qquad \mathrm{CRB}(\hat{n}_f | A_\epsilon) = \frac{\sigma_w^2}{4{A_\epsilon}^2 N_-}. \tag{24}$$

The CRB of $\hat{n}_f$ favors a receive correlation template that yields a large value for A_ϵ, an indicator of multipath energy capture. Note that both CRBs are inversely proportional

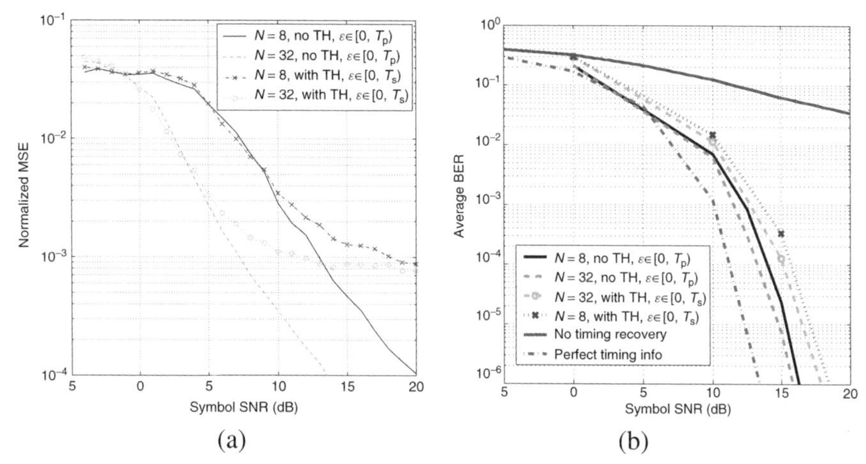

(a) (b)

Figure 4.2 Performance of CML timing acquisition: (a) MSE, (b) BER

to the numbers of training symbols in the respective subsets $\mathcal{G}_+$ and $\mathcal{G}_-$. On the basis of (24), we can optimize the optimum training pattern to minimize the timing estimation CRB, so that training resources in terms of placement, power, and number of training symbols are optimally allocated to $\mathcal{G}_+$ and $\mathcal{G}_-$ [26].

Figure 4.2 illustrates the performance of the CML timing algorithm. At the transmitter, $p(t)$ is the second-order derivative of the Gaussian pulse with pulse width $T_p = 1$ ns. Each symbol contains $N_f = 25$ frames, with frame duration $T_f = 100$ ns. The chip duration is $T_c = 1$ ns, and the TH code is chosen randomly over $[0, 90]$. The CM-1 multipath channel model in the IEEE 802.15.3a standard is used [27], according to which rays arrive in several clusters within an observation window. The cluster arrival times are modeled as Poisson variables with cluster arrival rate Λ, and returns within each cluster also arrive according to a Poisson process with arrival rate λ. On the basis of the doubly clustering structure, the amplitude of each arriving ray is a real-valued random variable having an exponentially decaying mean-square value with parameters Γ and γ. These channel parameters are set to $\Gamma = 30$ ns, $\gamma = 5$ ns, $\Lambda^{-1} = 2$ ns, and $\lambda^{-1} = 0.5$ ns. The channel delay spread is bounded to 99 ns.

The performance metrics of interest include the normalized acquisition MSE $\mathrm{E}\{|(\hat{n}_f - n_f)/N_f|^2\}$, and the BER when an optimum detector with perfect channel knowledge is available for symbol demodulation following timing. Figure 4.2 depicts MSE and BER versus SNR $\mathcal{E}_s/\sigma^2$ for two cases: one when TH is not employed and the pulse-level offset ϵ is small and the other case corresponding to large ϵ in the presence of TH. Two different sizes of training symbol sequence, $N = 8$ and $N = 32$, are considered.

Because the unknown offset ϵ has not been estimated during coarse timing acquisition, there is an error floor in the MSE curves in Figure 4.2(a) for the ϵ randomly generated in $[0, T_s)$. This floor is caused by the bounded unknown $\lambda_\epsilon \in [0, 1]$ in (20), which diminishes when ϵ approaches 0. Nevertheless, pulse-level offset ϵ only has a small impact on BER in dense multipath channels [6]. As Figure 4.2(b) illustrates, there are small BER gaps between the receiver using CML-based timing acquisition and an ideal receiver with perfect timing information. Operating at a practical sampling rate of $1/T_s$, the CML timing acquisition algorithm not only enjoys very low computational complexity but also offers good acquisition accuracy, since the MAT is able to collect sufficient multipath diversity gain.

4.6 Synchronization with Flexible Timing Resolution

After frame-level timing acquisition that identifies a coarse estimate of τ_0, we move on to search-based synchronizers that permit finer timing resolution and even excellent tracking performance. We will utilize the training patterns $(1, 1)$ and $(1, -1)$ to design TOE algorithms that search at flexible step sizes.

4.6.1 Timing-Offset Search via Sample Mean Square

Suppose that the MAT $\hat{p}_R(t; \tau_0)$ has been estimated from (16) as the mss average across a period of the periodic extension $\tilde{p}_R(t; \tau_0)$ in (14). An mss consistent estimate of $\tilde{p}_R(t; \tau_0)$ can be found by periodically extending $\hat{p}_R(t; \tau_0)$ to obtain $\hat{\tilde{p}}_R(t; \tau_0) = \sum_k \hat{p}_R(t + kT_s; \tau_0)$ [28]. During synchronization, the receiver generates symbol-rate samples at *candidate* time shifts $\tau \in [0, T_s)$ that serve as the search points for estimating τ_0:

$$y[n; \tau] = \int_{\tau+nT_s}^{\tau+(n+1)T_s} r(t)\hat{\tilde{p}}_R(t; \tau_0)\, dt = \int_{\tau}^{\tau+T_s} r(t + nT_s)\hat{\tilde{p}}_R(t; \tau_0)\, dt. \tag{25}$$

Here, the correlation template $\hat{\tilde{p}}_R(t; \tau_0)$ is independent of τ, as it always matches to the received waveform $r(t)$ that is determined by the unknown τ_0.

When $\tau = 0$, $y[n; 0]$ in (25) becomes $y[n]$ in (18), which entails up to two consecutive symbols within the corresponding T_s-long interval. Depending on τ, the two contributing symbols are $s[n]$ and $s[n-1]$ when $\tau < \tau_0$, while they become $s[n+1]$ and $s[n]$ when $\tau > \tau_0$. Let 1_τ denote a binary indicator that equals 1 when $\tau > \tau_0$, and 0 otherwise. Similar to (8), with $\tau = 0$, the symbol-rate correlation output, $y[n; \tau]$, $\forall \tau \in [0, T_s)$, can be derived from (25) as:

$$y[n; \tau] = A_\epsilon\big(s[n + 1_\tau](1 - \nu_\tau) + s[n - 1 + 1_\tau]\nu_\tau\big) + w[n; \tau], \tag{26}$$

where $A_\epsilon = \mathcal{E}_s\mathcal{E}_{\max}$ is the same as that in (18), $\nu_\tau = (1/\mathcal{E}_{\max}) \int_{T_s-\tau_\Delta}^{T_s} p_R^2(t)\, dt \in [0, 1]$ (recall $\tau_\Delta := [(\tau_0 - \tau) \bmod T_s]$), and $w[n; \tau]$ has the same variance σ_w^2.

The mean-square of the symbol-rate samples $y[n; \tau]$ in (26) can be deduced as:

$$\mathrm{E}\{y^2[n; \tau]\} = {A_\epsilon}^2 \left((1 - \nu_\tau)^2 + \nu_\tau^2\right) + \sigma_w^2 = {A_\epsilon}^2 \left[1 + (1 - 2\nu_\tau)^2\right]/2 + \sigma_w^2, \tag{27}$$

where, to cancel the cross-product terms in $y^2[n; \tau]$, we use $\mathrm{E}\{s[n]s[m]\} = 0$ and $\mathrm{E}\{s^2[n]s^2[m]\} = 1$ for $n \neq m$. It is obvious that $(1 - 2\nu_\tau)^2$ is maximized at $\tau_\Delta = 0$, by definition of ν_τ; hence, $\mathrm{E}\{y^2[n; \tau]\}$ is maximized when $\tau_\Delta = 0$, or equivalently, $\tau = \tau_0$. The peak value of $\mathrm{E}\{y^2[n; \tau]\}$ is $\mathcal{E}_s\mathcal{E}_{\max} + \sigma_w^2$, corresponding to the maximum energy capture. In practice, the ensemble mean-square $\mathrm{E}\{y^2[n; \tau]\}$ must be replaced by its consistent sample square average $z(\tau)$ formed by averaging N samples, giving rise to a TOE algorithm on the basis of SMS:

$$\hat{\tau}_0 = \arg \max_{\tau \in [0, T_s)} z(\tau), \qquad z(\tau) = \frac{1}{N} \sum_{n=1}^{N} y^2[n; \tau]. \tag{28}$$

This estimator in essence is a *discrete-time energy detector.*

It is worth stressing that the sample square average in (28) can be computed in either a DA mode or an NDA mode. However, this assumes that the MAT has been formed, which requires N_s all-one training symbols, as we saw earlier. The number of samples N required for reliably estimating τ_0 can be reduced markedly in the DA case in which we choose the training sequence consisting of binary symbols with alternating signs. When successive training symbol pairs are either $(1, -1)$ or $(-1, 1)$, they always have opposite signs in (26), leading to

$$y[n; \tau] = \pm A_\epsilon (1 - 2\nu_\tau) + w[n; \tau], \tag{29}$$

where the $\pm$ sign is determined by $s[n - 1 + 1_\tau]$ and plays no role in (28). Notice also that the noise term $w[n; \tau]$ does not alter the peak point of (28) because it just adds a constant noise floor to the objective in (28). In the noise-free case, when $\tau = \tau_0$, the integration window in (25) encompasses exactly one symbol, with its sample amplitude determined by the energy $\mathcal{E}_s \mathcal{E}_{\max}$ contained in the entire symbol duration. When $\tau \neq \tau_0$, $A_\epsilon |1 - 2\nu_\tau| < \mathcal{E}_s \mathcal{E}_{\max}$, reflecting the energy cancellation effect of this training pattern in the presence of mistiming. This explains why the peak amplitude of $y^2[n; \tau]$ yields the correct timing estimate for τ_0. It is the change of symbol signs exhibited in the alternating symbol pattern $(1, -1)$ that reveals the timing information in symbol-rate samples $y[n; \tau]$. The benefit of the DA approach is very rapid acquisition, since only two pairs $(N_s = 1, N = 1)$ of received symbol-long segments, carrying as few as 3 training symbols, are sufficient.

A complication to the TOE estimator in (28) arises when there is a noticeable gap between the symbol period T_s and the nonzero time-support of $p_R(t)$, defined as $T_R := \sup\{t : p_R(t) \neq 0\}$. The objective function $z(\tau)$ is maximized for any $\tau \in [\tau_0 - T_s + T_R, \tau_0]$, exhibiting a plateau around its peak. On the other hand, τ_0 is a unique maximum of $z(\tau)$ if we replace the integration range in (25) by $[\tau, \tau + T_R]$. Since $T_R = T_s - (c_1 - c_{N_f - 1})T_c - T_f + \tau_{L-1,0}$, it can be readily determined through channel sounding experiments that determine the delay spread $\tau_{L-1,0}$. Even when T_R is unknown, choosing any point on the plateau around τ_0 leads to maximum energy capture, which, as far as symbol detection is concerned, is what controls BER in demodulation. From this point of view, although nonunique, any point on the plateau can be considered as being equally good. In the following text, we assume that T_R is known to the receiver. The result on SMS-based TOE is summarized next.

Search-based SMS Synchronization with Flexible Timing Resolution *Consistent TOE can be accomplished in the absence of ISI, even when TH codes are present and the UWB multipath channel is unknown, as:*

$$\hat{\tau}_0 = \arg \max_{\tau \in [0, T_s)} \frac{1}{N} \sum_{n=1}^{N} \left(\int_{\tau}^{\tau + T_R} r(t + nT_s) \hat{\tilde{p}}_R(t; \tau_0)\, dt \right)^2, \tag{30}$$

where $\hat{\bar{p}}_R(t; \tau_0) = \hat{p}_R(t \bmod T_s; \tau_0)$. *Both training and blind modes exhibit excellent multipath energy capture and flexible timing resolution, determined conveniently by the search stepsize of the candidate time-shift* τ, *but the data-aided mode also enjoys rapid acquisition by relying on a training* $(1, -1)$ *symbol sequence.*

4.6.2 Timing-Offset Search via Cross-Correlation Mean Square

The TOE principle in (28) is based on discrete-time sample energy detection for correlators using the MAT as a template. This estimation principle can be established for cross-correlation outputs without using the MAT, which is known as TDT [29]. TDT permits a variety of implementation options, among which several notables ones are summarized below.

4.6.2.1 TDT Based on Cross-Correlation

This approach to estimating τ_0 hinges upon cross-correlating pairs of successive symbol-long segments of $r(t)$ at candidate time shifts $\tau \in [0, T_s)$, where one segment in each pair serves as the template for the other segment. Specifically, integrate and dump operations are performed on products of such segments to obtain $y[n; \tau]$, replacing those in (25) [29]:

$$y[n; \tau] = \int_0^{T_s} r(t + 2nT_s + \tau) r(t + (2n - 1)T_s + \tau)\, dt. \tag{31}$$

This cross-correlation–based TOE principle can be intuitively understood in the DA mode, in which the repeated training pattern $(1, 1, -1, -1)$ is particularly suitable. In this case, any two consecutive T_s-long segments, $r(t + 2nT_s + \tau)$ and $r(t + (2n - 1)T_s + \tau)$, for $t \in [0, T_s)$, will encounter exactly two distinct patterns of symbol pairs, $(1, 1)$ (or $(-1, -1)$) and $(1, -1)$ (or $(-1, 1)$), but not necessarily in this order. Therefore, in each cross-correlation, one segment corresponding to $\pm(1, 1)$ serves as the MAT without noise averaging, while the other segment corresponding to $\pm(1, -1)$ reveals timing. As a result, the expressions in (26) and (29) stand even without using an explicit MAT, proving the effectiveness of (31).

The timing estimator in (28) in combination with the cross-correlation samples $y[n; \tau]$ in (31) can also be applied in the NDA mode [3, 29]. In essence, the cross-correlation of successive symbol-long received segments reach its maximum if and only if these segments are scaled versions of each other, which is achieved uniquely at the correct timing $\tau = \tau_0$ if $T_s = T_R$. In its simplicity, this neat observation offers a distinct criterion for timing synchronization, which for years has relied on the idea that the autocorrelation of the noise-free template has a unique maximum at the correct timing; this idea has been the principle behind all existing narrowband timing schemes. For any $T_R \leq T_s$, TOE can be accomplished as follows [30]:

TDT-CC: TDT based on Cross-Correlation *Consistent data-aided and blind TOE can be accomplished in the absence of ISI, on the basis of cross-correlation among* T_s*-long* *segments of the received waveform, as follows:*

$$\hat{\tau}_0 = \arg \max_{\tau \in [0, T_s)} \frac{1}{N} \sum_{n=1}^{N} \left(\int_{nT_s}^{nT_s+T_R} r(t+\tau) r(t+\tau-T_s)\, dt \right)^2 . \tag{32}$$

In particular, the data-aided mode also enjoys rapid acquisition relying on as few as four training symbols $(1, 1, -1, -1)$.

4.6.2.2 TDT Based on First-Order Cyclic Mean

The timing estimation performance of TDT can be further improved in the DA mode if one opts to average analog waveforms (which may be more costly to implement). Specifically, by relying on the training pattern $(1, 1, -1, -1)$, one can average over every $2T_s$-long segments of $r(t)$ to yield

$$\bar{r}(t) = \frac{1}{N} \sum_{n=1}^{N} (-1)^n r(t + 2nT_s), \qquad t \in [0, 2T_s) \tag{33}$$

where $(-1)^n$ is used inside the summand because adjacent $2T_s$-long segments of $r(t)$ always entail the same training symbol pattern $\pm(1, 1, -1)$ (or $\pm(1, -1, -1)$), but of opposite signs. In essence, $\bar{r}(t)$ is the cyclic mean of $r(t)$ with period $2T_s$.

Without loss of generality, let us suppose that $\bar{r}(t)$ corresponds to the training pattern $(1, 1, -1)$. Accordingly, let us write explicitly the averaged symbol-long received segments, which entail symbol pairs of the patterns $(1, 1)$ and $(1, -1)$:

$$\bar{r}(t+\tau) = \sqrt{\mathcal{E}_s}\big(p_R(t + T_s - \tau_\Delta) + p_R(t - \tau_\Delta)\big) + \bar{w}(t+\tau), \tag{34a}$$

$$\bar{r}(t+\tau-T_s) = \sqrt{\mathcal{E}_s}\big(p_R(t + T_s - \tau_\Delta) - p_R(t - \tau_\Delta)\big) + \bar{w}(t+\tau-T_s), \tag{34b}$$

where $\bar{w}(t)$ is the averaged noise of PSD σ^2/N. Apparently, $\bar{r}(t+\tau)$ is the average noisy template playing the same role as the MAT $\hat{\tilde{p}}_R(t; \tau_0)$ in (25). Correlating it with the other segment $\bar{r}(t+\tau-T_s)$ containing symbols of opposite signs, we reach the same expression as (29):

$$\bar{y}(\tau) := \int_0^{T_s} \bar{r}(t+\tau)\bar{r}(t+\tau-T_s)\, dt = \mathcal{E}_s\mathcal{E}_{\max}\big(1 - 2\nu_\tau\big) + \bar{w}(\tau), \tag{35}$$

where $\bar{w}(\tau)$ is the zero-mean Gaussian distributed with the variance $\sigma^2_{\bar{w}} = \mathcal{E}_s\mathcal{E}_{\max}\sigma^2/N + BT_s\sigma^4/N^2$. As a consequence of (35), the following timing estimator arises [29]:

TDT-CM: TDT based on First-Order Cyclic Mean *With the training pattern* $(1, 1, -1, -1)$, *data-aided synchronization can be accomplished in the absence of ISI with as few as four training symbols, on the basis of cross-correlation among the* $2T_s$*-long segments of the cyclic mean of* $r(t)$ *in (33), as follows:*

$$\hat{\tau}_0 = \arg \max_{\tau \in [0, T_s)} z(\tau), \quad z(\tau) = \left[\int_0^{T_R} \bar{r}(t+\tau)\bar{r}(t+\tau-T_s)\, dt \right]^2 . \tag{36}$$

Compared with TDT based on cross-correlation, (36) results in better estimation performance because it averages noise more effectively.

4.6.3 *Comparative Study and Implementation Aspects*

All the timing algorithms in (30), (32), and (36) follow the estimation principle of energy detection using sample squares. The samples $y[n;\tau]$ all boil down to the form in (26), except that their noise variances are different. Accordingly, the mean values of all objective functions can be unified as

$$\bar{z}(\tau) := \mathrm{E}\{z(\tau)\} = \begin{cases} \mathcal{E}_s^2\mathcal{E}_{\max}^2(1-2\nu_\tau)^2+\sigma_w^2, & \text{DA;} \\ \mathcal{E}_s^2\mathcal{E}_{\max}^2(1-2\nu_\tau)^2/2+\mathcal{E}_s^2\mathcal{E}_{\max}^2/2+\sigma_w^2, & \text{NDA;} \end{cases} \tag{37}$$

where σ_w^2 is the noise variance contributed from the noise components in both $r(t)$ and the (implicit) MAT. The estimation performance of each TOE algorithm is determined by two metrics: one is the error distance $d(\bar{z}) := \max_\tau\{\bar{z}(\tau)\} - \min_\tau\{\bar{z}(\tau)\}$ that is preferably large, and the other is the variance $\sigma_z^2 := \mathrm{E}\{z^2(\tau)\} - \mathrm{E}^2\{z(\tau)\}$ that determines both the error performance and the convergence rate. It is straightforward to show that $d(\bar{z}) = \mathcal{E}_s^2\mathcal{E}_{\max}^2$ in the DA mode and $d(\bar{z}) = \mathcal{E}_s^2\mathcal{E}_{\max}^2/2$ in the NDA mode. This confirms that the DA algorithms are more effective than their NDA counterparts; moreover, the larger the receive-energy $\mathcal{E}_s\mathcal{E}_{\max}$, the better the TOE accuracy. The variances σ_z^2 of all the three algorithms are summarized in Table 4.1.

Note that the training pattern in SMS is $(1,-1)$, which is half the size of the $(1,1,-1,-1)$ pattern in TDT algorithms. Setting $N_s = N$ so that the total number of effective samples are the same for all algorithms, it can be deduced from Table 4.1 that

Table 4.1 Variances of the objective functions in TOE search algorithms

	Noise variance σ_w^2
SMS	$\sigma_{w,1}^2 = \mathcal{E}_s\mathcal{E}_{\max}\sigma^2(1+N_s)/N_s + BT_s\sigma^4/N_s$
TDT-CC	$\sigma_{w,2}^2 = 2\mathcal{E}_s\mathcal{E}_{\max}\sigma^2 + BT_s\sigma^4$
TDT-CM	$\sigma_{w,3}^2 = 2\mathcal{E}_s\mathcal{E}_{\max}\sigma^2/N + BT_s\sigma^4/N^2$
	Variance $\sigma_{z,\mathrm{DA}}^2$ in the DA mode
SMS	$\sigma_{z1,\mathrm{DA}}^2 = (2\sigma_{w,1}^2/N)\left[\sigma_{w,1}^2 + 2\mathcal{E}_s^2\mathcal{E}_{\max}^2(1-2\nu_\tau)^2\right]$
TDT-CC	$\sigma_{z2,\mathrm{DA}}^2 = (2\sigma_{w,2}^2/N)\left[\sigma_{w,2}^2 + 2\mathcal{E}_s^2\mathcal{E}_{\max}^2(1-2\nu_\tau)^2\right]$
TDT-CM	$\sigma_{z3,\mathrm{DA}}^2 = (2\sigma_{w,3}^2)\left[\sigma_{w,3}^2 + 2\mathcal{E}_s^2\mathcal{E}_{\max}^2(1-2\nu_\tau)^2\right]$
	Variance $\sigma_{z,\mathrm{NDA}}^2$ in the NDA mode ($\lambda_\tau := \nu_\tau - \nu_\tau^2$)
SMS	$\sigma_{z1,\mathrm{NDA}}^2 = \sigma_{z1,\mathrm{DA}}^2 + (4\mathcal{E}_s^2\mathcal{E}_{\max}^2/N)\left[2\sigma_{w,1}^2\lambda_\tau + \mathcal{E}_s^2\mathcal{E}_{\max}^2\lambda_\tau^2\right]$
TDT-CC	$\sigma_{z2,\mathrm{NDA}}^2 = \sigma_{z2,\mathrm{DA}}^2 + (4\mathcal{E}_s^2\mathcal{E}_{\max}^2/N)\left[2\sigma_{w,2}^2\lambda_\tau + \mathcal{E}_s^2\mathcal{E}_{\max}^2\lambda_\tau^2\right]$
TDT-CM	$\sigma_{z3,\mathrm{NDA}}^2 = \sigma_{z3,\mathrm{DA}}^2 + (4\mathcal{E}_s^2\mathcal{E}_{\max}^2)\left[2\sigma_{w,3}^2\lambda_\tau + \mathcal{E}_s^2\mathcal{E}_{\max}^2\lambda_\tau^2\right]$

the SMS and TDT-CM algorithms have competitive variances smaller than that of TDT-CC algorithms, thus yielding better estimation accuracy. In SMS, the separate parameters N_s and N make it possible to optimally allocate transmission resources between the (1, 1) and (1, −1) training patterns. On the other hand, both SMS and TDT-CM involve analog waveform averaging, which can be more expensive to implement. The TDT-CC algorithm is simple to implement, at the expense of timing accuracy. The SMS algorithm involves the MAT, which facilitates the recovery of the SAT needed as a template during data demodulation; both TDT algorithms, on the other hand, do not yield explicit MAT. Nevertheless, the use of cross-correlation samples in TDT is convenient during DA synchronization because it does not need any locally generated receive-template and only requires the knowledge of the training pattern rather than the actual symbol values.

The performance of the three search algorithms is depicted in Figure 4.3, using the same system setup as in Figure 4.2. As Figure 4.3(a) illustrates, the search stepsize Δ, affects the TOE accuracy at the high SNR region, where a smaller stepsize results in a lower error floor. In the low SNR region, the timing accuracy is primarily dictated by the multipath energy capture capability of the synchronizer, which is independent of the stepsize because of the asymptotically optimal template, $\hat{\tilde{p}}_R(t; \tau_0)$, used. Figure 4.3(b) corroborates the performance comparison of various search algorithms in Table 4.1. For this particular system step, the SMS algorithm outperforms the two TDT algorithms. On the other hand, when the time-bandwidth product BT_s is very large, the TDT-CM algorithm may yield the best TOE accuracy.

The estimators in (30), (32), and (36) enable timing synchronization at any desirable resolution, constrained only by the affordable complexity such as the following three types of timing synchronizations: (i) coarse timing with low complexity, for example, by picking the maximum over N_f candidate offsets $\{\tau = nT_f\}_{n=0}^{N_f-1}$ taken every T_f seconds in $[0, T_s)$; (ii) fine timing with higher complexity at the chip resolution; and (iii) adaptive timing (tracking) with voltage-controlled clock (VCC) circuits. An illustration of the VCC

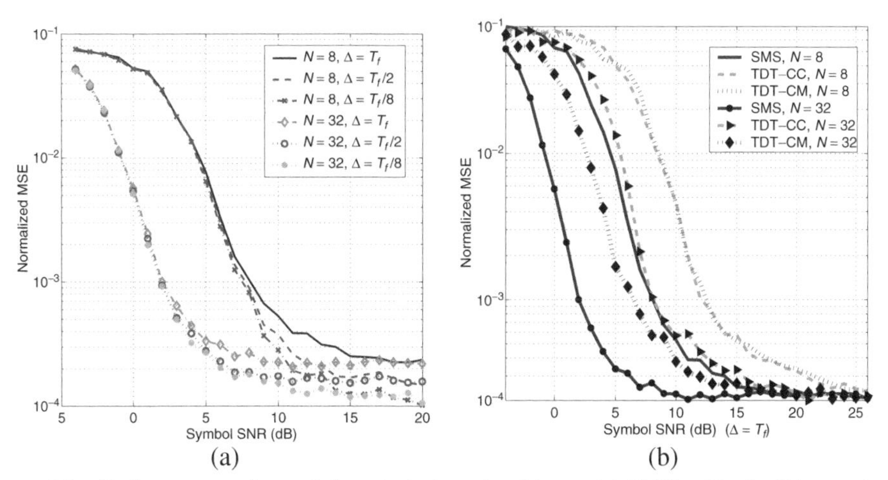

Figure 4.3 Performance of search-based timing algorithms: (a) SMS with flexible resolution, (b) comparison of SMS, TDT-CC, and TDT-CM

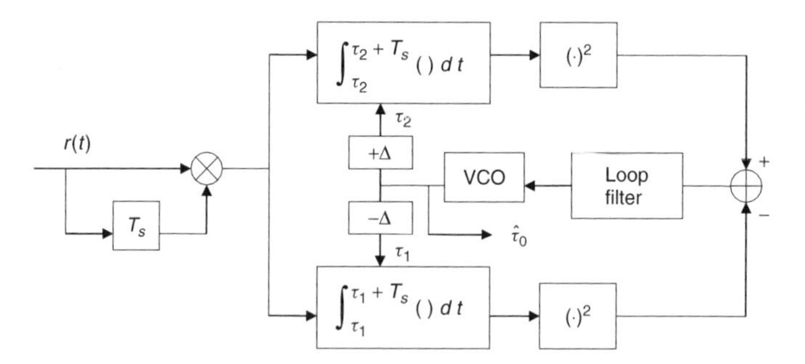

Figure 4.4 Voltage-controlled clock circuit for the TDT synchronizer based on cross-correlation

implementation is shown in Figure 4.4, using the TDT synchronizer based on cross-correlation as an example.

4.7 Timing Acquisition for Ad Hoc Multiple Access

So far, we have presented timing algorithms for a point-to-point link, where MUI is treated as noise. This is reasonable in a multiple-access setting, provided that user separation is accomplished through channelization or medium access control. Even so, user separation with despreading relies heavily on timing. In the section, we discuss the means of providing user separation prior to timing and present timing algorithms when MUI is present.

Replacing the point-to-point link model in (3), the received waveform in the presence of MUI becomes:

$$r(t) = \sqrt{\mathcal{E}_s} \sum_{k=0}^{\infty} s[k] p_R(t - kT_s - \tau_0) + \rho(t) + w(t), \tag{38}$$

where $\rho(t)$ represents the MUI and $w(t)$ is zero-mean additive noise. As in the single-user case, the issue here is how the desired template $p_R(t;\, \tau_0)$ can be obtained in the presence of unknown $\rho(t)$? We will address this question by relying on the first-order mean of $r(t)$. To this end, we make the following assumption:

A1. In the multiuser setup, there is only one user to be synchronized at a time, while all the interfering users communicate zero-mean independent and identically distributed information symbols, yielding $\mathrm{E}\{\rho(t)\} = 0$.

4.7.1 Training-Based Multiuser TOE

Let us consider the TDT algorithm that is based on the first-order cyclic mean. Suppose that a single user is in the DA synchronization phase and relies on the training pattern $(1, 1, -1, -1)$. Even if $\rho(t) \neq 0$, the average waveform $\bar{r}(t)$ still obeys the expressions in (34), simply because $\mathrm{E}\{\rho(t)\} = 0$ under A1. This observation suggests that the single-user TDT estimator in (36) is operational even in a multiuser environment. The price paid is an increase in the noise variance $\sigma_{\bar{w}}^2$ because of the MUI variance $\mathrm{E}\{\rho^2(t)\}$. Summarizing, we have that [29]:

Training-Based Multiuser TDT *With the desired user transmitting training symbols and other users transmitting zero-mean i.i.d. information symbols, MSS consistent data-aided TDT can be accomplished in the absence of ISI using (36) without requiring any knowledge of these users' channel or timing information.*

This multiuser TDT method relies on the first-order cyclic mean values of the transmitted symbols to separate the desired user from the interfering ones. This user separation principle can be exploited by any TOE method performing first-order mean as the first step immediately following the receiver's front-end. This category of timing algorithms include all DA versions of the flexible-rate timing algorithms in (30), (32), and (36), as well as the CML coarse timing algorithm in Section 4.5. Note that the mean-square operations in (30) and (32) need to be replaced by their first-order mean counterparts by substituting $(1/N)\sum_{n=1}^{N}\left(\int(\cdot)\,dt\right)^2$ with $\left((1/N)\sum_{n=1}^{N}(-1)^n\int(\cdot)\,dt\right)^2$. Although all these algorithms are operational in the multiuser setup, their TOE performances entail different levels of sensitivity to MUI.

4.7.2 Blind Synchronization for Multiuser Ad Hoc Access

In ad hoc multiple access, there is a need for simple and preferably blind synchronizers that are flexible to operate with transmissions of variable bandwidth over AWGN or multipath channels, in single- or multiuser settings. Given only $r(t)$ in (38), it is important to have transmission protocols and low-complexity receiver processors enabling blind estimation of the timing offset τ_0 and the desired correlation template $p_R(t)$, as well as coherent demodulators to detect $s[n]$, in the presence of noise, MUI, and even ISI.

4.7.2.1 NDA User Separation

For recovery of τ_0 with sufficient multipath energy capture, the first step is to (blindly) estimate the ideal correlation template $p_R(t;\tau_0)$ under mistiming. To this end, the key idea is to exploit occasional, yet periodic, nonzero mean (NZM) signaling, which provides a new dimension for signal separation in the presence of MUI [31].

For high–data-rate applications in multipath channels, we generalize the signal model in (38) to allow $p_R(t)$ to span over multiple symbol periods, that is, $T_R > T_s$, thus allowing for ISI to be present. During blind synchronization, the transmitter periodically sends an NZM symbol after every $M-1$ zero-mean symbols, where $M := \lceil T_R/T_s \rceil + 1$ ($\lceil \cdot \rceil$ denoting integer ceiling) is the number of symbol periods spanned by the ISI-inducing $p_R(t)$ plus one additional guard symbol time. The NZM property can be effected by asymmetric PAM (A-PAM), where equally probable binary symbols are transmitted at different amplitudes, that is, $s[n] = \theta > 1$ when transmitting bit '1'; and $s[n] = -1$ when transmitting bit '0'. The symbol mean value is thus $\mu_0 := \mathrm{E}\{s[n]\} = 0.5\theta + 0.5(-1) > 0$. The same mean arises if we superimpose a deterministic (training) constant $(\theta-1)/2$ to a symmetric PAM (S-PAM) modulation, as in [32]. However, power is wasted with superimposed training, and BER performance will degrade relative to the A-PAM if we maintain the detection threshold at zero, which is desirable because the same decoder is used for both zero-mean and NZM symbols.

Writing $n = kM + m$ with $m \in [0, M-1]$, the symbol stream $\{s[n]\}$ with periodic NZM signaling takes values from a finite alphabet equiprobably, yielding:

$$\mathrm{E}\{s[kM+m]\} = \mu_0 \delta(m), \qquad \mu_0 \neq 0. \tag{39}$$

If the receiver can only 'hear' a single transmitter broadcasting the NZM synchronization pattern, then the assumption A1 is satisfied regardless of how many zero-mean interfering signals from other communicating nodes are present. This is the case with star or clustered topologies of ad hoc networks, where a single (but not always the same) node undertakes the task of synchronizing neighbors. Taking into account (39), the mean of the received waveform in (38) is:

$$\mathrm{E}r(t) = \sqrt{\mathcal{E}_s}\mu_0 \sum_k p_R(t - kMT_s - \tau_0). \tag{40}$$

In the special case of $M = 1$, $\mathrm{E}r(t)$ reduces to its ISI-free counterpart $\tilde{p}_R(t; \tau_0)$ in (14). In its simple form, (40) has important implications: signaling in an NZM fashion among zero-mean interfering users *enables MUI suppression* after waveform averaging; meanwhile, raising the transmitter's voice only periodically with a period $MT_s > T_R$ *blindly avoids ISI* among NZM symbols, without the necessity of knowing the transmit-filter $p_T(t)$ or the channel $h(t)$. With both strategies, the blind recovery of the MAT is possible when both ISI and MUI are present.

Because $\mathrm{E}r(t)$ is periodic with period MT_s, it can be estimated over a period using the sample average across N segments of $r(t)$ each of size MT_s:

$$\bar{r}(t) = (1/N)\sum_{n=0}^{N-1} r(t + nMT_s), \qquad t \in [0, MT_s]. \tag{41}$$

Under mistiming, the average $\bar{r}(t)$ of length MT_s in the ISI case is the counterpart of the MAT $\hat{p}_R(t; \tau_0)$ in (16), which is of length T_s in the ISI-free case. We choose $MT_s \geq T_R + T_s$ to add a symbol-long zero-guard time that will be useful in the ensuing timing recovery based on simple energy detection.

4.7.2.2 Timing Estimation and Template Recovery

Having obtained $\mathrm{E}r(t)$, we then seek to recover the timing offset τ_0, as well as the optimal SAT $p_R(t)$.

Close examination of the relationship between $\mathrm{E}r(t)$ and $p_R(t)$ reveals that

$$p_R(t) = \frac{1}{\sqrt{\mathcal{E}_s}} \frac{1}{\mu_0} \mathrm{E}r(t + \tau_0), \qquad t \in [0, T_R]. \tag{42}$$

This implies that over any interval of size MT_s, the mean $\mathrm{E}r(t)$ in (40) contains a circularly shifted (by τ_0) copy of $p_R(t)$, which has a support of size $T_R \leq (M-1)T_s$. To determine τ_0, we will exploit the zero-guards of size $MT_s - T_R \geq T_s$ present in each period of $\mathrm{E}r(t)$. Specifically, let τ be a candidate shift, which we confine to $[0, MT_s)$, as per (40). With $\tau \in [0, MT_s)$, τ_0 can be uniquely acquired by [31, 33]

$$\tau_0 = \arg \max_{\tau \in [0, MT_s)} J(\tau) = \int_0^{T_R} [\mathrm{E}r(t+\tau)]^2 dt = \int_{\tau}^{\tau + T_R} [\mathrm{E}r(t)]^2 \, dt. \tag{43}$$

Indeed, for the correct timing $\tau = \tau_0$, the objective $J(\tau_0) = \mu_0^2 \mathcal{E}_s \mathcal{E}_{\max}$ extracts the whole energy of $p_R(t)$, which can be shown to be the unique maximum of $J(\tau)$ [31]. In practice, the ensemble mean $\mathrm{E}r(t)$ can be replaced by the sample mean $\bar{r}(t \bmod MT_s)$, where the modulo operation is involved to periodically extend the time-limited waveform $\bar{r}(t)$ to span over the integration window used in (43). In summary, blind synchronization and SAT recovery can be accomplished as follows [31]:

Blind Synchronization and SAT Recovery *The timing offset τ_0 and the SAT $p_R(t)$ can be estimated blindly in the presence of ISI and MUI using*

$$\hat{\tau}_0 = \arg \max_{\tau \in [0, MT_s)} \hat{J}(\tau) = \int_0^{T_R} \left[\bar{r}\big((t+\tau)\ mod\ MT_s\big)\right]^2 dt; \tag{44}$$

$$\hat{p}_R(t; \hat{\tau}_0) = \frac{1}{\sqrt{\mathcal{E}}} \frac{1}{\mu_0} \bar{r}(t + \hat{\tau}_0), \qquad t \in [0, T_R]. \tag{45}$$

It is worth emphasizing the universal applicability of this result to narrowband, wideband, and UWB regimes, in the presence or absence of ISI and/or MUI, with fixed or ad hoc access. Only readily available upper bounds on channel parameters are required for low-complexity blind estimation based on sample averaging and energy detection. In the ISI-free case, we have $T_R \leq T_s$, and $M = 2$ suffices. Notice that no available method can acquire the timing of UWB transmissions in the generic setting allowed by this NZM approach. There is no need to interrupt the information transmission for the training and transmit-filters, channels, or spreading codes need not be known, so long as they remain invariant while averaging in (41) takes place.

4.7.2.3 SAT-Based Demodulation

With the estimated SAT serving as the asymptotically optimal correlation template that captures the full multipath energy, we can form the decision statistic: $d[n] = \int_0^{T_R} \hat{p}_R(t; \hat{\tau}_0) r(t + \hat{\tau}_0 + nT_s)\, dt$, using $\hat{\tau}_0$ and $\hat{p}_R(t; \hat{\tau}_0)$ obtained as in (44) and (45). Substituting $r(t)$ from (38), we can rewrite $d[n]$ as [31]

$$d[n] = \sqrt{\mathcal{E}_s}\psi_{\hat{p}_R p_R}[0; \tilde{\tau}_0] s[n] + \sqrt{\mathcal{E}_s} \sum_{k=-2(M-1), k \neq 0}^{2(M-1)} \psi_{\hat{p}_R p_R}[k; \tilde{\tau}_0] s[n+k]$$
$$+ \rho[n; \hat{\tau}_0] + w[n; \hat{\tau}_0], \tag{46}$$

where $\tilde{\tau}_0 := \tau_0 - \hat{\tau}_0$, $\psi_{\hat{p}_R p_R}[k; \tilde{\tau}_0] := \int_0^{T_R} \hat{p}_R(t; \hat{\tau}_0) p_R(t - kT_s - \tilde{\tau}_0)\, dt$, $\rho[n; \hat{\tau}_0] := \int_0^{T_R} \hat{p}_R(t; \hat{\tau}_0) \rho(t + nT_s + \hat{\tau}_0)\, dt$, and likewise for $w[k; \hat{\tau}_0]$.

On the basis of (46), Viterbi's Algorithm (VA), sphere decoding, or linear equalization can be used depending on the application-specific trade-off between BER and affordable complexity [31]. In UWB receivers, in which (sub-)chip rate sampling is prohibitive, VA applied to $d[n]$ is the only ML optimal (in the absence of MUI) UWB receiver based on symbol-rate samples.

To further reduce the complexity, one can combine the ISI and MUI plus AWGN terms in (46) into a single colored noise term and proceed with a low-complexity (although suboptimal) slicer. In UWB single- or multiuser access with binary symbol transmissions, this amounts to demodulating symbols with the sign detector:

$$\hat{s}\,[n] = \text{sign}\left\{\int_0^{T_R} \hat{p}_R(t;\,\hat{\tau}_0) r(t + \hat{\tau}_0 + nT_s)\,dt\right\}. \tag{47}$$

In summary, both during and after the synchronization phase, ML optimal, linear equalization, and low-complexity matched-filter options are available for demodulating $s[n]$ from the decision statistic in (46) that is based on the MUI- and ISI-resilient SAT and the timing is estimated blindly.

Figure 4.5 depicts the results in (44) and (47) for the SAT-based synchronization and demodulation scheme, when both MUI and ISI are present [31]. Binary PAM symbols modulate the Gaussian monocycle $p(t)$ of pulse width $T_p = 1$ ns; in A-PAM, $\theta = 3$ is used to effect NZM. The IEEE 'CM-1' multipath channel is used [27], having a delay spread effectively upper bounded by 29 ns. The transmitter parameters are $T_f = 10$ ns and $N_f = 10$ to induce ISI, and $T_c = 1$ ns with randomly generated spreading code. The period of NZM symbols corresponds to $M = 3$. It is shown that with reasonable averaging (N), the SAT receiver will be only $1 - 2$ dBs away from the clairvoyant one with perfectly known timing and channel.

4.7.2.4 Timing Protocol for Ad Hoc Networks

To appreciate the merits of the blind synchronization and demodulation schemes in multiuser ad hoc access, we outline the overall protocol for establishing communication links within an ad hoc UWB network. As explained in Figure 4.6, the master node A broadcasts regularly to its cluster, the zero-mean S-PAM information symbols and occasionally A-PAM information symbols, obeying the NZM synchronization pattern. An idle neighbor node B 'wakes up' to link with node A, following the steps in Table 4.2 [31].

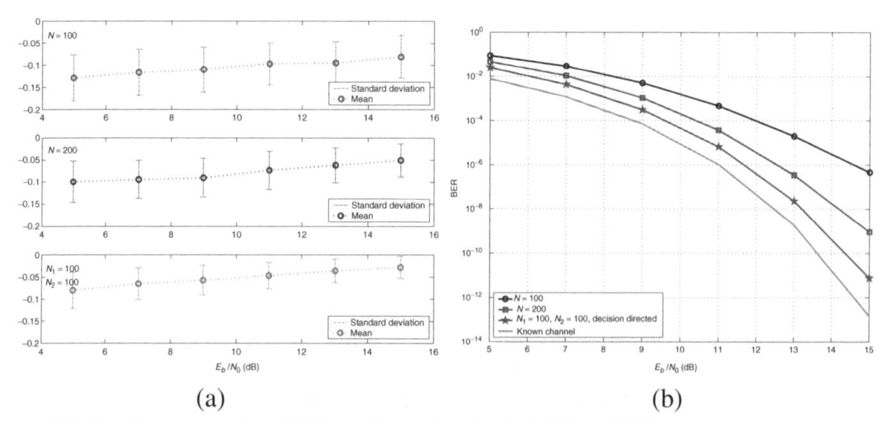

(a) (b)

Figure 4.5 Performance of an SAT receiver when both MUI and ISI are present: (a) normalized (with respect to T_s) mean $\pm$ half of the standard deviation of the timing estimation errors; (b) BER performance (the solid line with $\star$ denotes the decision-directed version of SAT [31])

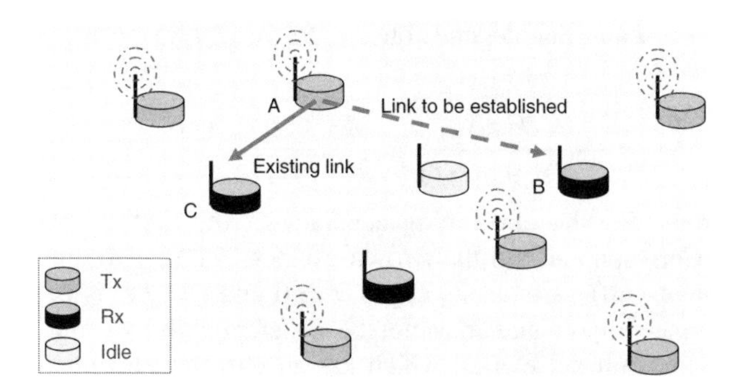

Figure 4.6 An ad hoc multiple-access configuration (cluster topology) Tx – Transmitter; Rx – Receiver [31]

Table 4.2 Blind timing protocol for ad hoc multiple-access networks

Step 1:	Node A switches intermittently from S-PAM to A-PAM;
Step 2:	Node B acquires timing and channel information in the A-PAM mode, using the blind estimation results in (44) and (45);
Step 3:	Node A returns to S-PAM and continues broadcasting information in an energy-efficient mode to all nodes, including node B;

While node B is synchronizing during Step 1, other nodes receiving information from (and already in sync with) node A do not have to interrupt communication; that is, the information received by the nodes in S-PAM before and after Step 2 is continuously received in the A-PAM format during synchronization. This implies that only occasionally does node A have to increase its power (while transmitting bit '1') to establish links with 'new' users (nodes), without interrupting or affecting the rate and BER performance of existing links.

4.7.3 TOE Performance Analysis

To benchmark timing estimation accuracy of MAT-based TOE algorithms, we take the SAT synchronizer in (44) as an example and derive its asymptotic estimation variance using first-order perturbation analysis [34]. Because of noise, the maximum of $\hat{J}(\tau)$ moves from τ_0 to $\hat{\tau}_0 = \tau_0 + \Delta\tau$, thus inducing an estimation error $\Delta\tau$. When N or $\mathcal{E}_s$ is sufficiently large, $\hat{\tau}_0$ will be close to the true timing offset τ_0 such that $|\Delta\tau| \leq \delta$ with probability close to 1, where $\delta \ll T_p$ is a small positive number.

Following perturbation analysis, let $\dot{J}(\tau)$ denote the derivative of the objective function $\hat{J}(\tau)$ with respect to τ. It follows from (44) that

$$\dot{J}(\tau) = \bar{r}^2(\tau + T_R) - \bar{r}^2(\tau). \tag{48}$$

Since $|\Delta\tau| \leq \delta$, we can use the mean value theorem to obtain:

$$\dot{J}(\hat{\tau}_0) = \dot{J}(\tau_0 + \Delta\tau) = \dot{J}(\tau_0) + \ddot{J}(\tau_0 + \mu\Delta\tau)\Delta\tau, \tag{49}$$

where $\mu \in (0, 1)$ is a scalar that depends on $\Delta\tau$. Because $\dot{J}(\hat{\tau}_0) = 0$, if follows from (48) and (49) that

$$\Delta\tau = -\frac{\dot{J}(\tau_0)}{\ddot{J}(\tau_0 + \mu\Delta\tau)} \approx \frac{1}{2\mathcal{E}_s} \frac{\dot{J}(\tau_0)}{p_R(\mu\Delta\tau)\dot{p}_R(\mu\Delta\tau)}. \tag{50}$$

It is evident from (50) that the timing accuracy of the SAT synchronizer is closely related to $p_R(t)$. Corresponding to the two cases, $\hat{\tau}_0 \in [\tau_0, \tau_0 + \delta]$ and $\hat{\tau}_0 \in [\tau_0 - \delta, \tau_0]$, the local behavior of $p_R(t)$ around its edges, $t = 0$ and $t = T_R$, is of interest. Because $p_R(t)$ has finite time-support and may not be differentiable, we impose the following conditions on the transmit-pulse $p(t)$, which has an ultrashort length T_p and determines the local behavior of $p_R(t)$.

C1: $p(t)$ is continuous and differentiable over $t \in [0, +\delta]$, and $p(t) \propto t^a$ with $a < 1/2$ for $t \in [0, +\delta]$;
C2: $p(t)$ is continuous and differentiable over $t \in [T_p - \delta, T_p]$, and $p(t) \propto t^b$ with $b < 1/2$ for $t \in [T_p - \delta, T_p]$.

Under conditions **C1** and **C2**, it can be shown from (50) that the MSE of the timing estimator is upper bounded for sufficiently large N by [34, 35]:

$$\mathrm{E}[\Delta\tau^2] \le \frac{1}{\min\{p_R^2(\delta)\dot{p}_R^2(\delta),\, p_R^2(T_R - \delta)\dot{p}_R^2(T_R - \delta)\}} \frac{B\sigma^4}{N^2\mathcal{E}_s^2}. \tag{51}$$

As (51) indicates, the SAT synchronizer is able to accurately recover the receive-time even in the presence of MUI, as its estimation variance decreases quadratically with N. This result also delineates the N required to achieve a desired level of timing accuracy. On the other hand, the asymptotic variance is only applicable to the small-error local region around τ_0, which requires the search time resolution to be sufficiently small and the SNR or N to be sufficiently large.

4.8 Demodulation and BER Sensitivity to Mistiming

Timing synchronization enables coherent data demodulation. In this section, we present a SAT-based demodulator and evaluate its performance. We focus on demodulation in the ISI-free case, which follows any of the synchronizers presented in Sections 4.5 and 4.6. The optimal SAT-based receiver in the presence of ISI has been presented in Section 4.7.2.3.

Let us focus on symbol-by-symbol optimum reception. Because mistiming gives rise to ISI, an optimal ML approach requires sequence detection, as in (46). On the other hand, when low-complexity demodulation is sought, symbol-by-symbol detection is typically used. The receiver we discuss here is an optimum symbol-by-symbol ML detector under mistiming, which reduces to a broad-sense optimal ML detector under perfect timing.

To gain full multipath diversity, the receiver may use the MAT to generate the decision statistic for $s[n]$, as in (25):

$$d[n] = \int_{\hat{\tau}_0}^{\hat{\tau}_0+T_s} r(t + kT_s)\hat{\bar{p}}_R(t; \tau_0)\, dt = \int_{\hat{\tau}_0}^{\hat{\tau}_0+T_s} r(t + kT_s)\hat{p}_R\big(t \bmod T_s; \tau_0\big)\, dt. \tag{52}$$

Equivalently, the receiver can recover the SAT $\hat{p}_R(t;\hat{\tau}_0) = \hat{p}_R\big((t+\hat{\tau}_0) \bmod T_s;\tau_0\big)$, as in (17) and use it as the template for detection [26, 31, 36]:

$$d[n] = \int_0^{T_s} r(t + kT_s + \hat{\tau}_0)\hat{p}_R(t;\hat{\tau}_0)\,dt. \tag{53}$$

Recalling (26), either (52) or (53) leads to:

$$d[n] = \mathcal{E}_s\mathcal{E}_{\max}\big[s[n]\nu(\hat{\tau}_0) + s[n+a](1-\nu(\hat{\tau}_0))\big] + w[n;\hat{\tau}_0], \tag{54}$$

where

$$\nu(\hat{\tau}_0) := \frac{1}{\mathcal{E}_{\max}}\int_{\max\{0,\hat{\tau}_0-\tau_0\}}^{\min\{T_s,T_s+\hat{\tau}_0-\tau_0\}} p_R^2(t)\,dt, \quad a = \begin{cases} 1, & \hat{\tau}_0 > \tau_0 \\ -1, & \hat{\tau}_0 < \tau_0 \end{cases},$$

and, as discussed in (26), the noise term $w[n;\hat{\tau}_0]$ can be well approximated as zero-mean Gaussian with variance σ_w^2 being the same as that in (19). It is evident from the presence of $\mathcal{E}_{\max}$ in (54) that sliding correlation with the SAT always leads to asymptotically full multipath energy capture regardless of $\hat{\tau}_0$. However, when there is a large discrepancy between $\hat{\tau}_0$ and τ_0, there will a considerable portion of symbol energy leaked to the ISI term $s[n+a]$ in (54). This portion depends on the integration range $[0, \hat{\tau}_0 - \tau_0]$ or $[T_s - (\hat{\tau}_0 - \tau_0), T_s]$, which has length $|\hat{\tau}_0 - \tau_0|$. Ignoring the ISI term, a symbol-by-symbol detector makes a decision on $s[n]$ directly from $d[n]$, using

$$\hat{s}\,[n] = \text{sign}\{d[n]\}. \tag{55}$$

Next, we analyze the BER performance of the detector in (55). In wireless fading channels, the taps $\{\alpha_l, \tau_{l,0}\}_{l=0}^{L}$ are treated as random, and possibly time-varying, which complicates performance analysis relative to the fixed channel case [37, 38]. To evaluate the BER in any random channel, we first investigate the instantaneous BER $P_e(\hat{\tau}_0|p_R)$ conditioned on $p_R(t)$, which depends on the underlying channel realization of $\{\alpha_l, \tau_l\}$. Depending on whether $(s[n], s[n+a])$ have the same or opposite signs, $P_e(\hat{\tau}_0|p_R)$ is given by [6]

$$\begin{aligned} P_e(\hat{\tau}_0|p_R) &= 0.5\ \text{Prob}\{d[n] > 0|(1,1)\} + 0.5\ \text{Prob}\{d[n] < 0|(1,-1)\} \\ &= \frac{1}{2}Q\left(\sqrt{\frac{\mathcal{E}_s\mathcal{E}_{\max}}{\sigma^2\frac{N_s+1}{N_s} + \frac{\sigma^4 BT_s}{N_s\mathcal{E}_s\mathcal{E}_{\max}}}}\right) + \frac{1}{2}Q\left(\sqrt{\frac{\mathcal{E}_s\mathcal{E}_{\max}(2\nu(\hat{\tau}_0)-1)^2}{\sigma^2\frac{N_s+1}{N_s} + \frac{\sigma^4 BT_s}{N_s\mathcal{E}_s\mathcal{E}_{\max}}}}\right), \end{aligned} \tag{56}$$

where N_s is the number of symbol pairs (1, 1) used to recover the MAT, and it also determines the noise averaging in the SAT. In comparison, the BER of an optimal ML detector with perfect channel and timing information is given by simply setting $\hat{\tau}_0 = \tau_0$ and $\sigma_w^2 = \sigma^2$ (or equivalently, $N_s \to \infty$), yielding

$$P_e(\tau_0|p_R) = Q\left(\sqrt{\frac{\mathcal{E}_s\mathcal{E}_{\max}}{\sigma^2}}\right). \tag{57}$$

It can be shown that the detector in (55) exploiting the noisy templates attains full energy capture $\mathcal{E}_s\mathcal{E}_{\max}$, while bypassing complicated estimation of $\{\alpha_l, \tau_l\}_{l=0}^{L-1}$. The price paid is an increase in the noise variance from σ^2 to σ_w^2 via the time-bandwidth product BT_s, which can be large for low–duty-cycle IR. Nevertheless, σ_w^2 approaches σ^2 asymptotically as the data size N_s and/or the receive-energy $\mathcal{E}_s\mathcal{E}_{\max}$ increase(s). In contrast, the detection performance of a RAKE receiver is limited by the number of RAKE fingers utilized, even in the noise-free case.

The BER expression in (56) also indicates that the detectors in (52) and (53) are quite robust to timing errors. A point worth emphasizing is the BER dependence on the percentage of energy capture $\nu(\hat{\tau}_0)$, which matters more than the accuracy of $\hat{\tau}_0$ itself. Indeed, even when the error $|\hat{\tau}_0 - \tau_0|$ is relatively large, if $\nu(\hat{\tau}_0) \in [0, 1]$ captures most of the $\mathcal{E}_{\max}$ energy such that $|2\nu(\hat{\tau}_0) - 1|$ is close to 1, then the resultant BER will be low. This is important, since, after all, the goal is reliable demodulation rather than 'superaccurate' synchronization.

The BER in (56) is expressed for a given channel realization $p_R(t)$ regardless of the channel type. For any random channel, the average BER can be numerically evaluated by first computing $\mathcal{E}_{\max}$ and $\nu(\tau)$ for each channel realization, followed by averaging (56) over a large number of realizations, yielding $P_e(\tau) = \mathrm{E}_{p_R}\{P_e(\tau|p_R)\}$; see, for example, [37, 38]. With a proper channel simulator, this BER evaluation can be performed offline instead of simulating the entire communication system.

4.9 Concluding Summary

Within the framework of *noisy templates*, we presented a suite of practical timing-offset estimators including a coarse CML synchronizer in closed form, as well as search-based SMS, TDT, and SAT synchronizers with flexible timing resolution, all achieving asymptotically optimal multipath energy capture. For practical timing synchronization, a receiver may start with a signal energy detector for symbol-level TOE, then use a low-complexity CML synchronizer to identify the frame-level timing offset, followed by a search algorithm, preferably carried out locally around the coarse timing estimate. When the search stepsize is reduced to the pulse level, the search algorithms automatically accomplish fine-scale timing tracking for both sparse and dense multipath channels.

Key to the noisy template concept is to observe that the aggregate receive-waveform $p_R(t)$, although unknown, is embedded in the noise-corrupted received waveform $r(t)$. It is thus possible to directly extract from $r(t)$ the MAT and SAT templates that are noisy yet perfectly matched to $p_R(t)$, completely avoiding cumbersome estimation of the channel taps, delays, and distorted shapes. That latter is needed in a RAKE receiver to capture the energy in $p_R(t)$ by an approximate L_r-finger template, $\hat{p}_R^{\mathrm{RAKE}}(t) = \sum_{l=0}^{L_r-1} \hat{\alpha}_l \hat{p}_l(t - \hat{\tau}_l)$; RAKE combiners are not only strictly suboptimal when $L_r < L$ limits multipath energy capture but also expensive because they require pulse-rate sampling and multiple ($L_r \gg 1$) RF chains. The advantages of MAT/SAT-based algorithms lie not only in their low complexity but also in their excellent multipath capture capability, which is indispensable for accurate timing accuracy, and even more critical to demodulation performance.

The TOE principles covered in this chapter can be epitomized as different forms of *energy detection*. Analog and discrete-time energy detectors are presented in Section 4.3 to recover the symbol-level timing-offset parameter n_s. The CML algorithm in Section 4.5 coarsely equals the offset ratio $(n_f T_f + \epsilon)/T_s$ to the amplitude scale factor ν_τ induced by the sample energy ratio of two different symbol pair patterns, $(1, -1)$ vs. $(1, 1)$. The class of search algorithms in Section 4.6 relies on discrete-time energy detection in the form of mean-square values of either digital samples (as in the SMS algorithm) or cross-correlation outputs (as in the TDT algorithm). A feature common to these algorithms in both DA and NDA modes is that the transmitted symbols effective for synchronization are related to the pattern $(1, 1)$ that generates the MAT and $(1, -1)$ that reveals timing. The SAT receiver in Section 4.7 recovers τ_0 by energy detection from the analog MAT waveform after it is blindly separated from both ISI and MUI via periodic NZM signaling. The design principles in these TOE techniques are applicable to general communication systems, including narrowband, wideband, and UWB systems. However, it is under the unique UWB signaling that the performance-complexity advantages of these algorithms are best illustrated, especially for practical UWB applications operating in dense multipath.

There are several notable means of equipping a timing synchronizer with the capability of suppressing ISI and MUI. The multiuser synchronizers in Section 4.7 do not rely on any spreading code, since timing precedes despreading. User separation is accomplished by user-unique patterns of the transmitted symbols, in the form of either NZM signaling or training. To protect the desired template-waveform from ISI, a guard time used in periodic transmissions turns out to be useful during the synchronization phase. These low-complexity strategies culminate in the SAT synchronizer, which periodically raises the transmitter's voice to synchronize blindly with resilience to both ISI and MUI and uses the SAT to demodulate with asymptotic optimality. Such salient properties of the SAT receiver enjoy universal applicability in transmissions of variable bandwidth over AWGN or multipath channels, in single- or multiuser settings designed for fixed or ad hoc access.

The last note of this chapter is with regard to the implementation of the MAT and SAT. Mathematically expressed as the average over multiple consecutive segments of the received waveform $r(t)$ [cf (16) and (41)], the MAT, and hence the SAT, can be computed either digitally or in the analog form. Analog approaches avoid the possibly high sampling rates needed in the UWB regime, but implementing the analog delay required to shift successive $r(t)$ segments by either T_s in (16) or MT_s in (41) can be challenging. Nonetheless, chips implementing analog delays from $20 - 1000$ ns are available [39].

References

[1] R. A. Scholtz, "Multiple Access with Time-Hopping Impulse Radio," *Proc. IEEE Milcom Conference*, Boston, MA, pp. 447–450, Oct. 1993.

[2] M. Z. Win and R. A. Scholtz, "Ultra Wide Band-width Time-Hopping Spread-Spectrum Impulse Radio for Wireless Multiple Access Communications," *IEEE Trans. Commun.*, vol. 48, no. 4, pp. 679–691, Apr. 2000.

[3] L. Yang and G. B. Giannakis, "Ultra-Wideband Communications: An Idea Whose Time Has Come," *IEEE Signal Process. Magn.*, vol. 21, no. 6, pp. 26–54, Nov. 2004.

[4] A. R. Forouzan, M. Nasiri-Kenari, and J. A. Salehi, "Performance Analysis of UWB Time-Hopping Spread Spectrum Multiple-Access Systems: Uncoded and Coded Schemes," *IEEE Trans. Wireless Commun.*, vol. 1, no. 4, pp. 671–681, Oct. 2002.

[5] W. M. Lovelace and J. K. Townsend, "The Effects of Timing Jitter and Tracking on the Performance of Impulse Radio," *IEEE J. Select. Areas Commun.*, vol. 20, no. 12, pp. 1646–1651, Dec. 2002.

[6] Z. Tian and G. B. Giannakis, "BER Sensitivity to Timing Offset in UWB Impulse Radios," *IEEE Trans. Signal Process.*, vol. 53, no. 4/5, pp. 1897–1907/pp. 1550–1560, Apr./May 2005.

[7] R. Blazquez, P. Newaskar, and A. Chandrakasan, "Coarse Acquisition for Ultra Wideband Digital Receivers," *Proc. International Conference on Acoustics, Speech, and Signal Processing*, Hong Kong, China, pp. IV. 137–140, Apr. 2003.

[8] E. A. Homier and R. A. Scholtz, "Rapid Acquisition of Ultra-Wideband Signals in the Dense Multipath Channel," *Proc. IEEE Conference on UWB Systems & Technologies*, Baltimore, MD, pp. 245–250, May 2002.

[9] R. Fleming, C. Kushner, G. Roberts, and U. Nandiwada, "Rapid Acquisition for Ultra-Wideband Localizers," *Proc. IEEE Conference on UWB Systems & Technologies*, Baltimore, MD, pp. 245–250, May 2002.

[10] J. Kusuma, I. Maravic, and M. Vetterli, "Sampling with Finite Innovation Rate: Channel and Timing Estimation in UWB and GPS," *Proc. International Conference on Communications*, Anchorage, Alaska, vol. 5, pp. 3540–3544, May 2003.

[11] I. Maravic, J. Kusuma, and M. Vetterli, "Low-Sampling Rate UWB Channel Characterization and Synchronization," *J. Commun. Netw.*, vol. 5, no. 4, pp. 319–327, Dec. 2003.

[12] R. D. Weaver, "Frequency Domain Processing of Ultra-Wideband Signals," *Proc. Asilomar Conference on Signals, Systems & Computers*, Pacific Grove, CA, pp. 1221–1224, Nov. 2003.

[13] "IEEE 802.15 WPAN High Rate Alternative PHY Task Group 3a (TG3a)," Internet on-line access at http://www.ieee802.org/15/pub/TG3a.html Aug. 2004.

[14] D. Cassioli, M. Z. Win, and A. F. Molisch, "The Ultra-Wide Bandwidth Indoor Channel: From Statistical Model to Simulations," *IEEE J. Select. Areas Commun.*, vol. 20, no. 6, pp. 1247–1257, Aug. 2002.

[15] J. Foerster, "The Effects of Multipath Interference on the Performance of UWB Systems in an Indoor Wireless Channel," *Proc. IEEE Vehicular Technology Conference*, Rhodes, Greece, pp. 1176–1180, 2001.

[16] V. Lottici, A. D'Andrea, and U. Mengali, "Channel Estimation for Ultra-Wideband Communications," *IEEE J. Select. Areas Commun.*, vol. 20, no. 12, pp. 1638–1645, Dec. 2002.

[17] M. Z. Win and R. A. Scholtz, "On the Energy Capture of Ultra-Wide Bandwidth Signals in Dense Multipath Environments," *IEEE Trans. Commun. Lett.*, vol. 2, no. 9, pp. 245–247, Sept. 1998.

[18] C. J. Le Martret and G. B. Giannakis, "All-Digital Impulse Radio for Wireless Cellular Systems," *IEEE Trans. Commun.*, vol. 50, no. 9, pp. 1440–1450, Sept. 2002.

[19] H. Urkowitz, "Energy Detection of Unknown Deterministic Signals," *Proc. IEEE*, vol. 55, pp. 523–531, Apr. 1967.

[20] I. I. Immoreev and D. V. Fedotov, "Detection of UWB Signals Reflected from Complex Targets," *Proc. IEEE Conference on UWB Systems & Technologies*, Baltimore, MD, pp. 193–196, May 2002.

[21] Z. Tian and G. B. Giannakis, "A GLRT Approach to Data-Aided Timing Acquisition in UWB Radios – Part I: Algorithms," *IEEE Trans. Wireless Commun.*, vol. 4, no. 6, pp. 2956–2967, Nov. 2005.

[22] S. M. Kay, *Fundamentals of Statistical Signal Processing: Vol. II – Detection Theory*, Prentice Hall, 1998.

[23] J. G. Proakis and M. Salehi, *Communication Systems Engineering*, 2nd edition, Section 7.8: Symbol Synchronization, Prentice Hall, Upper Saddle River, New Jersey, 2002.

[24] R. T. Hoctor and H. W. Tomlinson, "Delay Hopped Transmitted Reference Experimental Results," *Proc. IEEE Conference on Ultra Wideband Systems and Technologies*, Baltimore, MD, pp. 105–110, May 2002.

[25] L. Yang and G. B. Giannakis, "Optimal Pilot Waveform Assisted Modulation for Ultra-Wideband Communications," *IEEE Trans. Wireless Commun.*, vol. 3, no. 4, pp. 1236–1249, Jul. 2004.

[26] Z. Tian and G. B. Giannakis, "A GLRT Approach to Data-Aided Timing Acquisition in UWB Radios – Part II: Training Sequence Design," *IEEE Trans. Wireless Commun.*, vol. 4, no. 6, pp. 2994–3004, Nov. 2005.

[27] IEEE P802.15 Working Group for WPANs, *Channel Modeling Sub-Committee Report Final*, IEEE P802.15-02/368r5-SG3a, Nov. 2002.

[28] Z. Wang and X. Yang, "Channel Estimation for Ultra Wide-Band Communications," *IEEE Signal Process. Lett.*, vol. 12, no. 7, pp. 520–523, July 2005.

[29] L. Yang and G. B. Giannakis, "Timing Ultra-Wideband Signals with Dirty Templates," *IEEE Trans. on Communications*, vol. 53, no. 11, pp. 1952–1963, Nov. 2005.

[30] L. Yang and G. B. Giannakis, "Blind UWB Timing with a Dirty Template," *Proc. International Conference on Acoustics, Speech and Signal Processing*, Montreal, Quebec, Canada, vol. 4, pp. IV.509-512, May 2004.

[31] X. Luo and G. B. Giannakis, "Low-Complexity Blind Synchronization and Demodulation for (Ultra-) Wideband Multiuser Ad Hoc Access," *IEEE Trans. Wireless Commun.*, 2006, to appear.

[32] G. T. Zhou, M. Viberg, and T. McKelvey, "First-Order Statistical Method for Channel Estimation," *IEEE Signal Process. Lett.*, vol. 10, pp. 57–60, Mar. 2003.

[33] X. Luo and G. B. Giannakis, "Blind Timing Acquisition for Ultra-Wideband Multi-User Ad Hoc Access," *Proc. International Conference on Acoustics, Speech & Signal Proceeding*, Philadelphia, PA, vol. 3, pp. iii.313–316, Mar. 2005.

[34] Z. Tian, X. Luo, and G. B. Giannakis, "Cross-Layer Sensor Network Synchronization," *Proc. 38th Asilomar Conference on Signals, Systems, and Computers*, Pacific Grove, CA, pp. 1276–1280, Nov. 2004.

[35] S. Farahmand, X. Luo, and G. B. Giannakis, "Demodulation and Tracking with Dirty Templates for UWB Impulse Radio: Algorithms and Performance, " *IEEE Trans. Veh. Technol.*, vol. 54, no. 5, pp. 1595–1608, Sept. 2005.

[36] L. Wu and Z. Tian, "Capacity-Maximizing Resource Allocation for Data-Aided Timing and Channel Estimation in UWB Radios," *Proc. IEEE International Conference on Acoustics, Speech and Signal Proceeding*, Montreal, Canada, vol. 4, pp. 525–528, May 2004.

[37] H. Lee, B. Han, Y. Shin, and S. Im, "Multipath Characteristics of Impulse Radio Channels," *Proc. IEEE Vehicular Technology Conference*, Tokyo, pp. 2487–2491, Spring 2000.

[38] M. Z. Win and R. A. Scholtz, "Characterization of Ultra-Wide Bandwidth Wireless Indoor Channels: A Communication-Theoretic View," *IEEE J. Select. Areas Commun.*, vol. 20, pp. 1613–1627, Dec. 2002.

[39] RCD Components Inc., http://216.153.156.169:8080/rcd/rcdpdf/P1410-P2420.pdf, 2003

[40] Z. Tian and L. Wu, "Timing Acquisition with Noisy Template for Ultra-Wideband Communications in Dense Multipath," *EURASIP J. Appl. Signal Process.*, vol. 2005, no. 3, pp. 439–454, March 2005

5

Error Performance of Pulsed Ultra-wideband Systems in Indoor Environments

Huaping Liu

5.1 Introduction

The analysis of digital communications system performance over the ideal additive white Gaussian noise (AWGN) channels often starts with statistically independent, zero-mean, Gaussian noise samples corrupting data samples. It should be mentioned that since some practical communications receivers use some type of baseband filters which are not Nyquist filters before data demodulation, the noise samples at sampling points could become correlated. However, when averaged over a large number of data points, it is the variance of the zero-mean noise relative to the energy per data symbol, rather than the short correlation-window introduced by the receiver baseband filtering, that determines the system performance. Because of the Gaussian statistics of the noise samples, the probability of error can thus be written in terms of a Q-function.

The mechanisms that cause fading in communications channels were first studied and modeled in the 1950s, mostly based on extensive measurements of the channel frequency power transfer characteristics over time. Fading of terrestrial digital radio channels, owing to multipath reception when a large number of received rays are constructively or destructively added in the receiver, causes an additional multiplicative distortion, in addition to the distortion caused by the thermal noise generated in the receiver. This is typically called the *small-scale fading* and is a prime cause of performance degradation. Another aspect of fading, the large-scale fading owing to propagation path loss (signal attenuation), represents the average signal power loss over distance. Given a particular propagation terrain and a fixed transmission power, the average received signal power experiences very small changes over a number of data symbols for most existing mobile

Ultra-wideband Wireless Communications and Networks
Edited by Xuemin (Sherman) Shen, Mohsen Guizani, Robert Caiming Qiu and Tho Le-Ngoc

terminals. For this reason, large-scale fading has been important in system link budget analysis when the system radio frequency (RF) coverage is to be determined. On the other hand, the system error performance analysis should focus on modeling the small-scale fading.

For different communications environments, models of the small-scale fading could be different. For example, for mobile radio channels, the small-scale fading envelope typically satisfies the Rayleigh statistics when there are a large number of reflective paths in the absence of a line-of-sight (LOS) signal component. When there is a dominant LOS path, the small-scale fading envelope could be well described by a Ricean probability density function (pdf) or a Nakagami-m pdf. In an indoor fading environment, however, the fading envelope is more appropriately modeled as having a lognormal pdf.

Most of the existing analysis of wireless system performance has dealt with narrowband or spread-spectrum mobile communications environments for which the channel is characterized as Ricean or Nakagami-m fading. The focus of these analyses has been on Rayleigh fading channels whose fading coefficients are modeled as complex Gaussian random processes. There exists a set of systematic techniques for performance analysis when the system is distorted by AWGN and Rayleigh fading. As mentioned earlier, the fading statistics and models of indoor channels are different from those encountered in mobile radio channels, most notably, the fading magnitude statistics, which are more appropriately modeled as lognormal fading. Another difference is that the number of resolvable paths in pulsed UWB systems is typically much larger than those for narrowband mobile systems.

In this chapter, we focus on analyzing the error performance of pulsed UWB systems with commonly used data-modulation schemes such as pulse amplitude modulation (PAM) and pulse position modulation (PPM). Despite the aforementioned differences between pulsed UWB over indoor channels and narrowband or spread-spectrum systems in mobile communications scenarios, the existing models and analytical techniques are still the useful starting points for performance analysis of UWB systems in lognormal fading environments. We shall focus on the analytical techniques that are effective in addressing the characteristics unique to UWB systems, such as the lognormal fading statistics, the different modulation schemes, and the large number of resolvable paths available to the receiver. In order to underline the fundamental analysis techniques, we will focus on a single-user system (no cochannel interference) that employs binary signaling, and assume that there is no narrowband interference or intersymbol interference (ISI). In a realistic pulsed UWB system, the received pulses may overlap one another, causing interpulse interference (IPI). Performance analysis considering IPI is generally very complex and will not be studied in this chapter. Readers interested in this topic may read [1–3]. One of the most important features of UWB signaling is that tremendous bandwidth is exchanged for extremely low transmission power to provide reliable communications over short distances. As in any other wireless system, antenna elements could be introduced to UWB systems to extend system RF range through spatial diversity, to increase throughput that is limited by the channel excess delay through spatial multiplexing, and to cancel potential narrowband interference. Thus, we will analyze examples of simple cases of UWB signaling with multiple antennas.

5.2 System Model

The information bits converted into a nonreturn-to-zero form $b[i] \in \{1, -1\}$ are assumed to be independent and equiprobable. These bits are carried by a train of ultrashort pulses $p(t)$ of pulse width T_p and unit energy (i.e., $\int_{-\infty}^{\infty} p^2(t)\,dt = 1$). The bit interval T_b is much larger than T_p, resulting in a low-duty-cycle transmission form. With energy E_b per pulse, the transmitted UWB waveform is given by

$$s(t) = \sum_{i=-\infty}^{\infty} \sqrt{E_b}\, p\,(t - iT_b; b[i])\,, \tag{1}$$

where the unit-energy waveform $p\,(t - iT_b; b[i])$ carries information bit in either its amplitude or its position. Waveforms for three commonly used modulation schemes – PAM, PPM, and combined PAM and PPM – are expressed as

$$p\,(t - iT_b; b[i]) = \begin{cases} b[i]p(t - iT_b) & \text{PAM} \\ p\left(t - iT_b - \dfrac{1 - b[i]}{2}\Delta\right) & \text{PPM} \\ b_a[i]p\left(t - iT_b - \dfrac{1 - b_p[i]}{2}\Delta\right) & \text{Combined PAM and PPM,} \end{cases} \tag{2}$$

where Δ for PPM or combined PAM and PPM is the time shift used to represent different bits.

The PAM and PPM schemes are illustrated in Figure 5.1. For a single-user system with binary PPM signaling, bit '0' is represented by a pulse with a delay relative to the time reference. One of the major factors governing the performance of such a system is the choice of the time shift. The most commonly used PPM scheme is the orthogonal signaling scheme for which $p(t; 1)$ is orthogonal to $p(t; -1)$. There also exists an optimal

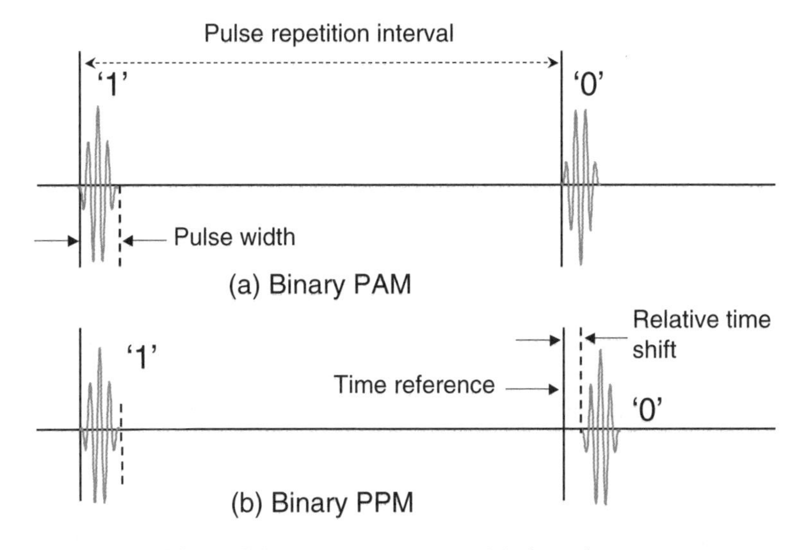

Figure 5.1 PAM and PPM modulation schemes

time shift for which Δ_{opt} is chosen to be such that the cross correlation between $p(t;\,1)$ and $p(t;\,-1)$ is minimized. Thus, Δ_{ortho} and Δ_{opt} are determined as

$$\Delta_{\text{ortho}} \triangleq \underbrace{\arg}_{\Delta(0<\Delta\leq T_p)} \int_{-\infty}^{\infty} p(t)p(t-\Delta_{\text{ortho}})\,dt = 0 \tag{3a}$$

$$\Delta_{\text{opt}} \triangleq \underbrace{\arg\min}_{\Delta(0<\Delta\leq T_p)} \int_{-\infty}^{\infty} p(t)p(t-\Delta_{\text{opt}})\,dt. \tag{3b}$$

Note that depending on the specific shape of waveform $p(t)$, the time shifts for orthogonal and optimum signaling could be less than T_p, that is, Δ is typically in the range $0 < \Delta \leq T_p$ with pass-band pulses. For binary PAM signaling, information bits modulate the pulse polarity. In the combined PAM and PPM scheme, the nth transmitted symbol is represented by two bits as $a[i] = \{b_a[i], b_p[i]\}$. This scheme is attractive in dense multipath environments where ISI caused by multipath delay limits the achievable data rate. Additionally, the combined PAM and PPM scheme does not have spectral lines as the PPM scheme. In the error performance analysis, we will focus on the orthogonal signaling only and use Δ, omitting the subscript 'ortho' for simplicity of notation, to represent the time shift. In the model given in (2), we assumed the simplest case that a single pulse is transmitted during each bit interval. In practical scenarios, however, multiple pulses could be used to carry one bit even in a single-user system. The generalization is quite straightforward and will not be discussed in the analysis in this chapter.

Pulsed UWB signaling gives rise to highly frequency-selective channels. Multipath components tend to arrive in clusters, and fading for each cluster as well as each ray within the cluster is independent. Neglecting the shadowing effects, we can express the impulse response of the channel as

$$h_c(t) = \sum_{l_c=0}^{L_c-1} \sum_{l_r=0}^{L_r-1} \alpha_{l_c,l_r} \delta(t - T_{l_c} - \tau_{l_c,l_r}), \tag{4}$$

where $\{\alpha_{l_r,l_c}\}$ are the lognormal multipath gain coefficients, T_{l_c} is the delay of the l_cth cluster, τ_{l_r,l_c} is the delay of the l_rth multipath component relative to the l_cth cluster arrival time T_{l_c}, and $\delta(t)$ is the Dirac delta function. This model can be simplified as

$$h_c(t) = \sum_{l=0}^{L_t-1} \alpha_l \delta(t - \tau_l), \tag{5}$$

where $l = l_c L_c + l_r$ and $L_t = L_c L_r$ are the number of resolvable multipath components, $\tau_l = T_{l_c} + \tau_{l_c,l_r}$ is the delay of the lth path relative to the first path (assuming $\tau_0 = 0$), and α_l is the fading coefficient for path l. As mentioned earlier in this chapter, we assume that adjacent paths do not overlap in time. It should be pointed out that although L_t in the model given by (6) is the total number of paths that the receiver experiences, the receiver does not necessarily or may not be capable of combining all L_t paths. We shall use L in this chapter to represent a subset of all possible paths that the receiver is interested in combining. The aggregated channel after convolving the channel impulse response with the transmitted pulse $p(t)$ is given by

$$h(t) = \sum_{l=0}^{L_t - 1} \alpha_l p(t - \tau_l). \tag{6}$$

The statistics of α_l are critical to the analysis. Let us examine the channel model proposed by the IEEE 802.15.3a working group [4]. In this model, the channel path gain α_l is modeled as $\alpha_l = \varepsilon \beta_l$, where ε, with an equal probability to take on the value of -1 and 1, accounts for random pulse polarity inversion that is resulted from reflection. The path magnitude β_l is real-valued and follows a lognormal distribution for indoor UWB channels. The standard deviation of fading amplitudes across $\{\beta_l\}$ is typically in the range of 3–5 dB. The power delay profile of the channel is modeled as exponentially decaying by rays and clusters, the relative power of the lth path to the first path can be expressed as $E\{|\alpha_l|^2\} = E\{|\alpha_0|^2\} e^{-T_{l_c}/\Gamma} e^{-\tau_{l_c,l_r}/\gamma}$ (note again $l = l_c L_c + l_r$, $\tau_l = T_{l_c} + \tau_{l_c,l_r}$), where Γ and γ are the power decay factors for the corresponding cluster and ray, respectively.

Perfect timing is assumed by setting $\tau_0 = 0$ and by knowing the exact values of τ_l. By keeping only the first L ($L \leq L_t$) paths, the received waveform is thus given by

$$\begin{aligned} r(t) &= \sum_{l=0}^{L-1} \alpha_l s(t - \tau_l) + n_o(t) \\ &= \sqrt{E_b} \sum_{i=-\infty}^{\infty} h\,(t - iT_b; b[i]) + n_o(t), \end{aligned} \tag{7}$$

where $n_o(t)$ is the additive noise typically treated as a zero-mean white Gaussian process with two-sided power spectral density (PSD) $N_0/2$. To avoid channel-induced ISI, the bit interval T_b is selected to be greater than the channel delay τ_{L-1}. In practice, the received signal $r(t)$ is first passed through a band-pass filter with center frequency f_0 and bandwidth B, which is chosen to be wide enough and larger than the signal's 10 dB bandwidth, to avoid severe filter-induced distortion. For simplicity of analysis, it is also assumed that B is an integer multiple of $1/(2T_b)$. The filtered noise $n(t)$ is no longer white, but its autocorrelation in time can be very small owing to the ultra-wide filter bandwidth B. If the band-pass filter is ideal, the autocorrelation function of $n(t)$ is

$$R_n(\tau) = E\{n(t)n(t + \tau)\} = N_o B \frac{\sin(2\pi B\tau)}{2\pi B\tau} \cos(2\pi f_0 \tau). \tag{8}$$

The received signal waveform can be described by the following one-shot model within each bit period $t \in [iT_b, (i + 1)T_b)$:

$$\begin{aligned} r_i(t) &= r(t + iT_b) \\ &= \sqrt{E_b} h\,(t; b[i]) + n_i(t), \qquad t \in [0, T_b), \end{aligned} \tag{9}$$

where $n_i(t) = n(t + iT_b)$, $t \in [0, T_b)$, is the corresponding T_b-long segment of the filtered noise $n(t)$.

In practical UWB systems, there are distortions that narrowband systems do not experience. These distortions could be introduced by hardware components such as UWB antennas or by the channel, which causes frequency-dependent per-path distortion [5]. In the basic analysis, we shall adopt an idealized model that the transmitted $p(t)$ will not be distorted by the transmit and receive chain. Under this condition, the traditional matched filter-based approach can be used for data detection. The optimum receiver for the binary PAM or PPM scheme consists of a matched filter that is matched to

$$\phi(t) = h(t;\, 1) - h(t;\, -1), \tag{10}$$

which is called the *template waveform*. This receiver processing is equivalent to correlating the received signal with $\phi(t)$ followed by an integrator, which integrates over one bit duration. This choice of the template implies perfect channel estimation in the receiver, that is, channel fading coefficients and relative delays for all L paths are perfectly known.

Note that when $L > 1$, this scheme could be realized using the traditional rake receiver [6]. In this case, the received signal is passed through a matched filter, which is matched to the transmitted pulse $p(t)$, and then sampled according to the delays of each resolvable path. For coherent detection, the channel fading amplitudes $\{\alpha_l\}$ need to be estimated and then passed to a combiner. In the presence of AWGN distortion only, the optimum combiner remains to be a maximal ratio combiner (MRC).

Alternatively, in a slowly fading channel, the template waveform for PAM signaling can be adaptively derived by using a sliding-window method combined with decision feedback. In such a scheme, the template waveform for detection of the ith bit is derived as

$$\phi_i(t) = \frac{1}{\mathcal{N}} \sum_{j=1}^{\mathcal{N}} \hat{b}[i-j] r_{i-j}(t), \tag{11}$$

where $\mathcal{N}$ is the window length and $\hat{b}[i-j]$ is the decision to bit $b[i-j]$. The process of deriving the template waveform and the detection stage using the derived template is illustrated in Figure 5.2.

A simpler method to obtain the template waveform is to use the transmitted-reference (TR) scheme in which pulses are transmitted in pairs, where the first pulse of each pair

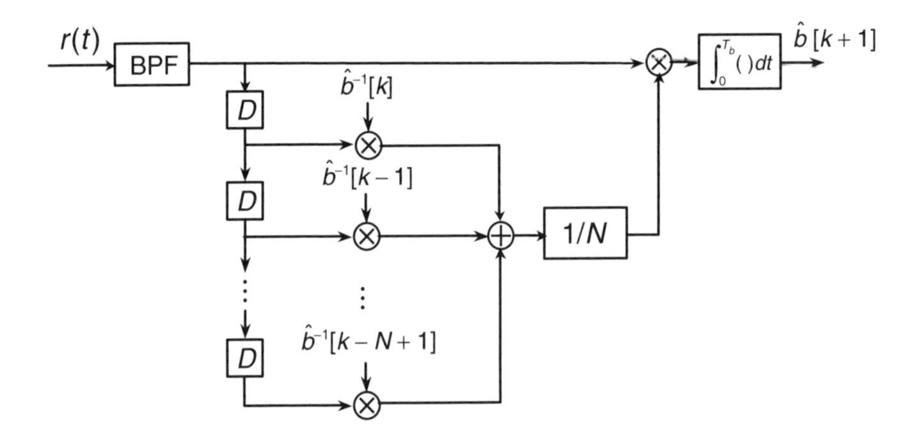

Figure 5.2 Sliding-window-based method for receiver template estimation

serves as the reference signal and the second pulse is data modulated. In the receiver, the reference signal is used as the template waveform. The main drawbacks of the TR scheme are performance degradation due to the use of a noisy template waveform and significant transmission overhead due to the reference signals. If the channel is time invariant or slowly fading so that the channel coefficients and relative delays remain unchanged over many bit intervals, the effect of noise in the template waveform can be mitigated by averaging over multiple reference periods.

In the following section, we will first derive the error performance of PAM and PPM schemes, assuming an ideal template waveform $\phi(t)$ (i.e., perfect fading coefficient and path delay estimation). The performances of the system with an ideal template and that with a matched filter, matched to $p(t)$, followed by a sampler and an MRC are identical. After this, we analyze the case of an adaptively generated template using the decision-directed sliding-window scheme described by (11). Error performance of the common TR scheme is a special case of this scheme with a window length of one and perfect feedback decisions ($\hat{b}[i-1] = b[i-1]$).

5.3 Error Performance in Indoor Environments

Having discussed the mathematical models of the received signals and the matched filter receiver, we now move to our main topic, analysis of error performance of pulsed UWB systems in lognormal fading channels. We shall first examine the case when the template waveform given in (10) is perfectly known in the receiver. Since the multipath resolution of the receiver is equal to T_p, there are typically a large number of resolvable paths. As in other spread-spectrum systems, a rake receiver could be exploited to capture multipath energy and to provide multipath diversity. Such a receiver structure could be implemented by using a bank of correlators ($p(t-\tau_l), l = 0, 1, \ldots, L-1$), corresponding to multipath delays $\tau_0, \tau_1, \ldots, \tau_{L-1}$, followed by an MRC to capture multipath energy. Both the multipath delays and fading magnitude for each path must be estimated so that an appropriate integration window and path scaling factors can be determined. The more paths the rake receiver combines, the higher is the system complexity.

An idealization that we make in the analysis is the absence of IPI when a rake receiver is employed. For typical indoor environments, measurements have shown that the average multipath arrival rate is in the range of 0.5–2 ns. Thus, the relative path delays $\tau_0, \ldots, \tau_{L-1}$ are random variables (RVs) whose distribution will most likely depend on the propagation environment. The pulse width T_p depends on the signal bandwidth occupied. In dense multipath environments, the received multipath components may experience severe time overlap (i.e., when $\tau_{l+1} - \tau_l < T_p$), especially when the signaling bandwidth is close to the minimum bandwidth of 500 MHz. As mentioned in Section 5.1, the analysis considering partial correlation between adjacent received paths becomes very complex and will not be conducted in this chapter.

Another assumption that we make is that of slow fading, which implies that the fading process (fading amplitude and delays) varies very slowly compared with signaling rates so that the fading process is considered to be effectively constant during one-bit interval. For cases when we need to adaptively derive the receiver template waveforms, the slow fading condition is assumed to hold true even when considering a period of many bit intervals.

5.3.1 Pulse Amplitude Modulation and Pulse Position Modulation

5.3.1.1 PAM Signaling

For binary PAM signaling, the template waveform given in (10) is simplified as $\phi(t) = 2h(t;\,1)$, or equivalently for performance analysis, it can be simplified as $\phi(t) = h(t)$. The decision variable for the ith bit is given as

$$d_i = \int_{iT_b}^{(i+1)T_b} r_i(t)\phi(t)\,dt. \tag{12}$$

This results in the traditional rake receiver with maximal ratio combining. Let the noise component corresponding to the lth received pulse be $n_l = \int_{iT_b}^{(i+1)T_b} n_i(t)p(t-\tau_l)\,dt$. Since the energy of $p(t)$ is normalized to unity, the decision variable can be written as the sum of the signal term, $\sqrt{E_b}\sum_{l=0}^{L-1}\alpha_l^2$, and the noise term, $\sum_{l=0}^{L-1}\alpha_l n_l$. The bit-error-rate (BER) expression can then be derived using the traditional method by finding the conditional probability of error conditioned on a set of channel coefficients $\{\alpha_l\}$. Then the conditional error probability is averaged over the probability density function of the $\{\alpha_l\}$. The only difference between the analysis of system performance over Rayleigh fading channels and over lognormal fading channels is that the distribution of the channel attenuation factors is different, causing a different pdf of $\{\alpha_l\}$.

For a fixed set of fading coefficients $\{\alpha_l\}$, d_i is a Gaussian RV. Variance of the noise component associated with the lth finger (conditioned on α_l) is $\sigma_n^2 = \alpha_l^2 N_0/2$. The instantaneous signal-to-noise ratio (SNR) per bit, γ_b, is obtained as

$$\gamma_b = \frac{2E_b}{N_0}\sum_{l=0}^{L-1}\alpha_l^2 = \sum_{l=0}^{L-1}\gamma_l, \tag{13}$$

where $\gamma_l = \frac{2E_b}{N_0}\beta_l^2$ ($\alpha_l = \varepsilon_l\beta_l$ and $\varepsilon_l \in \{\pm 1\}$). Because β_l is a lognormal RV, $\gamma_l = \frac{2E_b}{N_0}\beta_l^2$ is also a lognormal RV. Thus, γ_b is a sum of independent lognormal RVs.

For the scenario being analyzed ($\varepsilon_l \in \{\pm 1\}$), conditioning on $\{\alpha_l\}$ is equivalent to conditioning on $\{\beta_l\}$. We shall then average the conditional BER over the PDF of γ_b. The conditional BER for a fixed set of $\{\beta_l\}$ is given as

$$p(\gamma_b) = \frac{1}{2}\,\mathrm{erfc}\left(\sqrt{\frac{\gamma_b}{2}}\right). \tag{14}$$

Let $\beta_l = e^{u_l}$, where u_l is a normal RV, that is, $u_l \sim N(\mu_{u_l}, \sigma_{u_l}^2)$. It follows that

$$\gamma_l = e^{c_0 + 2u_l}, \tag{15}$$

where $c_0 = \ln\left(\frac{2E_b}{N_0}\right)$. The kth moment of β_l is given as

$$E\{\beta_l^k\} = e^{k\mu_{u_l} + k^2\sigma_{u_l}^2/2}. \tag{16}$$

Although an exact closed-form expression of the PDF of a sum of independent lognormal RVs does not exist, such a sum can be approximated by another lognormal RV [7].

The approximation can be obtained by using a number of methods, one of which is the Wilkinson's method [8].

Let $\gamma_b = e^x$, where x, $x \sim N(\mu_x, \sigma_x^2)$, is a normal RV. In Wilkinson's method, the two parameters μ_x and σ_x are obtained by matching the first two moments of γ_b with the first two moments of $\sum_{l=0}^{L-1} \gamma_l$. These two parameters are given as

$$\mu_x = \ln\left(E_{L1}^2/\sqrt{E_{L2}}\right) \tag{17a}$$

$$\sigma_x = \ln\left(E_{L2}/E_{L1}^2\right) \tag{17b}$$

where E_{L1} and E_{L2} are related to μ_{u_l} and $\sigma_{u_l}^2$ as

$$E_{L1} = \sum_{l=0}^{L-1} e^{(c_0+2\mu_{u_l}+2\sigma_{u_l}^2)} \tag{18a}$$

$$E_{L2} = \sum_{l=0}^{L-1} e^{2(c_0+2\mu_{u_l}+4\sigma_{u_l}^2)} + 2\sum_{m=1}^{L-1}\sum_{n=0}^{m-1} e^{2(c_0+\mu_{um}+\mu_{un}+\sigma_{um}^2+\sigma_{un}^2)}. \tag{18b}$$

The approximated pdf of γ_b is given as

$$f(\gamma_b) = \frac{1}{\gamma_b\sqrt{2\pi\sigma_x^2}} \exp\left[-\frac{(\ln(\gamma_b)-\mu_x)^2}{2\sigma_x^2}\right]. \tag{19}$$

The average BER can be calculated by averaging the conditional BER $p(\gamma_b)$ over $f(\gamma_b)$ as

$$P_b = \int_0^\infty p(\gamma_b) f(\gamma_b) d\gamma_b. \tag{20}$$

5.3.1.2 PPM Signaling

For orthogonal binary PPM signaling, the template waveform is expressed as $\phi(t) = h(t) - h(t-\Delta)$. For any bit, the signal energy per bit in the correlator output is the same as PAM signals. However, the variance of the total noise component is twice that of PAM signals, yielding a 3-dB SNR disadvantage compared with PAM signaling. Other than this scaling factor in SNR, all the BER related expressions for PAM signaling are applicable for PPM signaling.

5.3.1.3 Combined PAM and PPM Signaling

In the combined PAM and PPM scheme, the received signal is correlated by using two template waveforms

$$\phi_1(t) = h(t) + h(t-\Delta), \tag{21a}$$

$$\phi_2(t) = h(t) - h(t-\Delta), \tag{21b}$$

where Δ was defined in (2). The time shift Δ is chosen to be such that $\int_{-\infty}^{\infty} p(t)p(t-\Delta)\,dt = 0$. We shall also make an assumption, or make appropriate choices for Δ so that

$\int_{-\infty}^{\infty} p(t-\Delta)p(t-T_p)\,dt \approx 0$. The correlator output is integrated over one bit interval, which results in effective maximal ratio combining of L paths. Decisions are made independently for each correlator output and then multiplexed, forming the estimate of the transmitted symbols. If the input symbols $a[i] \in \{11, 1-1, -11, -1-1\}$ are precoded in the transmitter by interchanging the two symbols $\{-11\}$ and $\{-1-1\}$, the two outputs of the two correlators with waveform $\phi_1(t)$ and $\phi_2(t)$ correspond respectively to the first and the second bit of each pair of input bits. With independent input bits, the symbol error rate (SER) can be easily calculated using the BER.

The decision variables for the ith symbol are expressed as

$$d_{i,k} = \int_{iT_b}^{(i+1)T_b} r_i(t)\phi_k(t)\,dt, \quad k = 1, 2, \tag{22}$$

where $\phi_1(t)$ and $\phi_2(t)$ are the template waveforms applied. Due to the orthogonality between $\phi_1(t)$ and $\phi_2(t)$, the Gaussian noise components at the output of the two correlators for the lth path can be easily shown to be independent of each other. Thus, symbol decisions can be made by passing $d_{i,1}$ and $d_{i,2}$ independently through a decision device with a threshold zero. The precoding scheme described earlier ensures that the decisions based on $d_{i,1}$ and $d_{i,2}$ correspond respectively to bits $b_a[i]$ and $b_p[i]$ before the precoding. Because decisions for bits $b_a[i]$ and $b_p[i]$ of each symbol can be made independently, the BER can be analyzed based on the statistics of either $d_{i,1}$ or $d_{i,2}$. From the decision variables for both bits of each pair, the signal part is the same as the PAM case, but noise variance per path doubles. Thus, the performance of the combined PAM and PPM scheme is the same as that of the PPM scheme.

5.3.2 Receiver with Self-Derived Template Waveforms

We assume a slowly fading environment where the channel does not change over $\mathcal{N}$ bit intervals. When the template is self-derived, using the sliding-window scheme given by (11), $\phi_i(t)$ consists of two terms; the desired waveform $h(t)$ scaled by a template-estimation efficiency factor δ_i and $\sqrt{E_b}$, and a noise term $\xi_i(t)$ as

$$\phi_i(t) = \delta_i\sqrt{E_b}h(t) + \xi_i(t), \tag{23}$$

where

$$\delta_i = \frac{1}{\mathcal{N}}\sum_{j=1}^{\mathcal{N}} \hat{b}[i-j]b[i-j] \tag{24a}$$

$$\xi_i(t) = \frac{1}{\mathcal{N}}\sum_{j=1}^{\mathcal{N}} \hat{b}[i-j]n_{i-j}(t). \tag{24b}$$

For binary PAM, $\xi_i(t)$ is a zero-mean Gaussian process with a power spectral density (PSD) of $\frac{1}{\mathcal{N}}(N_0/2)$. In the ideal case when feedback decisions are correct, the channel is time invariant, and the window length $\mathcal{N} \to \infty$, $\phi_i(t) = \sqrt{E_b}h(t)$.

To analyze the BER performance, we will first assume that feedback decisions are correct (thus $\delta_i = 1$) in the template-derivation process, which is valid in the high SNR region. We will then incorporate the potential erroneous feedback decisions for BER analysis in the low-SNR region. To provide a general result, we allow the time-window of the correlator to have a flexible length $T_L \leq T_b$, which corresponds to collecting the first arrived L paths. By correlating the filtered signal waveform $r_i(t)$ with the self-derived template waveform $\phi_i(t)$, the decision variable for the ith bit is obtained as

$$d_i = \int_{iT_b}^{iT_b+T_L} r_i(t)\phi_i(t)\,dt = b[i]\theta + z_1 + z_2 + z_3, \tag{25}$$

where the first term $b[i]\theta$ in (25) is the desired signal term, while z_1, z_2 and z_3 are noise terms. These four terms are expressed as

$$\theta = E_b \int_{iT_b}^{iT_b+T_L} h^2(t)\,dt \tag{26a}$$

$$z_1 = \sqrt{E_b} \int_{iT_b}^{iT_b+T_L} h(t)\xi_i(t)\,dt \tag{26b}$$

$$z_2 = \sqrt{E_b} \int_{iT_b}^{iT_b+T_L} h(t)n(t)\,dt \tag{26c}$$

$$z_3 = \int_{iT_b}^{iT_b+T_L} n(t)\xi_i(t)\,dt. \tag{26d}$$

Note that in (26b) bit $b[i] \in \{\pm 1\}$ has been absorbed into the noise term $\xi_i(t)$.

In order to determine the analytical BER expression for binary PAM signals, we will derive the conditional BER from (25) conditioned on a given realization of the random channel coefficients $\boldsymbol{\alpha} = \{\alpha_0, \ldots, \alpha_{L-1}\}$. This instantaneous performance will then be integrated over the joint pdf of the random parameters to obtain the average BER.

5.3.2.1 Distribution of $z = z_1 + z_2 + z_3$

Evaluation of the distribution of the composite noise $z = z_1 + z_2 + z_3$ is a complex procedure [9–11]. Here, we derive a BER expression that can approximately predict the performance reliably while keeping the complexity reasonable.

Because both $\xi_i(t)$ and $n(t)$ are zero-mean Gaussian random processes independent of $h(t)$, the noise terms z_1 and z_2, when conditioned on $\boldsymbol{\alpha}$, can be regarded as zero-mean Gaussian RVs. When BT_b is large, the noise product term z_3 can also be well approximated as a Gaussian RV for SNR regions of practical interest [9–12]. The mean value of z_3 is zero when the bandwidth B is an integer of $1/(2T_b)$, as we have assumed earlier. By using $R_n(T_b) = 0$, it is straightforward to show that all three noise terms are uncorrelated, that is, $E\{z_1 z_2\} = E\{z_1 z_3\} = E\{z_2 z_3\} = 0$ [9–12].

By using (8) and (26), and employing the assumption that $B \gg 1/T_L$, the noise variance can be approximately evaluated as [10, 11]

$$\sigma_z^2 \cong \theta \frac{N_0}{2}\left(1 + \frac{1}{\mathcal{N}}\right) + \frac{1}{\mathcal{N}} B T_L N_0^2. \tag{27}$$

5.3.2.2 Distribution of θ

Let $R_h(\tau) = \int_0^\tau h^2(t)\,dt$ denote the channel energy as a function of the integration window length τ. From (26a), the effective sample gain θ is given by $\theta = E_b R_h(T_L)$, whose statistical characteristics can be evaluated via Monte Carlo methods for any channel type. Suppose, for simplicity, that all channel paths arrive separately in time, that is, there is no pulse overlapping. With this assumption, the received sample gain θ can be simplified to $\theta = E_b \sum_{l=0}^{L-1} \alpha_l^2$. Apparently, it is the sum of squared multipath fading coefficients, which are independent lognormal RVs, and the pdf of θ, $f(\theta)$, can be approximated using the same method as given in Section 5.3.1 as

$$f(\theta) = \frac{1}{\theta\sqrt{2\pi\sigma_x^2}} \exp\left[-\frac{(\ln(\theta) - \mu_x)^2}{2\sigma_x^2}\right], \tag{28}$$

where μ_x and σ_x are given by (17a) and (17b), respectively.

5.3.2.3 Error Performance in High SNR Region

Conditioned on θ, the BER expression of the binary PAM scheme is given by

$$P(\theta) = \frac{1}{\sqrt{2\pi\sigma_z^2}} \int_{-\infty}^{0} \exp\left[-\frac{(\eta - \theta)^2}{2\sigma_z^2}\right] d\eta = \frac{1}{2} \operatorname{erfc}\left(\sqrt{\frac{\theta}{2\sigma_z^2}}\right). \tag{29}$$

The average BER can be calculated by averaging the conditional BER $P(\theta)$ over the probability density function $f(\theta)$ as

$$P_b = \int_0^\infty P(\theta) f(\theta)\,d\theta. \tag{30}$$

Numerical methods can be used to evaluate the BER in (30), which provides the performance lower bound under the assumption of perfect feedback decisions.

5.3.2.4 Error Performance in Low-SNR Region

Albeit fairly accurate at the high SNR region, the BER expression in (30) cannot precisely describe the receiver error performance for low SNRs, in which case the number of erroneous past decisions becomes nontrivial. Here we examine the low-SNR case by incorporating the probability of erroneous feedback decisions.

In the presence of erroneous feedback decisions, the noise term $\xi_i(t)$ in the template $\phi_i(t)$ given in (23) remains unaffected, but the template-estimation efficiency factor δ_i is lowered. Each wrong bit decision will cause its corresponding waveform to counteract a correctly demodulated waveform. As a result, δ_i does not equal 1, but should be replaced by $1 - 2P(\theta)$ to account for the effect of wrong feedback decisions. It is reasonable to assume that $P(\theta) \leq 0.5$. Corresponding to (23), the low-SNR expression for $\phi_i(t)$ can be written as

$$\phi_i(t) = (1 - 2P(\theta))\sqrt{E_b}h(t) + \xi_i(t). \tag{31}$$

The decision statistic d_i in (25) should be adjusted accordingly. The effective sample amplitude θ should be replaced by $\theta' = (1 - 2P(\theta))\theta$ and the second noise term in (26c) replaced by $z_2' = (1 - P(\theta))z_2$, while z_1 and z_3 remain the same. The composite noise variance is now $\sigma_z'^2 \cong \theta(1 - 2P(\theta) + 1/\mathcal{N})(N_0/2) + BT_L N_0^2/\mathcal{N}$. Substituting θ' and $\sigma_z'^2$ for θ and σ_z^2 in (29), we obtain an expression for $P(\theta)$ in the low-SNR region as

$$P(\theta) = \frac{1}{2}\,\mathrm{erfc}\left(\sqrt{\frac{\theta'}{2\sigma_z'^2}}\right). \tag{32}$$

Equation (32) does not immediately lead to a manageable closed-form solution to $P(\theta)$. Nevertheless, some of the one-dimensional numerical methods may be used to find an approximate solution to $P(\theta)$ when θ is given.

5.3.3 System with Multiple Antennas

5.3.3.1 Transmitted Signal

When there are multiple transmit and receive antennas, the system model described in Section 5.2 needs to be modified. We shall only focus on PAM signaling for analysis in this subsection. Consider a spatial multiplexing communication system with N transmit antennas and M receive antennas ($M \geq N$), which operates over a UWB channel. For simplicity, we focus again on binary PAM signaling. Input bits are serial-to-parallel converted into N streams without space-time encoding. Each of the N streams of bits is then used to modulate the pulse amplitude of a short-duration UWB pulse. Finally, the N streams of waveforms are distributed to N transmit antennas for simultaneous transmission. Each receive antenna responds to each transmit antenna through statistically independent fading processes. The received signals are further corrupted by additive white Gaussian noise that is statistically independent among the M receive antennas, and across different bit periods as well as different resolvable paths of the same bit. The transmitted binary PAM signal from the nth transmit antenna is expressed as

$$s_n(t) = \sum_{i=-\infty}^{\infty} \sqrt{E_b}\, b_n[i]\, p(t - iT_b), \tag{33}$$

where $b_n[i]$ is the ith information bit from the nth transmit antenna, E_b scales the energy of the basic unit-energy pulse $p(t)$ or is equivalently the energy per bit, and T_b, as in the single-antenna system, is the bit interval for each transmitted bit stream. Again, we assume that T_b is sufficiently large compared with the channel excess delay experienced by the system to avoid severe ISI.

5.3.3.2 Received Signal

The transmitted signal from each transmit antenna goes through a highly frequency-selective channel modeled by (6), resulting in multiple delayed and independently faded copies of the same signal aggregated at the receiver. The reccived signal $r_m(t)$ at the mth receive antenna is a sum of signals from all N transmit antennas expressed as

$$r_m(t) = \sum_{n=1}^{N} \sum_{l=0}^{L-1} h_{mn,l} s_n(t - \tau_l) + \nu_m(t), \quad m = 1, \ldots, M, \tag{34}$$

where $h_{mn,l}$ represents the fading coefficient of the lth path for the signal from the nth transmit antenna to the mth receive antenna, $\nu_m(t)$ is a real zero-mean white Gaussian noise process with a two-sided PSD of $N_0/2$, and $s_n(t)$ was given in (33).

The receiver separates the N simultaneously transmitted signals on a path-by-path basis. Thus, the model is different from the single-antenna system that uses the template given in (10). Instead, we model the receiver as consisting of a matched filter per receive antenna, which is matched to the UWB pulse $p(t)$ and then sampled at each path interval. The output of the matched filter is processed by an array processing unit to separate the signals from the N transmit antennas on a path-by-path basis. A temporal rake receiver is then used to combine the array-processed paths that are carrying information of the same bit and form the decision statistics for a bit-by-bit detection.

Mathematical models for such a detection procedure can be briefly described as follows. Over each bit period, the N bits sent over N transmit antennas form an $N \times 1$ transmit signal vector $\boldsymbol{s} = [b_1, b_2, \ldots, b_N]^{\mathrm{T}}$, where $[\cdot]^{\mathrm{T}}$ denotes transpose, and bit time index is omitted. Corresponding to the lth path, the received $M \times 1$ spatial signal vector at the matched filter output, $\boldsymbol{r}_l$, is represented by

$$\boldsymbol{r}_l = [r_{1,l}, r_{2,l}, \ldots, r_{M,l}]^{\mathrm{T}}, \; l = 0, \ldots, L-1. \tag{35}$$

Within $\boldsymbol{r}_l$, the received noise vector across M receive antennas, $\boldsymbol{\nu}_l$, is expressed as

$$\boldsymbol{\nu}_l = [\nu_{1,l}, \; \nu_{2,l}, \ldots, \nu_{M,l}]^{\mathrm{T}}, \quad l = 0, \ldots, L-1. \tag{36}$$

For the lth delayed path, the discrete-time channel gains among all pairs of transmit–receive antennas can be organized into an $M \times N$ matrix as

$$\boldsymbol{H}_l = \left[\boldsymbol{h}_{1,l} \; \boldsymbol{h}_{2,l} \; \cdots \; \boldsymbol{h}_{N,l}\right] = \begin{bmatrix} h_{11,l} & h_{12,l} & \cdots & h_{1N,l} \\ h_{21,l} & h_{22,l} & \cdots & h_{2N,l} \\ \vdots & & & \vdots \\ h_{M1,l} & h_{M2,l} & \cdots & h_{MN,l} \end{bmatrix}, \tag{37}$$

where the $M \times 1$ column vector $\boldsymbol{h}_{n,l}$ is given as $\boldsymbol{h}_{n,l} = [h_{1n,l}, h_{2n,l}, \ldots, h_{Mn,l}]^{\mathrm{T}}$.

With the assumption as described in Section 5.3, there is no interpath interference for data from the same transmit antenna and the received signal can be modeled on a path-by-path basis. Assuming perfect timing and synchronization, the received spatial signal vector $\boldsymbol{r}_l$ in (35) can now be written in a vector-matrix form as

$$\boldsymbol{r}_l = \sqrt{E_b}\boldsymbol{H}_l\boldsymbol{s} + \boldsymbol{\nu}_l, \quad l = 0, \ldots, L-1, \tag{38}$$

where the noise vector $\boldsymbol{\nu}_l$ is independent of channel fading processes.

5.3.3.3 Detection

Suppose that the zero-forcing (ZF) scheme is adopted to separate data streams from N transmit antennas on a path-by-path basis. Performance of the ZF scheme approaches that of the minimum mean-square error (MMSE) scheme at high SNRs [13], and it is convenient to analyze the ZF structure. In the ZF receiver, the received signal vector in (38) is premultiplied by $\boldsymbol{H}_l^+$, where $(\cdot)^+$ denotes the pseudoinverse. Thus, the zero-forced $N \times 1$ signal vector $\boldsymbol{y}_l$ for the lth path of a particular bit can be written as

$$\begin{aligned} \boldsymbol{y}_l &= [y_{1,l}, y_{2,l}, \ldots, y_{N,l}]^{\mathrm{T}} \\ &= \boldsymbol{H}_l^+ \boldsymbol{r}_l = \sqrt{E_b}\boldsymbol{s} + \boldsymbol{\xi}_l, \end{aligned} \tag{39}$$

where $\boldsymbol{\xi}_l = \boldsymbol{H}_l^+ \boldsymbol{v}_l$ is the zero-mean noise vector. The *instantaneous* noise power (conditioned on a fixed realization of channel coefficients) on the lth path of the nth data stream is obtained as

$$[E\{\boldsymbol{\xi}_l \boldsymbol{\xi}_l^H\}]_{nn} = (N_0/2)\left[\boldsymbol{H}_l^+ \boldsymbol{H}_l^{+H}\right]_{nn},$$

where $(\cdot)^H$ denotes conjugate transpose and $[\cdot]_{nn}$ represents the (n, n)th element of a matrix. Note that for the channel model adopted, all channel coefficients are real. The notation of conjugate transpose is still kept when operating on real vectors or matrices because of its mathematical convenience. If $\boldsymbol{H}_l^H \boldsymbol{H}_l$ is a full rank matrix, which is usually satisfied when signals from the transmit and receive antennas are independent or have low correlations, the following relation can be easily obtained

$$[E\{\boldsymbol{\xi}_l \boldsymbol{\xi}_l^H\}]_{nn} = (N_0/2)\left[\left(\boldsymbol{H}_l^H \boldsymbol{H}_l\right)^{-1}\right]_{nn}. \tag{40}$$

For notational simplicity, let us introduce a new variable

$$\kappa_{n,l} = \frac{1}{\sqrt{\left[(\boldsymbol{H}_l^H \boldsymbol{H}_l)^{-1}\right]_{nn}}}. \tag{41}$$

It was shown in [13] that $\kappa_{n,l}^2 = 1/\left[(\boldsymbol{H}_l^H \boldsymbol{H}_l)^{-1}\right]_{nn}$ can be written in a quadratic form as

$$\kappa_{n,l}^2 = \boldsymbol{h}_{n,l}^H \boldsymbol{G}_l \boldsymbol{h}_{n,l}, \tag{42}$$

where $\boldsymbol{G}_l$ is an $M \times M$ positive semidefinite Hermitian matrix formed from channel coefficients defined in (37) $\boldsymbol{h}_{1,l}, \ldots, \boldsymbol{h}_{n-1,l}, \boldsymbol{h}_{n+1,l}, \ldots, \boldsymbol{h}_{N,l}$, and thus is independent of $\boldsymbol{h}_{n,l}$. For example, with $N = 2$ and $n = 1$, $\boldsymbol{G}_l$ is determined to be $\boldsymbol{G}_l = \boldsymbol{h}_{2,l}^{\mathrm{T}} \boldsymbol{h}_{2,l} \boldsymbol{I}_2 - \boldsymbol{h}_{2,l} \boldsymbol{h}_{2,l}^{\mathrm{T}}$, where $\boldsymbol{I}_2$ is the 2×2 identify matrix.

Mathematically, it is possible to detect the transmitted signal $\boldsymbol{s}$ from the spatial-processed, received signal in (39) that corresponds to a single path. However, for better performance, the receiver should combine the signal energy contained in many resolvable paths corresponding to a particular bit. From (39), we collect the zero-forced signal elements from all L paths carrying the same bit of the nth data stream and write them in an $L \times 1$ temporal vector as

$$\begin{aligned} \boldsymbol{y}_n &= [y_{n,0}, y_{n,1}, \ldots, y_{n,L-1}]^{\mathrm{T}} \\ &= \left[\sqrt{E_b} s_n + \xi_{n,0}, \ldots, \sqrt{E_b} s_n + \xi_{n,L-1}\right]^{\mathrm{T}}. \end{aligned} \tag{43}$$

The decision variable for a bit from the nth transmit antenna can be obtained by combining the L elements of $\boldsymbol{y}_n$. The MRC output for a particular bit of the nth transmitted data stream is expressed as a single decision variable in the form

$$d_n = \sum_{l=0}^{L-1} \kappa_{n,l}^2 \left[\sqrt{E_b} s_n + \xi_{n,l}\right]. \tag{44}$$

5.3.3.4 Error Performance

Let us examine the error behavior of the transmitted bit b_n. For notational simplicity, the subscript 'n' indexing the nth data stream for some variables (e.g., in $\kappa_{n,l}$ and $\boldsymbol{h}_{n,l}$) will be omitted whenever it does not cause a confusion. Notice that with bipolar PAM signaling and the channel model adopted, both the signal component and the noise component are real-valued. For a fixed set of $\{\kappa_l\}$, or equivalently, a fixed realization of channel coefficients $\{\boldsymbol{H}_l\}$, the *instantaneous* SNR γ_p of the decision variable in (44) can be written as

$$\gamma_p = \frac{2E_b}{N_0} \sum_{l=0}^{L-1} \kappa_l^2. \tag{45}$$

The conditional BER expression conditioned on a fixed set of channel coefficients is given as

$$P(\gamma_p) = Q\left(\sqrt{\gamma_p}\right), \tag{46}$$

where $Q(x) = (1/\sqrt{2\pi}) \int_x^\infty e^{-y^2/2} dy$ is the Q-function.

In order to derive the average BER, we need the pdf of γ_p. The challenge for the analysis is on obtaining the statistics of $\{\kappa_l\}$ when the magnitude of the channel coefficients has lognormal statistics and when there are multiple transmit and receive chains. Even with a ZF receiver for spatial de-multiplexing, we must resort to some approximations. Substituting (42) into (45) yields

$$\gamma_p = \gamma_0 \sum_{l=0}^{L-1} \boldsymbol{h}_l^H \boldsymbol{G}_l \boldsymbol{h}_l,$$

where $\gamma_0 = 2E_b/N_0$. Let us form an $LM \times LM$ matrix $\boldsymbol{G}$ and an $LM \times 1$ vector $\boldsymbol{h}$ as

$$\boldsymbol{G} = \mathrm{diag}\left[\boldsymbol{G}_0\ \boldsymbol{G}_1\ \ldots\ \boldsymbol{G}_{L-1}\right] \tag{47}$$

$$\boldsymbol{h} = [\boldsymbol{h}_0^{\mathrm{T}}\ \boldsymbol{h}_1^{\mathrm{T}}\ \ldots\ \boldsymbol{h}_{L-1}^{\mathrm{T}}]^{\mathrm{T}}, \tag{48}$$

where $\boldsymbol{h}_l = \left[h_{1,l}, h_{2,l}, \ldots, h_{M,l}\right]^{\mathrm{T}}$, $l = 0, \ldots, L-1$. The instantaneous SNR γ_p can now be written in a quadratic form as

$$\gamma_p = \gamma_0 \boldsymbol{h}^H \boldsymbol{G} \boldsymbol{h}, \tag{49}$$

where $\boldsymbol{G}$, being a Hermitian matrix, can be diagonalized in the form [14]

$$\boldsymbol{G} = \boldsymbol{U} \boldsymbol{\Lambda} \boldsymbol{U}^H. \tag{50}$$

Here, $\boldsymbol{U}$ is a unitary matrix (i.e., $\boldsymbol{U}\boldsymbol{U}^H = \boldsymbol{U}^H\boldsymbol{U} = \boldsymbol{I}_{LM}$) and

$$\boldsymbol{\Lambda} = \mathrm{diag}[\lambda_1, \ldots, \lambda_{LM}] \tag{51}$$

is a diagonal matrix whose diagonal elements $\lambda_1, \ldots, \lambda_{LM}$ are the eigenvalues of matrix $\boldsymbol{G}$. Therefore, γ_p can be expressed as a diagonal Hermitian form with independent variates as

$$\gamma_p = \gamma_0 \boldsymbol{f}^H \boldsymbol{\Lambda} \boldsymbol{f} = \gamma_0 \gamma_p', \tag{52}$$

where

$$\boldsymbol{f} = [f_1, \ldots, f_{LM}]^{\mathrm{T}} = \boldsymbol{U}^H \boldsymbol{h}. \tag{53}$$

Since elements of the zero-mean vector $\boldsymbol{h}$ are independent of elements of $\boldsymbol{G}$, $\boldsymbol{f}$ is obviously also a zero-mean vector.

The key in evaluating the analytical performance of a MIMO UWB system is to derive the statistics of $\boldsymbol{f}$, which is difficult to obtain for a lognormal fading channel. For practical implementation constraints, the number of paths to be combined (L) is typically small (e.g., less than 10). Hence, among all paths combined by the MRC, the average power of the weakest path is approximately equal to the average power of the strongest path. To simplify the analysis, we assume that all L paths combined by the receiver have equal power. Thus, $E\{\boldsymbol{h}\boldsymbol{h}^H\} = \mathrm{diag}[1, 1, \ldots, 1]$, and each element of $\boldsymbol{f}$ is a linear combination of zero-mean, independent and identically distributed RVs (i.e., elements of $\boldsymbol{h}$), with the combining weights determined by the unitary matrix $\boldsymbol{U}$. Although this key observation is deduced based on the assumption of equal power for different paths, it also holds true for practical UWB indoor channels. As a visual proof, Figure 5.3 compares the simulated histogram of an element of $\boldsymbol{f}$ with the pdf of a zero-mean, unit-variance Gaussian RV, for $L = 2$, $M = 3$ ($LM = 6$), and a typical standard deviation of 5 dB for $20\log_{10}|\alpha_l|$ across $l = 0, \ldots, L-1$. It can be seen from this figure that $\boldsymbol{f}$ can be indeed modeled as zero-mean Gaussian. As the value of LM increases ($LM > 6$), the approximation is even more accurate. Therefore, $\boldsymbol{f}$ will be treated as a Gaussian vector with a mean $\boldsymbol{\mu}_f = \boldsymbol{0}_{LM\times 1}$ and a covariance matrix $\boldsymbol{R}_f = \boldsymbol{I}_{LM}$, where $\boldsymbol{0}$ is a vector with all zero elements. Clearly, with such an approximation on the pdf of $\boldsymbol{f}$, γ_p' is equal to a weighted (by the eigenvalues of $\boldsymbol{G}$) sum of independent central chi-square variables, yielding

$$\gamma_p' = \sum_{i=1}^{LM} \lambda_i f_i^2, \tag{54}$$

where $f_i, i = 1, \ldots, LM$, are zero-mean, unit-variance Gaussian RVs.

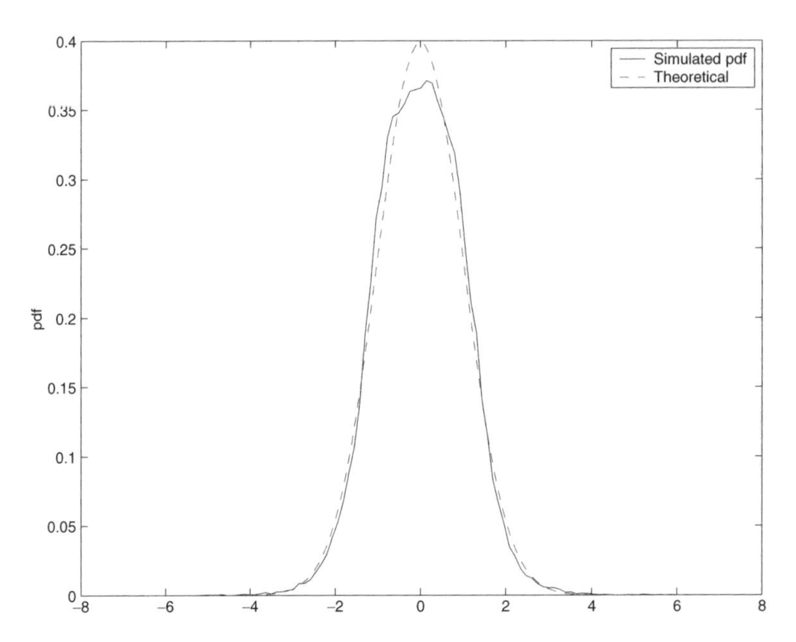

Figure 5.3 Approximated probability density function of f: $L = 2$, $M = 3$, and the standard deviation of $20 \log_{10} |\alpha_l|$ equals 5 dB

Because the matrix $\boldsymbol{G}$ depends on the channel fading coefficients, one would expect that its eigenvalues, λ_i, $i = 1, \ldots, LM$, are random, and thus difficult to obtain analytically. It was shown in [13] that the eigenvalues of $\boldsymbol{G}_l$ are either 1 or 0 with exactly $d = M - N + 1$ eigenvalues equaling 1. Therefore, the matrix $\boldsymbol{G} = \text{diag}[\boldsymbol{G}_0 \ \ldots \ \boldsymbol{G}_{L-1}]$ has LM eigenvalues, $L(M - N + 1)$ of which are equal to 1 and the other $L(N - 1)$ are equal to 0. If these eigenvalues $(\lambda_1, \ldots, \lambda_{LM})$ are arranged in a descending order, we can write

$$[\lambda_1, \ldots, \lambda_{LM}] = [\ \underbrace{1 \ldots 1}_{L(M-N+1)}\ \underbrace{0 \ldots 0}_{L(N-1)}]. \tag{55}$$

With (55), (54) reduces to $\gamma_p' = \sum_{i=1}^{d} f_i^2$, where $d = L(M - N + 1)$. Since f_i is distributed as $f_i \sim \mathcal{N}(0, 1)$, the variable γ_p' is a central chi-square distributed RV with d degrees of freedom and has a pdf expressed as [6]

$$p_{\gamma_p'}(y) = \frac{1}{2^{d/2}\Gamma(\frac{1}{2}d)} y^{d/2-1} e^{-y/2}, \tag{56}$$

where $\Gamma(\cdot)$ is the gamma function defined as $\Gamma(a) = \int_0^{\infty} \tau^{a-1} e^{-\tau} d\tau$, $a > 0$.

Because $\gamma_p = \gamma_0 \gamma_p'$, the conditional probability of error $P(\gamma_p)$ in (46) is also a function of γ_p'. The average BER in a lognormal fading channel can be computed by averaging $P(\gamma_p')$ over $p(\gamma_p')$ as

$$P_b = \int_0^{\infty} P(y) p_{\gamma_p'}(y)\, dy. \tag{57}$$

This BER expression captures any MIMO structure (M, N, L) by a single degree-of-freedom parameter d, which is exactly the diversity order of the corresponding MIMO system.

References

[1] X. Chu and R. Murch, "Performance Analysis of DS-MA Impulse Radio Communications Incorporating Channel-Induced Pulse Overlap," *IEEE Trans. Wireless Commun.*, 2006, to appear.

[2] S. Zhao and H. Liu, "On the Optimum Linear Receiver for Impulse Radio System in the Presence of Pulse Overlapping," *IEEE Commun. Lett.*, vol. 9, pp. 340–342, Apr. 2005.

[3] S. Zhao and H. Liu, "Prerake Diversity Combining for Pulsed UWB Systems Considering Realistic Channels with Pulse Overlapping and Narrow-Band Interference," *Proc. IEEE Globecom'05*, St. Louis, MO, Nov. 2005, to appear.

[4] A. F. Molisch, J. R. Foerster, and M. Pendergrass, "Channel Models for Ultrawideband Personal Area Networks," *IEEE wireless Commun.*, vol. 10, no. 6, Dec. 2003.

[5] R. Qiu, "A Study of the Ultra-Wideband Wireless Propagation Channel and Optimum UWB Receiver Design," *IEEE J. Select. Areas Commun.*, vol. 20, pp. 1628–1637, Dec. 2002.

[6] J. G. Proakis, *Digital Communications (Chapter 14)*, 3rd edition, McGraw-Hill, New York, 1995.

[7] N. C. Beaulieu, A. A. Abu-Dayya, and P. J. Mclane, "Estimating the Distribution of a Sum of Independent Lognormal Random Variable," *IEEE Trans. Commun.*, vol. 43, pp. 2869–2873, Dec. 1995.

[8] P. Cardieri and T. S. Rappaport, "Statistics of the Sum of Lognormal Variables in Wireless Communications," *Proc. 51st IEEE VTC (Spring)*, Tokyo, Japan, vol. 3, pp. 1823–1827, 2000.

[9] J. D. Choi and W. E. Stark, "Performance of Ultra-Wideband Communications with Suboptimal Receiver in Multipath Channels," *IEEE J. Select. Areas Commun.*, vol. 20, no. 9, pp. 1754–1766, Dec. 2002.

[10] M. K. Simon, S. M. Hinedi, and W. C. Lindsey, *Digital Communication Techniques*, Prentice Hall, Englewood Cliffs, NJ, 1995.

[11] J. H. Park, "On Binary DPSK Detection," *IEEE Trans. Commun.*, vol. COM-26, pp. 484–486, Apr. 1978.

[12] R. T. Hoctor and H. W. Tomlinson, "An Overview of Delay-Hopped, Transmitted-Reference RF Communications," Technical Information Series 2001CRD198, G.E. Research and Development Center, pp. 1–29, Jan. 2002.

[13] J. H. Winters, J. Salz, and R. D. Gitlin, "The Impact of Antenna Diversity on the Capacity of Wireless Communications Systems," *IEEE Trans. Commun.*, vol. 42, pp. 1740–1751, Feb./Mar./Apr. 1994.

[14] M. Schwartz, W. R. Bennett, and S. Stein, *Communication Systems and Techniques (Appendix B)*, IEEE Press, Piscataway, NJ, 1995.

6

Mixed-Signal Ultra-wideband Communications Receivers

Sebastian Hoyos and Brian M. Sadler

6.1 Introduction

The extremely high bandwidth signals used in Ultra-wideband (UWB) communication systems impose several challenging tasks for the implementation of both the transmit and the receive ends of the system. In particular, when the receiver is implemented in an all-digital topology, the extremely high sampling rates required to comply with Nyquist criteria have serious implications in the complexity and power consumption of the analog-to-digital converter (ADC) that provides the interface between the received continuous-time signal and the digital receiver. Although analog architectures are also possible, they do not allow for some of the important desired features such as reprogrammability of the receiver. In addition, UWB is inherently an overlay technology, and must deal with significant in-band interference [1]. Strong interferers may swamp the UWB receiver front end and cause ADC saturation, especially when only a few bits of ADC resolution are available over the entire bandwidth. Consequently, robust ADC approaches that lower the sampling rate, the bit resolution, or the power consumption requirements are of great interest.

The currently available architectures used in the fabrication of ADCs include the *flash* architecture, which is based on parallel techniques that use $2^b - 1$ comparators to achieve b bits of resolution. All comparators sample the analog input signal simultaneously, making the flash ADC inherently fast. Because of the parallelism of this architecture, the number of comparators grows exponentially with b, thus increasing the power consumption and also the circuitry area. This in turn increases the input capacitance, limiting the system bandwidth and making it difficult to match components. Some variations of the flash architecture such as the folded-flash [2–4], pipelined [5, 6], and time-interleaved [7] architectures have been proposed in order to overcome some of these problems. Among the difficulties that have slowed the evolution of ADCs is the aperture jitter or aperture

Ultra-wideband Wireless Communications and Networks
Edited by Xuemin (Sherman) Shen, Mohsen Guizani, Robert Caiming Qiu and Tho Le-Ngoc

uncertainty, which is the sample-to-sample variation of the instant in time at which sampling occurs. Moreover, the speed of sampling is limited by the frequency characteristic of the device used in the design, which limits the ability of the comparators to make an unambiguous decision about the input voltage.

In order to overcome these problems, techniques that aim to relax the operational conditions of the ADCs have been proposed. In general, these techniques perform multiband signal processing [8] in which the spectrum of the signal is channelized into several bands by means of a bank of band-pass filters. Analog-to-digital (A/D) conversion thus occurs at a much reduced speed for each one of the resultant band-pass signals. Further, a bank of frequency modulators can be used to shift the signal spectrum so that the center frequency of each subband becomes the zero frequency [9], allowing the use of a bank of identical low-pass filters. Sigma-delta modulation [10] has also been proposed since it enables A/D conversion with low-resolution. The noise penalty associated with the use of fewer bits in the quantization process is overcome in the sigma-delta scheme by using either signal oversampling or multiband processing techniques [11, 12]. In particular, when a single bit is used, the implementation is greatly simplified and practical mono-bit digital receivers have significant potential [13–15]. Since all these techniques are based on time-domain A/D conversion, they suffer from the aforementioned high-speed problems, making it desirable to channelize the signal spectrum into several subbands. However, the implementation of the bank of band-pass filters needed in the multiband processing schemes can be potentially troublesome; problems such as spectrum sharing due to the nonideal characteristics of the band-pass filters will affect the overall system performance. The design of analog filters with sharp roll-off needed in the multiband ADC approaches is not only very difficult but also suffers from heavy power consumption and requirement of large circuitry area to accommodate the passive elements (i.e., inductors and capacitors).

This chapter overviews the approaches to the A/D conversion problem and its application for the design of UWB communications receivers that can potentially overcome some of the limitations encountered during the implementation of time-domain A/D converters. These approaches exploit the signal representation in domains other than the classical time domain, which reduces the speed of the comparators that perform the signal quantization. This potentially improves the distortion versus average bit rates of the A/D conversion, which implies lower bit resolution requirements, which in principle can lead to lower power consumption.

First, the reduction of the ADC's speed is achieved because the quantization of the coefficients is carried out at the end of a time-window during which the signal is projected over basis functions. Thus, the ADCs run at a speed that is inversely proportional to the time-window duration, which can be properly designed to meet the speeds allowed by the technology used in the implementation. The speed reduction comes at the cost of the implementation of the local basis function generators, mixers, and integrators needed to project the continuous-time signal onto the set of basis functions. This introduces a trade-off between sampling speed reduction and system complexity, which is studied in this chapter. Similar reductions in the speed of the quantizers could be achieved in the time domain by using a time-interleaved bank of quantizers [7, 16, 17]. However, the synchronization problems, the very fine time resolution in high-speed applications, and the fact that all the ADCs see the full bandwidth of the input signal in a time-interleaved architecture, make it difficult to design the sample/hold circuitry and cause the overall design to require significant power. In addition, the signal expansion approach avoids the sharp-roll-off filter bank needed in multiband ADC architectures. In fact, mixing with basis functions, followed by integration over a time-window required to project the signal, synthesizes a filter bank with overlapping spectra and smooth transitions. The relaxed implementation requirements are a key motivation for the ideas presented here.

Second, the potential lower bit resolution requirements can be achieved by optimally allocating the available number of bits in the quantization of the coefficients, obtained through the projection of the continuous-time signal over the basis set. The possibility of efficiently allocating the available resources, in terms of number of bits per sample, is a powerful feature that is not available in conventional time-domain ADC. Optimal bit allocation is possible in the proposed A/D conversion scheme because some signal characteristics that are hidden in the time domain, such as power distribution, can now be explored by projecting the continuous-time signal into a more appealing domain.

As a particular case, we consider A/D conversion in the frequency domain [18–21] in which samples of the signal spectrum are taken at a rate that guarantees no aliasing in the discrete-time signal domain. The discrete frequency samples are then quantized by a set of quantizers operating at a rate that is a fraction of the Nyquist rate needed in the sampling of the time domain signal. Other domains, such as those provided by the Hadamard, Walsh, Walsh-Fourier, and Haar wavelet transforms, are also potential candidates.

This chapter presents the design of mixed-signal communications receivers based on the ADC framework obtained via signal expansion. A generalization of the mixed-signal receiver problem is first presented, and then two frequency-domain receiver design examples are discussed based on (i) multicarrier and (ii) transmitted-reference signaling.

6.2 Analog-to-Digital Conversion via Signal Expansion

The block diagram depicted in Figure 6.1 shows the basic signal expansion principle of the proposed A/D conversion. The received signal $s(t)$ is decomposed every T_c seconds into N components, which are obtained through the projection over the basis $\Phi_l(t)|_{l=0}^{N-1}$. The coefficients $s_l^{(m)}|_{l=0}^{N-1}$ are found as

$$s_l^{(m)} = \langle s(t), \Phi_l(t) \rangle_{m,T_c} = \int_0^{T_c} s(t + mT_c)\Phi_l^*(t)\, dt. \tag{1}$$

If the mean-square error (MSE) criterion is used to reconstruct the received signal $s(t)$ in the interval $mT_c \leq t \leq (m+1)T_c$ through a linear combination of the basis functions $\Phi_l(t)|_{l=0}^{N-1}$, in general, the coefficients $s_l^{(m)}|_{l=0}^{N-1}$ will have to be linearly transformed. We define the $N \times N$ matrix Γ, which contains the correlation coefficients of the basis functions,

$$\gamma_{n,l} = \langle \Phi_n(t), \Phi_l(t) \rangle_{T_c} = \int_0^{T_c} \Phi_n(t)\Phi_l^*(t)\, dt, \quad n, l = 0, \ldots, N-1. \tag{2}$$

The coefficients $a_l^{(m)}$ that provide the best MSE approximation are found by solving the linear equation

$$s^{(m)} = a^{(m)}\Gamma, \tag{3}$$

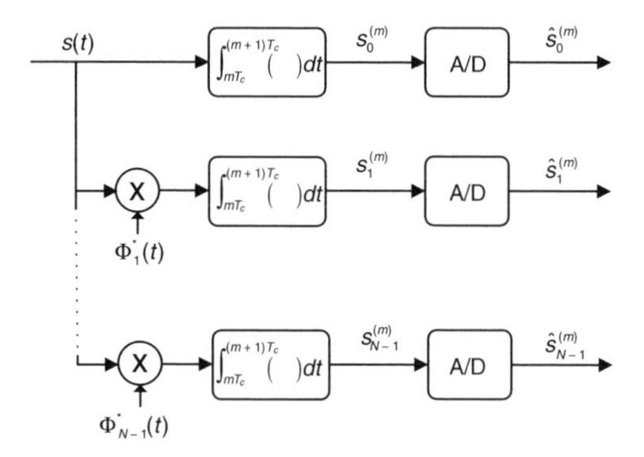

Figure 6.1 Block diagram of the analog-to-digital converter that expands the received signal using a set of basis functions

where the vectors $s^{(m)}$ and $a^{(m)}$ are defined as $s^{(m)} = [s_0^{(m)} \ldots s_{N-1}^{(m)}]$, and $a^{(m)} = [a_0^{(m)} \ldots a_{N-1}^{(m)}]$. Solving (3) requires invertibility of the matrix Γ. If the basis functions are orthonormal, then $a^{(m)} = s^{(m)}$. The best MSE approximation is given by

$$\tilde{s}^{(m)}(t) = \sum_{l=0}^{N-1} a_l^{(m)} \Phi_l(t), \qquad 0 \leq t \leq T_c, \tag{4}$$

where the signal $\tilde{s}^{(m)}(t)$, $0 \leq t \leq T_c$, is the best MSE approximation of the input signal $s(t)$, $mT_c \leq t \leq (m+1)T_c$. At the end of the conversion-time T_c, the coefficients $s_l^{(m)}|_{l=0}^{N-1}$ reach a constant value that is fed to a set of quantizers $Q_l^{(m)}|_{l=0}^{N-1}$, one for each coefficient, which return the digital words $\hat{s}_l^{(m)}|_{l=0}^{N-1}$. The lth quantizer $Q_l^{(m)}$ has 2^{b_l} output levels, where $b_l|_{l=0}^{N-1}$ is the number of bits used to obtain the quantized set of coefficients $\hat{s}_l^{(m)}|_{l=0}^{N-1}$. These values represent the output of the ADC for the input signal in a T_c second interval. Notice that the signal $s(t)$ is being segmented by a rectangular window for simplicity; however, windows with preferable characteristics can be used instead. The number of coefficients N used in the A/D conversion is intimately related to the conversion-time T_c, and will affect the degree of the approximation indicated in (4), up to the point where the signal $s(t)$ is represented with zero error energy with a sufficient number of coefficients[1] N_*. When only a limited number of coefficients ($N \leq N_*$) are used in the A/D signal conversion, some distortion is introduced. This distortion plus the distortion introduced in the quantization process constitute the major sources of distortion

[1] The existence of the number N_* that makes the MSE zero, assumes that the basis functions $\Phi_l(t)|_{l=0}^{N_*-1}$ span the input signal $s(t)$. It is also possible that N_* tends to infinity, as, for example, happens with signals with infinite spectral support (non-band-limited signals) when they are projected in the frequency domain. However, for simplicity in the analysis, it is assumed that the input signal $s(t)$ is a smooth, well-behaved signal (ideally a band-limited signal) that can be represented with a finite number of coefficients N_*. The particular conditions that $s(t)$ must satisfy for the existence of N_* will depend on the domain chosen for the A/D conversion.

of the proposed A/D conversion, which are analyzed in [18, 19, 22]. Timing and frequency offset distortion are considered in [22].

6.3 Mixed-Signal Communication Receivers Based on A/D Conversion via Signal Expansion

We will investigate the design of mixed-signal communications receivers based on the ADC ideas presented here. They are mixed-signal receivers in the sense that in their analog front end, signal projection over basis functions is performed before the parallel ADCs are applied. Additionally, the information bits are detected through a discrete matched filter operation that takes place in the domain on which the received signal has been expanded.

6.3.1 Transmitted Signal and Channel Model

To elaborate, assume that the signal $s(t)$ is transmitted over a linear communication channel with impulse response $h(t)$

$$r(t) = s(t) * h(t) + z(t), \quad 0 \leq t \leq T. \tag{5}$$

where '$*$' indicates continuous-time convolution and $z(t)$ is additive white Gaussian noise (AWGN). In conventional all-digital linear communication receivers, the received continuous-time signal is first passed through a time-domain ADC running at Nyquist rate, and the discrete-time samples are then demodulated by performing a discrete-time linear filtering operation. The following subsection presents a fundamentally different approach for the implementation of all-digital linear receivers, based on the coefficients obtained from A/D conversion via signal expansion.

6.3.2 Digital Linear Receivers Based on ADC via Signal Expansion

Assume that the transmitted signal $s(t)$ conveys the information symbol a. In order to obtain an estimate of the transmitted symbol from the set of coefficients provided by the ADC of the signal expansion, we begin by expressing the receiver structure as a linear filtering problem in the time domain

$$\hat{a} = r(t) * p(t)|_{t=T} = \int_0^T r(\tau)p(T-\tau)\,d\tau, \tag{6}$$

where $\hat{a}$ is the symbol estimate, $p(t)$ is the impulse response of the linear filter demodulator, which can be a simple matched filter, a RAKE receiver, an MMSE receiver, and so on. The output of this filter is sampled at $t = T$. For convenience, (6) is expressed as

$$\hat{a} = \int_0^T r(\tau)p(T-\tau)\,d\tau = \int_0^T r(\tau)g^*(\tau)\,d\tau, \tag{7}$$

where we define $g^*(\tau) = p(T - \tau)$. Now, we proceed to segment the signal duration time T into M time-slots of duration T_c. We define the following signals

$$w_m(t) = \begin{cases} 1 & mT_c \leq t \leq (m+1)T_c \\ 0 & \text{elsewhere} \end{cases} \tag{8}$$

$$r_m(t) = r(t)w_m(t) \tag{9}$$

$$g_m(t) = g(t)w_m(t), \tag{10}$$

for $m = 0, \ldots, M-1$, and $w_m(t)$ is a rectangular window for simplicity of the analysis, although other windows with desired characteristics could be used instead.

Using these definitions, the linear receiver output in (6) can be expressed as

$$\begin{aligned} \hat{a} &= \sum_{m=0}^{M-1} \int_{mT_c}^{(m+1)T_c} r(\tau)g^*(\tau)\,d\tau \\ &= \sum_{m=0}^{M-1} \int_{mT_c}^{(m+1)T_c} r_m(\tau)g_m^*(\tau)\,d\tau \\ &= \sum_{m=0}^{M-1} \int_{-\infty}^{\infty} r_m(\tau)g_m^*(\tau)\,d\tau, \end{aligned} \tag{11}$$

in which the integral in (6) has been segmented into M integrals that run over intervals of duration T_c each, such that $T = MT_c$.

In order to express the matched filter operations in the new conversion domain, the signal expansions over the basis functions $\Phi_l(t)$ and $\Psi_l(t)$ are used to represent the segmented received signal and the segmented receive filter respectively, leading to

$$\begin{aligned} \hat{a} &= \sum_{m=0}^{M-1} \int_{-\infty}^{\infty} r_m(\tau)g_m^*(\tau)\,d\tau \\ &= \sum_{m=0}^{M-1} \int_{-\infty}^{\infty} \sum_{n=0}^{\infty} R_m(n)\Phi_n(\tau) \sum_{l=0}^{\infty} G_m^*(l)\Psi_l^*(\tau)\,d\tau \\ &= \sum_{m=0}^{M-1} \sum_{n=0}^{\infty} \sum_{l=0}^{\infty} R_m(n)G_m^*(l) \int_{-\infty}^{\infty} \Phi_n(\tau)\Psi_l^*(\tau)\,d\tau \\ &= \sum_{m=0}^{M-1} \sum_{n=0}^{\infty} \sum_{l=0}^{\infty} R_m(n)G_m^*(l)\zeta_{n,l} \\ &\approx \sum_{m=0}^{M-1} \sum_{n=0}^{N-1} \sum_{l=0}^{N-1} R_m(n)G_m^*(l)\zeta_{n,l}, \end{aligned} \tag{12}$$

where $\zeta_{n,l} = \int_{-\infty}^{\infty} \Phi_n(\tau)\Psi_l^*(\tau)\,d\tau$, $R_m(n)|_{n=0}^{N-1}$, and $G_m(l)|_{l=0}^{N-1}$ are the best MSE coefficients representation as explained in (4), which requires reversing the linear transformation

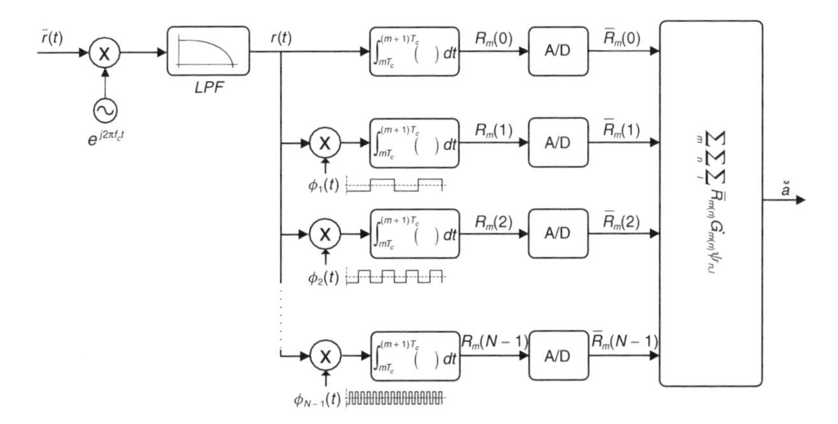

Figure 6.2 Mixed-signal receiver block diagram with ADC via signal expansion. A binary basis case is illustrated

of (3). Note that the series expansion in (12) has been truncated, leading to some degree of error. Although this truncation error should in principle degrade the receiver performance, we will show in the examples that any desired performance can be achieved if the trade off between complexity in terms of number of coefficients N, and sampling speed $\Delta F_c = 1/T_c = M/T$, is adequately set up. Note that if the basis functions are orthonormal to each other, (12) reduces to

$$\hat{a} \approx \sum_{m=0}^{M-1} \sum_{n=0}^{N-1} R_m(n) G_m^*(n), \tag{13}$$

which reduces the complexity of detection. The trade off between the choice of the basis functions, complexity of the detection formula, and the degree of truncation error is fundamental in the receiver design. Figure 6.2 illustrates the mixed-signal receiver architecture.

6.4 Analog-to-Digital Conversion in the Frequency Domain

The frequency domain represents a fundamental and well understood domain where A/D conversion can be carried out. Figure 6.3 shows the block diagram of the frequency domain ADC in which the complex exponential functions that constitute the orthogonal basis allow sampling of the continuous-time signal spectrum at the frequencies $F_l|_{l=0}^{N-1}$, leading to the set of frequency coefficients

$$c_l = \int_0^{T_c} s(t) e^{-j2\pi F_l t}\, dt, \qquad l = 0, \ldots, N-1. \tag{14}$$

These coefficients are then quantized by a set of quantizers $Q_l|_{l=0}^{N-1}$, which in turn produce the ADC output digital coefficients $\hat{c}_l|_{l=0}^{N-1}$. The frequency sample spacing $\Delta F = F_l - F_{l-1}$

complies with $\Delta F \leq \frac{1}{T_c}$ in order to avoid aliasing in the discrete-time domain [23]. Thus, the optimal number of coefficients N_* necessary to fully sample the signal spectrum with bandwidth W^2, without introducing time aliasing, is proportional to the time-bandwidth product

$$N_* = \left\lceil \frac{W}{\Delta F} + 1 \right\rceil \geq \lceil WT_c + 1 \rceil, \tag{15}$$

where the operator $\lceil\ \rceil$ is used to ensure that N_* is the closest upper integer that avoids discrete-time aliasing. When $\Delta F = \frac{1}{T_c}$, (15) turns into an equality and the discrete-time alias-free condition is satisfied without oversampling of the frequency spectrum.

Note that the frequency-domain ADC is fundamentally different from the ADC architectures based on filter bank theory [8, 9]. The difference lies in the fact that the frequency-domain ADC samples the spectrum of time-segments of the received signal, whereas the filter bank approach performs frequency channelization, which is a continuous-time signal filtering operation. The implication of this simple but fundamental difference is that the mixing of the received signal with the local oscillator, followed by integration over the projection time T_c, synthesizes a filter with very smooth transitions. In fact, the frequency response of the branches in Figure 6.3 overlap with each other, but this overlapping does not introduce ADC distortion. On the other hand, the filter bank approach requires filters with very sharp roll-off, and any leakage between the channels will seriously degrade the ADC performance. This is perhaps the most important motivation for the implementation of the frequency-domain ADC. We note that the frequency-domain ADC can suffer from the typical implementation impairments of any parallel architecture; these are considered in [22].

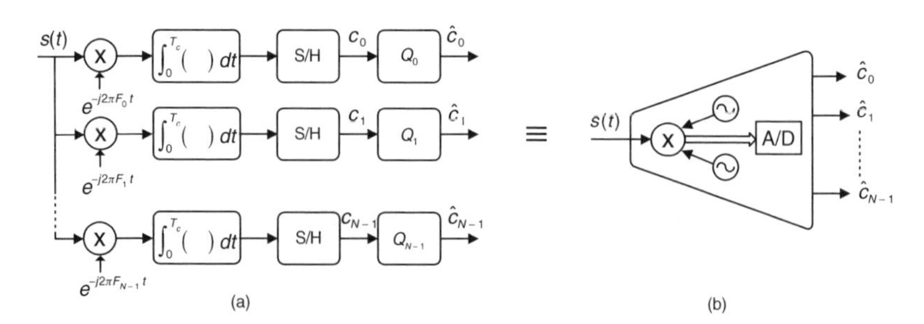

Figure 6.3 (a) Block diagram of the analog-to-digital converter in the frequency domain. (b) Block representation of the ADC in the frequency domain

[2] Because signals found in applications are time-limited, the term bandwidth here refers to the range of frequencies in which the signal power is larger than some defined power level, for instance, many signal bandwidths are defined at 3 dB of attenuation, although more conservative attenuation could be desirable for some applications such as A/D conversion. Moreover, the bandwidth W here is the bandwidth of the time-segmented signal, which in general is larger than the bandwidth of the signal $s(t)$, as the segmentation introduces sidelobes that should be sampled in order to obtain lower distortion error.

6.5 Frequency-Domain Mixed-Signal Receivers

The frequency-domain ADC architecture allows the implementation of linear and nonlinear receivers with potentially lower sampling rates and lower bit resolution requirements [24]. In this section we describe two such receivers.

6.5.1 Multicarrier Communication Systems Based on A/D Conversion in the Frequency Domain

6.5.1.1 A/D Conversion of Multicarrier Signals in the Frequency Domain

The A/D conversion of multicarrier signals in the frequency domain has interesting features that are investigated here before going into the details of the new receiver structure. In particular, it is important to understand how the number of coefficients N changes with the conversion-time T_c for the specific case of multicarrier signals. Let us consider the complex envelope of a multicarrier signal $x(t)$ composed of the sum of S complex exponentials with complex amplitudes $a_s|_{s=0}^{S-1}$

$$x(t) = \sum_{s=0}^{S-1} a_s e^{j2\pi f_s t}, \qquad 0 \le t \le T, \tag{16}$$

where the intercarrier frequency spacing is given by $\Delta F = f_s - f_{s-1} = \frac{1}{T}\ \forall s$, which is the minimum carrier separation that guarantees orthogonality between the carriers. We first notice that by making the conversion time T_c equal to the multicarrier signal duration T, it is evident that a total of $N = S$ samples of the spectrum of $x(t)$ are needed to recover the coefficients $a_s|_{s=0}^{S-1}$. In such a case, the frequency-domain ADC is just a simple correlator bank. The problem with choosing $T_c = T$ is that when a large number of carriers are used, twice this number of multiply and integrate devices are needed in the implementation of the correlator bank, which easily makes the system impractical. Therefore, it is of great interest to investigate cases in which the conversion time satisfies $T_c < T$. In particular, we chose $T_c = T/M$, with M an integer. In order to reflect the effect of segmenting the signal duration time T into M time-slots of duration T_c, we define the following signals

$$x_m(t) = x(t)w_m(t), \tag{17}$$

where $m = 0, \ldots, M-1$, and the window $w_m(t)$ was defined in (10). It is easy to verify that the Fourier transform ($\mathcal{F}$) of $w_m(t)$ is given by

$$\mathcal{F}\{w_m(t)\} = W_m(F) = \frac{\sin(\pi F T_c)}{\pi F} e^{-j\pi(2m+1)T_c F}. \tag{18}$$

The Fourier transform of $x_m(t)$, denoted as $X_m(F)$, can be expressed as

$$\begin{aligned} X_m(F) &= \mathcal{F}\{x_m(t)\} \\ &= \mathcal{F}\{x(t)\} * \mathcal{F}\{w_m(t)\} \\ &= \mathcal{F}\left\{\sum_{s=0}^{S-1} a_s e^{j2\pi f_s t}\right\} * W_m(F) \end{aligned}$$

$$= \left(\sum_{s=0}^{S-1} a_s \delta(F - f_s) \right) * \left(\frac{\sin(\pi F T_c)}{\pi F} e^{-j\pi(2m+1)T_c F} \right)$$

$$= \sum_{s=0}^{S-1} a_s \frac{\sin(\pi T_c(F - f_s))}{\pi(F - f_s)} e^{-j\pi T_c(2m+1)(F-f_s)}. \tag{19}$$

The spectrum in (19) is sampled every $\Delta F_c = \frac{1}{T_c}$ Hz. A total of N samples are taken based on (15). The bandwidth W changes depending on the segmentation ratio $M = T/T_c$. Figure 6.4 (a) illustrates the bandwidth expansion that $X_m(F)$ experiences owing to time windowing. The figure shows that with no windowing, the multicarrier signal occupies a bandwidth equal to $S\Delta F$[3], while windowing introduces a bandwidth expansion equal to $\frac{T-T_c}{T_c}\Delta F$. Thus, the bandwidth W is given by

$$W = S\Delta F + \frac{T - T_c}{T_c} \Delta F$$
$$= S\Delta F + \left(\frac{\Delta F_c}{\Delta F} - 1 \right) \Delta F$$
$$= (S + M - 1)\, \Delta F. \tag{20}$$

Additionally, on the basis of (15), the number of frequency samples N is given by

$$N = \left\lceil (S + M - 1) \frac{\Delta F}{\Delta F_c} \right\rceil$$
$$= \left\lceil \frac{S}{M} + \frac{M - 1}{M} \right\rceil. \tag{21}$$

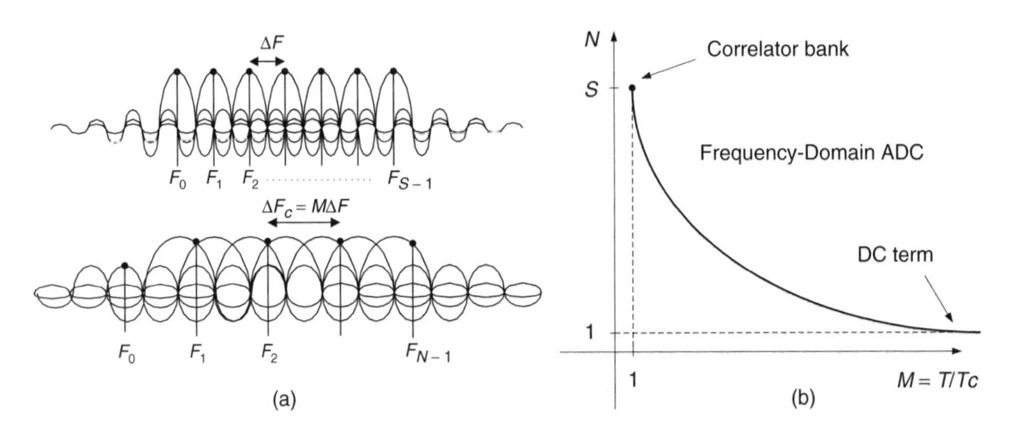

Figure 6.4 Characteristics of A/D conversion of multicarrier signals in the frequency domain. (a) Effect of time segmentation in the bandwidth of the multicarrier signal and in the samples frequency spacing ΔF_c. (b) Number of coefficients N versus the symbol-period to segmentation-time ratio $(M = T/T_c)$

[3] This bandwidth accounts only for the range of frequencies occupied by the main lobe of the carriers.

Note also that if no bandwidth expansion is taken into account, (21) reduces to $N = S/M$. These equations show that although the bandwidth of all the individual carriers that compose $x(t)$ is inversely proportional to T_c, only the changes in the bandwidth of the carriers that lie at the edges of the spectrum of the multicarrier signal will affect the overall bandwidth. Although some truncation error will occur here, N and M can be adequately chosen to achieve some desired performance. The inner carriers' spectrum will also spread out, overlapping the neighbor carriers, but will not increase the bandwidth of $x(t)$. Although the carriers are no longer orthogonal within each segment for a conversion time $T_c < T$, we show in Section 6.5.1.2 that no intercarrier interference (ICI) is introduced when the receiver processes all the segments to calculate the symbol estimates.

This simple result represents a fundamental concept in the development of the ideas presented here. The result shows us that although further segmentation of the multicarrier signal increases the bandwidth of the signal to be A/D converted, an even larger reduction in the number of coefficients N is obtained. The favorable trade-off between N and T_c is very important in the implementation of the ADC in the frequency domain for multicarrier signals, since it indicates that a practical number of frequency samples N can be obtained by adequately selecting a symbol-period to conversion-time ratio $M = T/T_c$.

Additionally, it is interesting to notice that as $M = T/T_c$ is increased, the number of frequency samples N will eventually reach the value 1, which is just the point where the ADC in the frequency domain turns into a zero-frequency term of the transform. At this point of operation, the ADC samples the DC frequency, which is just the average of the signal over the integration time T_c. If T_c is decreased further, the signal in the interval T_c approaches a flat level, which turns the ADC into a conventional time-domain ADC with some degree of oversampling. Figure 6.4(b) illustrates the three regions of operation of the ADC in Figure 6.3, which are (1) $T_c = T$, which leads to $N = S$, a straightforward correlator bank, (2) $T_c < T$ such that $N < S$, which is the proposed ADC in the frequency domain, and (3) $T_c \ll T$ such that $N = 1$, the DC term in the Fourier transform, which can turn into a conventional time-domain ADC if a sufficient oversampling rate is allowed, that is, T_c smaller than the Nyquist sampling period.

Assume that the available channel bandwidth B_W[4] is divided into S subchannels of bandwidth B_W/S with center frequencies $f_s|_{s=0}^{S-1}$. Thus, a block of S symbols $a_s|_{s=0}^{S-1}$ is simultaneously transmitted through the channel over a signal block time T. The transmitted signal for a block of symbols can be expressed as in (16), where the symbols $a_s|_{s=0}^{S-1}$ are often chosen from a QAM or PSK constellation. In OFDM systems, the generation of the transmit signal $x(t)$ is achieved by performing an inverse DFT operation over the block of symbols $a_s|_{s=0}^{S-1}$, which are then passed through a digital-to-analog (D/A) converter to generate a continuous-time signal that is then frequency shifted to some desired frequency band.

The signal $x(t)$ is transmitted over a linear communication channel with impulse response $h(t)$ and frequency response $H(f)$, as in (5). Assuming that the channel is

[4] Notice that the bandwidth B_W is the bandwidth of the multicarrier signal with duration equal to the symbol-period T, whereas the bandwidth W is the bandwidth of the segmented multicarrier signal with duration T_c. In general $W \geq B_W$.

flat in each of the transmission subbands, the received signal can be expressed as

$$r(t) = x(t) * h(t) + z(t) \approx \sum_{s=0}^{S-1} a_s e^{j2\pi f_s t} H(f_s) + z(t), \quad 0 \le t \le T \tag{22}$$

where '$*$' indicates continuous-time convolution and $z(t)$ is AWGN. In conventional OFDM systems, the received continuous-time signal is first passed through a time-domain A/D converter running at Nyquist rate, and the discrete-time samples are then demodulated by performing a DFT operation. The following subsection presents a fundamentally different approach for the implementation of the multicarrier receiver, based on A/D conversion in the frequency domain.

6.5.1.2 Multicarrier Receiver Based on Analog-to-Digital Conversion in the Frequency Domain

Figure 6.5 illustrates the proposed receiver architecture. The received signal is selected by a front-end band-pass filter tuned to the frequency range of interest, leading to the signal $r(t)$. The ADC in the frequency domain provides the set of full resolution samples[5] $R_m(F_n)|_{n=0}^{N-1}$ every T_c seconds, where $m = 0, \ldots, M-1$, and the information symbol block period T is related with the A/D conversion time T_c as $T = MT_c$. Notice that the receiver could either directly sample the received signal, or shift the band-pass multicarrier signal to its baseband[6] equivalent before the ADC in frequency domain takes place. The latter scenario would relax the generation of the local oscillators needed in the ADC in the frequency domain by lowering their frequency. The analysis presented here is valid for both sampling of the passband received signal or its baseband equivalent.

In order to obtain the estimates of the transmitted symbols $a_s|_{s=0}^{S-1}$ from the set of spectrum samples $R_m(F_n)|_{n=0}^{N-1}|_{m=0}^{M-1}$, we begin by expressing the calculation as a matched filter problem in time domain

$$\begin{aligned}\bar{a}_s &= r(t) * p_s(t)|_{t=T} \\ &= \int_0^T r(\tau) p_s(t-\tau)\, d\tau|_{t=T} \\ &= \int_0^T r(\tau) p_s(T-\tau)\, d\tau,\end{aligned} \tag{23}$$

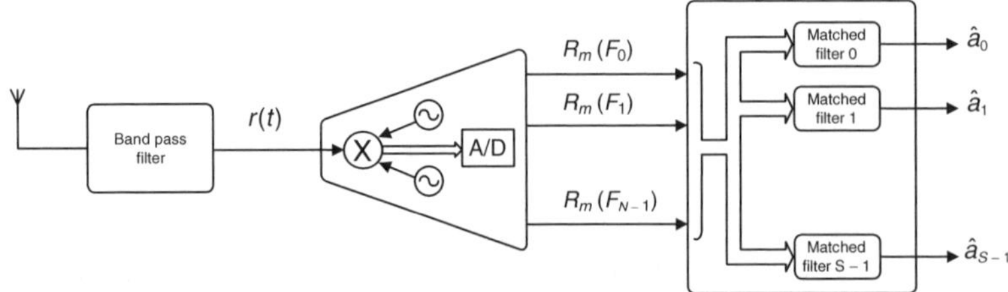

Figure 6.5 Block diagram of multicarrier receiver based on ADC in the frequency domain

[5] To facilitate the analysis, quantization error is ignored in this section.

[6] The received signal could be shifted to any convenient intermediate frequency.

where $\bar{a}_s$ is the estimated symbol associated to the carrier at frequency f_s, $p_s(t)$ is the impulse response of the matched filter for the sth carrier, and the output of this filter is sampled at $t = T$. Assuming that the channel is flat within each subband, the matched filter impulse response is given by

$$\begin{aligned} p_s(t) &= H^*(f_s)x_s^*(T-t), \\ &= H^*(f_s)e^{-j2\pi f_s(T-t)}, \\ &= H^*(f_s)e^{j2\pi f_s t}, \quad 0 \le t \le T \end{aligned} \tag{24}$$

where $x_s(t) = e^{j2\pi f_s t}$ is the carrier at frequency f_s, the result $e^{-j2\pi f_s T} = 1$ is used[7], and $(\cdot)^*$ indicates complex conjugate. Further,

$$\begin{aligned} p_s(T-t) &= H^*(f_s)e^{j2\pi f_s(T-t)}, \\ &= H^*(f_s)e^{-j2\pi f_s t}, \\ &= g_s^*(t). \quad 0 \le t \le T. \end{aligned} \tag{25}$$

For convenience, (23) is expressed as

$$\begin{aligned} \bar{a}_s &= \int_0^T r(\tau)p_s(T-\tau)\,d\tau \\ &= \int_0^T r(\tau)g_s^*(\tau)\,d\tau. \end{aligned} \tag{26}$$

In order to reflect the effect of segmenting the signal duration time T into M time-slots of duration T_c, we define the following signals

$$r_m(t) = r(t)w_m(t) \tag{27}$$

$$g_{s,m}(t) = g_s(t)w_m(t), \tag{28}$$

where $m = 0, \ldots, M-1$, and $w_m(t)$ is a rectangular window for simplicity of the analysis, although as mentioned earlier other windows with desired characteristics could be used instead.

Using these definitions, the matched filter output in (23) can be expressed as

$$\begin{aligned} \bar{a}_s &= \sum_{m=0}^{M-1} \int_{mT_c}^{(m+1)T_c} r(\tau)g_s^*(\tau)\,d\tau \\ &= \sum_{m=0}^{M-1} \int_{mT_c}^{(m+1)T_c} r_m(\tau)g_{s,m}^*(\tau)\,d\tau \\ &= \sum_{m=0}^{M-1} \int_{-\infty}^{\infty} r_m(\tau)g_{s,m}^*(\tau)\,d\tau, \end{aligned} \tag{29}$$

[7] This condition requires that $f_s T = k$, with k being some integer. This condition holds if, for example, the multicarrier signal has been downconverted to baseband frequencies where $f_s = \pm k\Delta F = \pm k/T$.

in which the integral in (23) has been segmented into M pieces of duration T_c each, such that $T = MT_c$.

In order to express the matched filter operations in the frequency domain, we use Parseval's theorem in (11), leading to

$$\bar{a}_s = \sum_{m=0}^{M-1} \int_{-\infty}^{\infty} r_m(\tau) g_{s,m}^*(\tau)\, d\tau$$
$$= \sum_{m=0}^{M-1} \int_{-\infty}^{\infty} R_m(F) G_{s,m}^*(F)\, dF, \quad (30)$$

where $R_m(F) = \mathcal{F}\{r_m(t)\}$ and $G_{s,m}(F) = \mathcal{F}\left\{g_{s,m}(t)\right\}$. Therefore, the exact calculation of the matched filter output in the frequency domain requires the Fourier transforms of all the segmented received signals, namely, $R_m(F)|_{m=0}^{M-1}$, and the Fourier transform of the segmented matched filters, $G_{s,m}(F)|_{m=0}^{M-1}$. However, since only N samples of the received signal spectrum are provided by the ADC in the frequency domain, (30) is approximated as

$$\bar{a}_s = \sum_{m=0}^{M-1} \int_{-\infty}^{\infty} R_m(F) G_{s,m}^*(F)\, dF$$
$$\approx \sum_{m=0}^{M-1} \Delta F_c \sum_{n=0}^{N-1} R_m(F_n) G_{s,m}^*(F_n), \quad (31)$$

where $R_m(F_n)|_{n=0}^{N-1}$ and $G_{s,m}(F_n)|_{n=0}^{N-1}$ are the samples from the spectrum of $r_m(t)$ and $g_{s,m}(t)$, respectively. If the signal has been downconverted to baseband frequencies, I and Q components need to be combined first before the ADC samples the spectrum of the signal. Note also that when N is even, for symmetry reasons, no DC component from the spectrum is collected even if the signal has been downconverted to baseband.

If the signal is A/D converted in its original frequency band (i.e. no downconversion performed), the signal spectrum is symmetric around the DC frequency and only the samples associated to the positive frequencies need to be evaluated, because the negative frequency samples are just complex conjugate of the positive frequency samples.

Note that truncation of the integrals in (31) has been used, leading to some degree of estimation error. However, a good approximation is obtained if the conversion frequency spacing ΔF_c avoids discrete-time aliasing and samples are taken only in the frequency band of interest where most of the signal energy is concentrated. The samples of the spectrum of the segmented matched filter, $G_{s,m}^*(F_n)|_{n=0}^{N-1}$, need to be estimated in order to calculate the information symbol estimates in (31). In particular, it is necessary to estimate the channel frequency response $H(F)$, or equivalently, if the channel is assumed flat within each subchannel, the samples $H(f_s)|_{s=0}^{S-1}$. To this end, training can be used to estimate the coefficients leading to the set $\hat{G}_{s,m}(F_n)|_{n=0}^{N-1}$. Thus, the final estimates of the information symbols are calculated as

$$\hat{a}_s = \sum_{m=0}^{M-1} \int_{-\infty}^{\infty} R_m(F) G_{s,m}^*(F)\, dF$$
$$\approx \sum_{m=0}^{M-1} \Delta F_c \sum_{n=0}^{N-1} R_m(F_n) \hat{G}_{s,m}^*(F_n). \quad (32)$$

The accuracy of the estimates $\hat{a}_s$ will depend on how close the channel estimates $\hat{G}_{s,m}(F_n)|_{n=0}^{N-1}$ are to their true values. When the channel estimates are perfect, the signal is free of intercarrier or intersymbol interference, and the number of spectrum samples N satisfies (15), which avoids discrete-time aliasing and makes the error in (31) negligible. Then, the probability of error associated with this receiver will be the same probability of error of a conventionally implemented multicarrier or multichannel communication system [25].

6.5.2 Relationship to the Fourier Series Coefficients

The presentation of the analog-to-digital converter in the frequency domain, which has been provided here, is general in the sense that the frequency samples $R_m(\pm F_n)|_{n=0}^{N-1}$ can be conveniently allocated anywhere in the signal spectrum. One particular sample allocation is provided by the Fourier series coefficients. Assuming that the signal $r(t)$ is a baseband signal, the Fourier coefficients provide samples at the frequencies $F_k = \pm k \Delta F_c$, and $k = 0, 1, \ldots, K-1$. Then, the signal $r_m(t)$ can be represented as

$$\tilde{r}_m(t) = \sum_{k=-K+1}^{K-1} R_m(F_k) e^{-j2\pi F_k t}, \qquad mT_c \le t \le (m+1)T_c, \quad (33)$$

where $\tilde{r}_m(t)$ is the best MSE approximation of $r_m(t)$.

Now, we will derive the data detection formula starting from (29)

$$\bar{a}_s = \sum_{m=0}^{M-1} \int_{-\infty}^{\infty} r_m(\tau) g_{s,m}^*(\tau)\, d\tau$$
$$= \sum_{m=0}^{M-1} \int_{-\infty}^{\infty} \sum_{k=-\infty}^{\infty} R_m(F_k) e^{-j2\pi F_k \tau} \sum_{l=-\infty}^{\infty} G_m^*(F_l) e^{j2\pi F_l \tau}\, d\tau$$
$$= \sum_{m=0}^{M-1} \sum_{k=-\infty}^{\infty} R_m(F_k) \sum_{l=-\infty}^{\infty} G_m^*(F_l) \int_{-\infty}^{\infty} e^{-j2\pi F_k \tau} e^{j2\pi F_l \tau}\, d\tau$$
$$= \sum_{m=0}^{M-1} \sum_{k=-\infty}^{\infty} R_m(F_k) G_m^*(F_k)$$
$$\approx \sum_{m=0}^{M-1} \sum_{k=-K+1}^{K-1} R_m(F_k) G_m^*(F_k), \quad (34)$$

where the orthogonality between the complex exponentials has been used, the approximation follows from truncation of the Fourier series coefficients. This is exactly the same result obtained in (31) through Parseval's theorem, where $N = 2K + 1$, and the DC term is included in the frequency allocation.

6.5.2.1 UWB Example

In this example, we consider the design of a mixed-signal multicarrier receiver based on the ideas presented here, which are summarized in the detection formula of (12). In order to provide a practical example, we use the system specifications of the multiband UWB OFDM receiver presented in the standard draft IEEE P802.15-03/268r1 [26]. The multicarrier signal is composed of 128 tones, with a frequency spacing of 4.125 MHz and a central frequency that lies in any of the UWB subbands. The receiver proposed in [26] requires two conventional time-domain ADCs (for I/Q paths) operating at a Nyquist rate of 528 Ms/s. Additionally, up to seven times this speed will be required to fully exploit all the UWB spectrum.

For our receiver design we consider a single 528 MHz subband. As illustrated in Figure 6.2, the received signal is downconverted to baseband frequencies and then is projected onto the set of basis functions. We use the traditional complex exponential functions as the basis functions, with $N = 3$ frequency samples (3 Fourier coefficients). Figure 6.6 shows the signal-to-noise ratio (SNR) at the output of the discrete frequency matched filter (AWGN power plus truncation error power) versus the sampling rate ΔF_c, for different values of $\text{SNR}_{\text{AWGN}} = 10 * \log_{10}(E_b/N_o)$, where E_b is the signal energy and $N_o/2$ is the two-sided power spectral density of the AWGN. The figure shows the required sampling rate to make negligible the truncation error compared with AWGN. For practical values of SNR, the truncation error is indeed negligible. For example, in a UWB multiband multicarrier application, the SNR_{AWGN} ranges between 4.0 and 4.9 dB or even lower if coding is taken into account, which makes this receiver an interesting possibility to lower the sampling rate requirements, especially as more subbands are included in the design. However, for applications operating at larger SNR_{AWGN} values the truncation error could become the dominant impairment and needs to be taken into consideration.

6.5.3 Mixed-Signal Transmitted-Reference Receiver

The general result provided in (12) is only applicable to linear receiver structures. However, the ideas presented here can also be used for the design of nonlinear receivers. Since generalization with nonlinear receivers is not straightforward, we instead present a nonlinear receiver design example based on transmitted-reference signaling.

6.5.3.1 Signal Model for Transmitted-Reference Receiver

A transmitted-reference (TR) communications system may be implemented by transmitting both an unmodulated and modulated pulse with an appropriate time-delay separation

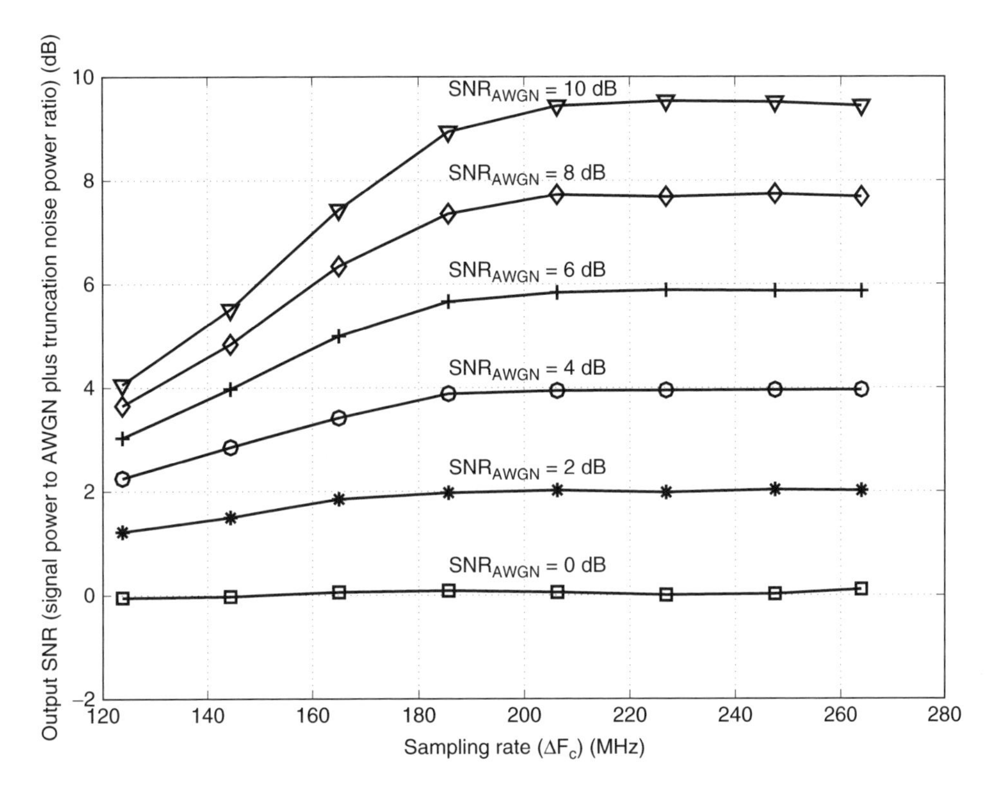

Figure 6.6 Output SNR of the mixed-signal multicarrier receiver implemented with $N = 3$ coefficients, versus the sampling rate for different values of $\text{SNR}_{\text{AWGN}} = 10 * \log_{10}(E_b/N_o)$

[27, 28] given by[8]

$$s_{tx}(t) = \sum_{j=-\infty}^{\infty} \sqrt{\frac{E_s}{2}} \left(w_{tr}(t - jT) + b_j w_{tr}(t - jT - D) \right), \tag{35}$$

where the pulse $w_{tr}(t)$ has unit energy, and therefore the total transmitted energy is E_s. The modulated pulse carries the binary information in its polarity through the variable b_j, and it is delayed D seconds with respect to the transmitted-reference pulse. The received signal can be represented as the convolution of the transmitted signal and the linear channel model

$$\begin{aligned} s_{rx}(t) = s_{tx}(t) * h(t) &= \sum_{j=-\infty}^{\infty} \sqrt{E_s/2} \left(w_{rx}^{j}(t - jT) + b_j w_{rx}(t - jT - D) \right) + z(t) \\ &= \sum_{j=-\infty}^{\infty} \sqrt{E_s/2} \sum_{l=0}^{L_j - 1} \beta^{j,l} \left(w_{rx}^{j,l}(t - jT - \tau^{j,l}) \right. \\ &\quad \left. + b_j w_{rx}(t - jT - D - \tau^{j,l}) \right) + z(t), \end{aligned} \tag{36}$$

[8] Various modifications are possible, such as employing a direct sequence like approach. In a static channel, fewer training pulses may be utilized.

where $*$ means convolution and, for the jth transmitted pulse, the received pulse waveform $w_{rx}^{j}(t - jT) = w_{tx}(t - jT) * h(t) = \sum_{l=0}^{L_j - 1} \beta^{j,l} w_{rx}^{j,l}(t - jT - \tau^{j,l})$ is composed of L_j paths components arriving at the receive antenna with associated attenuation $\beta^{j,l}$ and time-delay $\tau^{j,l}$. The received multipath components are denoted as $w_{rx}^{j,l}(t)$, which differ from the transmitted pulse $w_{tx}(t)$ due to waveform distortion introduced by the frequency selectivity of the medium, as well as the differentiation induced by the receiving antenna. Additionally, it has been assumed in (36) that both the reference and the modulated pulse have been equally distorted as they pass through the channel. The TR receiver uses the received reference to calculate the decision variable. This is done by band-pass filtering the received signal $s_{rx}(t)$, leading to the band-pass filtered received signal $r(t)$, and then estimating the autocorrelation at lag D.

6.5.3.2 Transmitted-Reference Scheme in Frequency Domain

The samples provided by the frequency-domain ADC can be used in the implementation of the autocorrelation receiver. To see this, we begin expressing the conventional autocorrelation operation used in the calculation of the decision variable estimate $\hat{a}$,

$$\hat{a} = \int_0^T r(t) r(t - D)\, dt, \tag{37}$$

which can also be expressed in frequency domain,

$$\begin{aligned}
\hat{a} &= \int_0^T r(t) r(t - D)\, dt \\
&= \int_{-\infty}^{\infty} R(F) R^*(F) e^{j2\pi DF}\, dF \\
&= \int_{-\infty}^{\infty} |R(F)|^2 e^{j2\pi DF}\, dF \\
&\approx \Delta F \sum_{n=0}^{N-1} |R(F_n)|^2 e^{j2\pi DF_n} + \Delta F \sum_{n=0}^{N-1} |R(-F_n)|^2 e^{-j2\pi DF_n},
\end{aligned} \tag{38}$$

where the second line follows from the Parsevals' theorem and the error in the approximation of the last line is negligible if the frequencies $F_n|_{n=0}^{N-1}$ satisfy the conditions stated in 15. Equation (38) can be further simplified by noting that the received signal is real-valued, which leads to $R(F_n) = R^*(-F_n)$. We have that

$$\begin{aligned}
\hat{a} &\approx \Delta F \sum_{n=0}^{N-1} |R(F_n)|^2 e^{j2\pi DF_n} + \Delta F \sum_{n=0}^{N-1} |R(-F_n)|^2 e^{-j2\pi DF_n} \\
&= \Delta F \sum_{n=0}^{N-1} |R(F_n)|^2 \left(e^{j2\pi DF_n} + e^{-j2\pi DF_n}\right) \\
&= 2\Delta F \sum_{n=0}^{N-1} |R(F_n)|^2 \cos(2\pi DF_n).
\end{aligned} \tag{39}$$

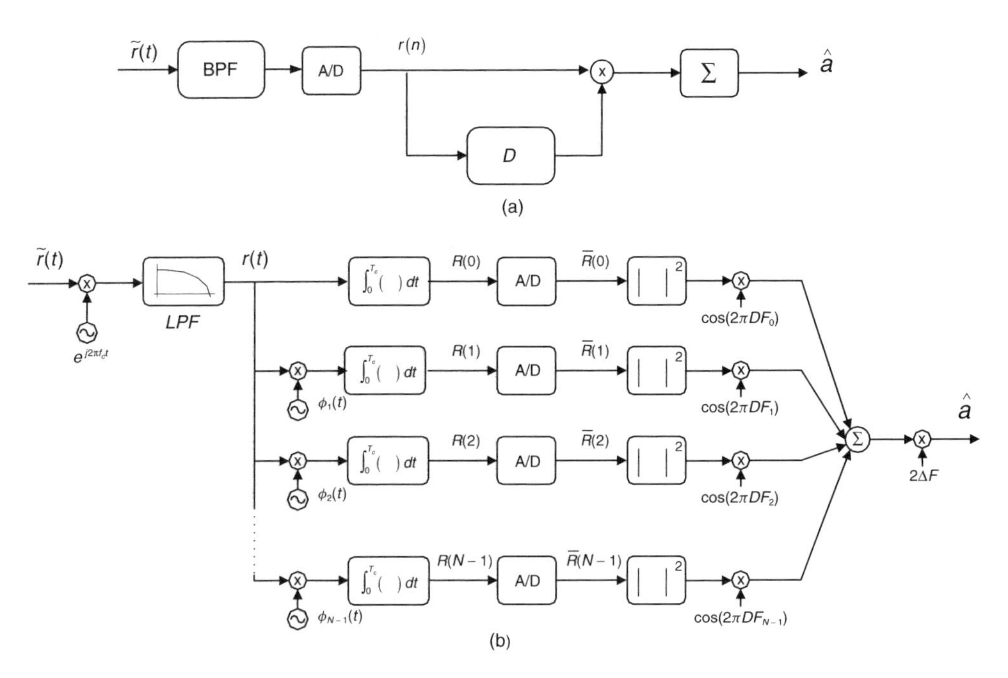

Figure 6.7 Block diagram of the transmitted-reference receiver. (a) Continuous–time-domain version and (b) mixed-signal implementation

The frequency-domain dual of the autocorrelation receiver reveals some of its most important characteristics. The absolute value square in the frequency samples indicates the nonlinear nature of the receiver structure and the weights $w_n = \cos(2\pi DF_n)|_{n=0}^{N-1}$ account for the time-delay D at which the autocorrelation of the received signal is evaluated. If the samples of signal spectrum comply with (15), the probability of error of the information estimates in (39) is the same as the one from its time-domain equivalent [14, 27].

The receiver structure of the transmitted-reference receiver is shown in Figure 6.7 where the conventional time-domain implementation is presented in part (a), and the new mixed-signal counterpart is presented in part (b). Figure 6.8 shows a performance comparison between the discrete time-domain and discrete frequency-domain version (with $N = 4$ coefficients) of the receivers implemented with single-bit ADCs. The plot shows that the $N = 4$ mono-bit discrete frequency-domain receiver not only reduces by four times the ADC sampling speed but also performs better than its time-domain counterpart. Thus, in this specific example, quantization noise can be better handled with a frequency-domain implementation than with the conventional time-domain implementation. Both receivers have the same performance if full resolution is used, which is also indicated in the figure for comparison purposes.

6.5.3.3 Synchronization Issues of the Proposed Frequency-Domain Transmitted-Reference Receiver

One of the most important advantages of the conventional time-domain autocorrelation receiver is that the synchronization is highly relaxed since the only parameter needed

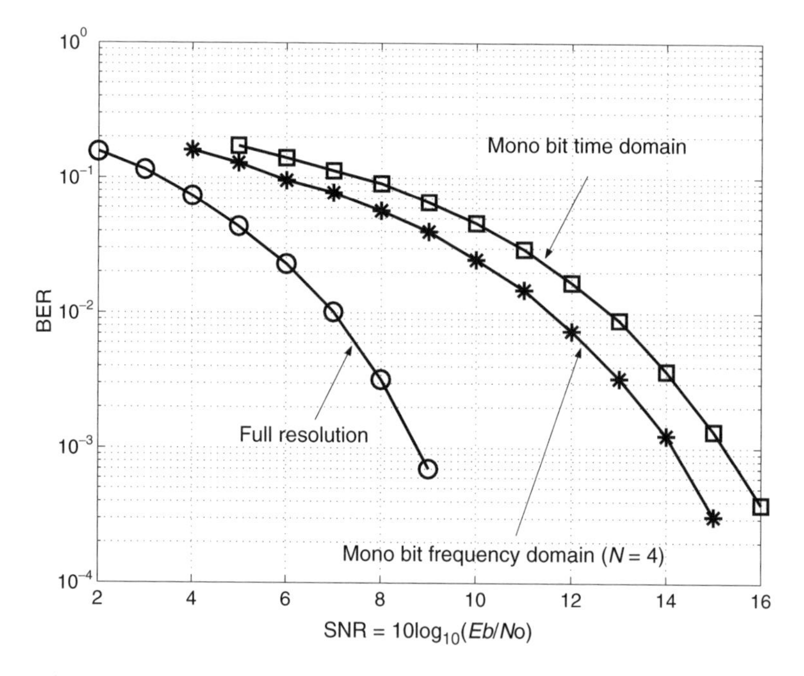

Figure 6.8 BER curves for the mixed-signal transmitted-reference receiver, comparing with mono-bit and full-precision cases

to perfectly align the reference signal and the modulated pulse is the delay D, which is known by the receiver. Therefore, only synchronization at the symbol level and not at the pulse level is needed, which can be easily accomplished. It is important that the new implementation of the receiver structure preserves the synchronization advantage of the original receiver. However, we noticed that the frequency samples $R(F_n)$ are obtained in the conversion-time window T_c, which first should satisfy $T_c > T_w + D$ in order to capture a whole doublet and second, should be synchronized with the doublet in order to avoid missing either the beginning of the transmitted-reference pulse or the end of the modulated pulse, which would result in performance degradation. Although this synchronization issue seems to be problematic with the structure of Figure 6.7, we solve this problem next by a simple modification that exploits the digital representation of the received signal provided by the frequency-domain ADC.

It is desired to make the performance of the new receiver structure independent of any unknown time offset of the received signal. To accomplish this, we reduce the conversion time to a fraction of the time delay D, that is, we make $T_c = D/M$, where M is an integer. As illustrated in Figure 6.9, this simple choice of T_c ensures that these segments that are $D = MT_c$ seconds apart align perfectly with respect to the doublet signal, independently of any time offset. So, the receiver structure reduces to simple correlation between the sample frequencies $R_m(F_n)$ with the sample frequencies $R_{m-M}(F_n)$ followed by a summation of M consecutive correlation segments. In mathematical terms, the received signal is segmented with the window $w_m(t)$ as defined in (10).

$$r_m(t) = r(t)w_m(t). \tag{40}$$

This allows us to express the autocorrelation receiver as

$$\begin{aligned}\hat{a} &= \int_0^T r(t)r(t-D)\,dt \\ &= \sum_{m=0}^{M-1} \int_{mT_c}^{(m+1)T_c} r_m(t)r_{m-M}(t)\,dt.\end{aligned} \tag{41}$$

The frequency samples $R_m(F_n)$ associated to the mth segment can be used in the all-digital implementation of the receiver as

$$\begin{aligned}\hat{a} &= \sum_{m=0}^{M-1} \int_{mT_c}^{(m+1)T_c} r_m(t)r_{m-M}(t)\,dt \\ &= \sum_{m=0}^{M-1} \int_{mT_c}^{(m+1)T_c} r_m(t)r_{m-M}(t)\,dt \\ &= \sum_{m=0}^{M-1} \int_{mT_c}^{(m+1)T_c} r_m(t)r_{m-M}(t)\,dt \\ &= \sum_{m=0}^{M-1} \int_{-\infty}^{\infty} R_m(F)R_{m-M}^*(F)\,dF \\ &\approx \sum_{m=0}^{M-1} \Delta F \left(\sum_{n=0}^{N-1} R_m(F_n)R_{m-M}^*(F_n) + \sum_{n=0}^{N-1} R_m(-F_n)R_{m-M}^*(-F_n) \right) \\ &= \sum_{m=0}^{M-1} \Delta F \sum_{n=0}^{N-1} 2\mathrm{Re}\left\{ R_m(F_n)R_{m-M}^*(F_n) \right\},\end{aligned} \tag{42}$$

where $Re\{\cdot\}$ indicates real part.

As illustrated in Figure 6.9, the receiver is not sensitive to a possible time offset in the received signal as the corresponding segments are still separated D seconds apart, which is exactly the time delay between the transmitted-reference and the modulated pulse. However, the receiver still needs to figure out the range of segments corresponding to a particular information symbol, which is equivalent to determining the initial and final integration times in the conventional continuous-time approach. This is just the symbol level synchronization. One potential advantage of the proposed scheme is that, because the digital data obtained from the frequency-domain ADC would probably be stored in a buffer before further processing, the symbol level synchronization can be enhanced by having an observation set of segments that is larger than the number of segments needed for the estimation of every symbol. In fact, the larger the number of stored segments, the better the chances of obtaining a precise symbol level synchronization, which is achieved in a practical implementation by having a large enough memory size and a synchronization algorithm that takes advantage of it. On the other hand, in time-domain analog implementation, the symbol level synchronization is still challenging, and adding

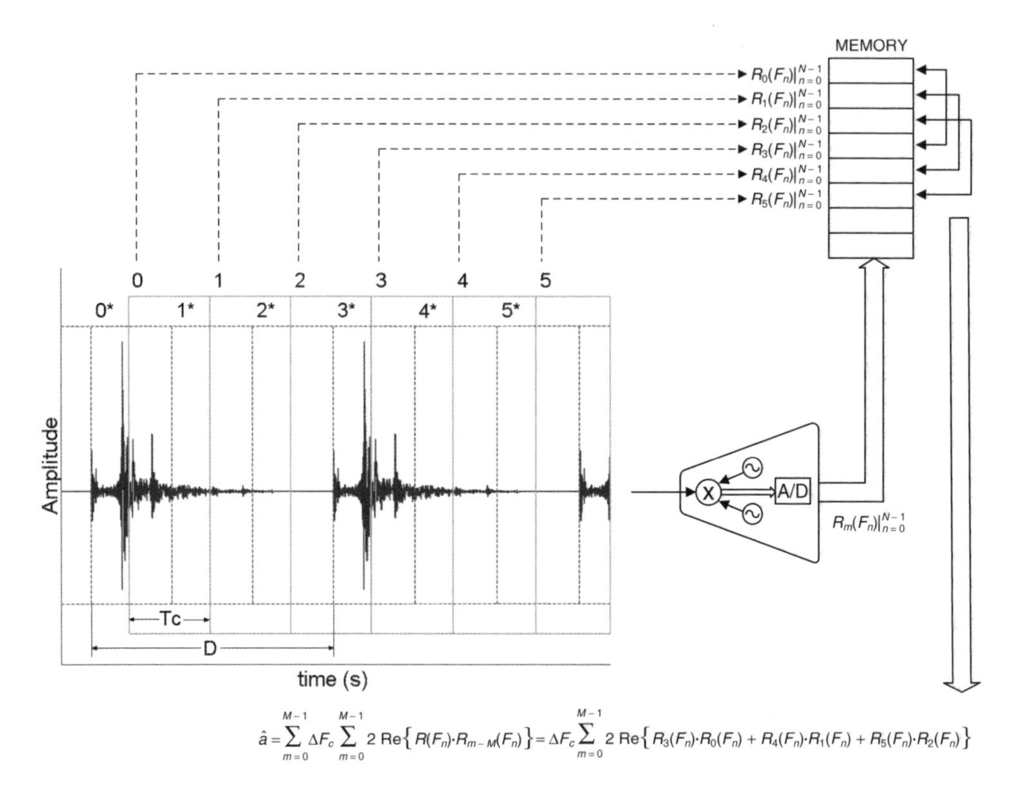

Figure 6.9 Block diagram of frequency-domain autocorrelation receiver with time synchronization that is achieved by choosing $T_c = D/M$. The figure shows two sets of segments that are separated by some arbitrary offset. However, time offset in the received signal does not affect the receiver performance

analog hardware to precisely determine the initial and final integration times is a very expensive and nonflexible solution.

6.6 Conclusions

This chapter presented a family of mixed-signal receivers based on A/D conversion via signal expansion. The receivers are aimed to function at large signal bandwidth as well as high information rates. The parallel architecture reduces the sampling speed, which makes it attractive for broadband applications. The performance of the receivers depends on both the number of coefficients obtained from the continuous-time signal projection over the basis functions and the sampling rate, which in turn imposes an important trade-off between the complexity and sampling rate of the proposed ADC structure. Any linear and nonlinear receiver can be designed with the architecture introduced here. We have presented a general result for linear receivers, an example with a multicarrier receiver, as well as an example with a nonlinear receiver based on transmitted-reference signaling.

References

[1] A. Swami and B. M. Sadler, "On the Coexistence of Ultra-wideband and Narrowband Radio Systems," *Chapter 7 in an Introduction to Ultra Wideband Communication Systems*, Prentice Hall, New Jersey, 2005.

[2] K. Nary, R. Nubling, S. Beccue, W. Colleran, J. Penney, and K. Wang, "An 8-bit, 2-gigasample per Second Analog to Digital Converter," *GaAs IC Symp. Tech. Dig.*, vol. 17, pp. 246–303, Oct. 1995.

[3] K. Poulton, K. L. Knudsen, J. Corcoran, K. C. Wang, R.B. Nubling, R. L. Pierson, M. C. F. Chang, and P. M. Asbeck, "An 8-bit 650 MHz folding ADC," *GaAs IC Symp. Tech. Dig.*, vol. 27, pp. 1662–1666, Dec. 1992.

[4] J. van Valverg and R. J. van de Plassche, "An 8-bit 650 MHz Folding ADC," *IEEE J. Solid State Circuits*, vol. 27, pp. 1662–1666, Dec. 1992.

[5] W. T. Colleran, T. H. Phan, and A. A. Abidi, "A 10b 100 Ms/s pipelined A/D converter," *GaAs IC Symp. Tech. Dig.*, vol. 36, pp. 68–69, Feb. 1993.

[6] K. Sone, N. Naotoshi, Y. Nishida, M. Ishida, Y. Sekine, and M. Yotsuyanagi, "A 10b 100 Ms/s Pipelined Subranging BiCMOS ADC," *IEEE ISSCC Dig. Tech. Papers*, vol. 36, pp. 66–67, Feb. 1993.

[7] C. Schiller and P. Byrne, "A 4-GHz 8-b ADC System," *IEEE J Solid State Circuits*, vol. 26, pp. 1781–1789, Dec. 1991.

[8] S. R. Velazquez, T. Q. Nguyen, and S. R. Broadstone, "Design of Hybrid Filter Banks for Analog/Digital Conversion," *IEEE Trans. Signal Process.*, vol. 4, pp. 956–967, April 1998.

[9] W. Namgoong, "A Channelized Digital Ultra-wideband Receiver," *IEEE Trans. Wireless Commun.*, vol. 3, pp. 502–510, May 2003.

[10] R.M. Gray, "Oversampled Sigma-delta Modulation," *IEEE Trans. Commun.*, vol. COM-35, pp. 481–489, May 1987.

[11] P. Aziz, H.Sorensen, and J. Van der Spiegel, "Multiband Sigma-delta Modulation," *Electron. Lett.*, vol. 29, no. 9, pp. 760–762, Apr. 1993.

[12] I. Galton and H.T. Jensen, "Delta-sigma Modulator Based A/D Conversion without Oversampling," *IEEE Trans. Circuits Syst. II: Analog and Digital Signal Process.*, vol. 42, no. 12, pp. 773–784, Dec. 1995.

[13] S. Hoyos, B. M. Sadler, and G. R. Arce, "Dithering and Sigma-delta Modulation in Mono-bit Digital Receivers for Ultra-wideband Communications," *Proc. UWBST 2003, IEEE Conference on Ultra Wideband Systems and Technologies*, Reston, VA, 2003, Nov. 2003.

[14] S. Hoyos, B. M. Sadler, and G. R. Arce, "Mono-bit Digital Receivers for Ultra-wideband Communications," *IEEE Trans. Wireless Commun.*, vol. 4, no. 4, pp. 1337–1344, July 2005.

[15] I. D. O'Donnell, M. S. W. Chen, S. B. T, Wang, and R. W. Brodersen, "An Integrated, Low Power, Ultra-wideband Transceiver Architecture for Low-rate, Indoor Wireless Systems," *IEEE CAS Workshop on Wireless Communications and Networking*, Pasadena, CA, Sept. 2002.

[16] R. Khoini-Poorard, "Mismatch Effects in Time-interleaved Oversampling Converters," *Proc. IEEE International Symposium on Circuits and Systems*, London, UK, pp. 429–432, May 1994.

[17] A. Petraglia and S. Mitra, "Analysis of Mismatch Effects Among A/D Converters in Time-interleaved Waveform Digitizer," *IEEE Trans. Intstrum. Meas.*, vol. 40, pp. 831–835, Oct. 1991.

[18] S. Hoyos, B. M. Sadler, and G. R. Arce, "Analog to Digital Conversion of Ultra-wideband Signals in Orthogonal spaces," *Proc. UWBST 2003, IEEE Conference on Ultra Wideband Systems and Technologies*, Reston, VA, 2003, Nov. 2003.

[19] S. Hoyos, B. M. Sadler, and G. R. Arce, "High-speed A/D Conversion for Ultra-wideband Signals Based on Signal Projection Over Basis Functions," *Proc. ICASSP 2004, International Conference on Acoustics, Speech and Signal Processing*, Montreal, Quebec, Canada, 2004.

[20] H.-J. Lee and D. S. Ha, "Frequency Domain Approach for CMOS Ultra-wideband Radios," *Proc. IEEE Computer Society Annual Symposium on VLSI*, pp. 236–237, Feb. 2003.

[21] H.-J. Lee, D. S. Ha, and H.-S. Lee, "A Frequency-domain Approach for All-digital CMOS Ultra-wideband Receivers," *Proc. UWBST 2003, IEEE Conference on Ultra Wideband Systems and Technologies*, Reston, VA, 2003, Nov. 2003.

[22] S. Hoyos and B. M. Sadler, "Ultra-wideband analog-to-digital conversion via signal expansion," *IEEE Transactions on Vehicular Technology, (invited)*, vol. 54, no. 5, pp. 1609–1622, Sept. 2006.

[23] A. V. Oppenheim and R. W. Shafer, *Discrete-Time Signal Processing*, 1st edition, Prentice Hall, Englewood Cliffs, NJ, p. 07632, 1989.

[24] S. Hoyos, B. M. Sadler, and G. R. Arce, "Broadband Multicarrier Communications Receiver based on Analog to Digital Conversion in the Frequency domain," *Accepted at IEEE Transactions on Wireless Communications*, to appear, 2006.

[25] J. G. Proakis, *Digital Communications*, 4th edition, Mc Graw Hill, New York, 2001.

[26] A. Batra and *et al.*, "Multi-band OFDM physical layer proposal for IEEE 802.15 task group 3a. Sept. 2003.

[27] J. D. Choi and W. E. Stark, "Performance of Ultra-wideband Communications with Suboptimal Receivers in Multipath Channels," *IEEE J. Select. Areas Commun.*, vol. 20, pp. 1754–1766, Dec. 2002.

[28] R. Hoctor and H. Tomlinson, "Delay-hopped Transmitted-reference RF Communications," *IEEE Conference on Ultra Wideband Systems and Technologies*, Baltimore, MD, pp. 265–269, May 2002.

7

Trends in Ultra-wideband Transceiver Design

Zhengyuan Xu

7.1 Introduction

Ultra-wideband (UWB) communication technology has attracted considerable attention owing to its appealing features [1, 2] and the recent release of a large spectral mask [3]. A UWB system can communicate by short-duration pulses with a very low duty cycle, and it easily supports multiuser communication. It offers low power transmission, easier signal penetration to obstructs, and fine path resolution [4]. Thus, it has become an ideal candidate technology for not only short to medium range high rate data communication [5] but also for low rate sensor networks [6]. However, various issues still remain unsolved in low-complexity channel acquisition and transmitter and receiver (transceiver) design. The lack of effective solutions to these problems clearly hinders the practical application of UWB technology.

This chapter suggests different solutions to tackle the aforementioned issues by considering characteristics and design constraints unique to UWB channel, namely, severe multipath distortion, low power operation, and low-complexity implementation. Several transceiver design methods are proposed. They are classified into two primary categories – digital and mixed analog/digital solutions. Solutions belonging to the first category directly cope with severe design constraints of UWB channels in the digital domain, being aware of the current device technology [7, 8]. In order to maintain high path resolution, those digital approaches utilize analog-to-digital converters (ADCs)[1] at a high sampling rate, such as the Nyquist rate. But a quantizer in ADC may have very low resolution, such as a few bits, or even mono bit. After the received analog signals are digitized, various advanced digital processing techniques can be applied to estimate the multipath components and information-bearing symbols. Approaches in this category thus offer

[1] ADC may also represent analog-to-digital conversion in this chapter.

Ultra-wideband Wireless Communications and Networks
Edited by Xuemin (Sherman) Shen, Mohsen Guizani, Robert Caiming Qiu and Tho Le-Ngoc

abundant flexibility in digital design. Trade-offs between complexity and performance constitute an important topic of study. The overall complexity stems from both signal processing algorithms and ADCs. Complexity from the former is reduced by designing computationally less intensive detection and estimation algorithms. Although complexity reduction for the latter is constrained by current microprocessor and ADC CMOS technology, it will be further enhanced as the device technology evolves. Practical ADC performance appears to be a concern, but numerical study shows that ADC at a sampling rate of Giga samples per second and a few bit resolution suffices to provide detection solutions as good as full resolution counterparts. This observation is consistent with findings by leading companies using UWB technology, such as Intel Corporation, Motorola Inc., and Staccato Communications. Thus, digital design and implementation of UWB transceivers will become a promising direction to pursue. Alternatively, transceiver design solutions in the second category may utilize existing analog components, or mixed analog/digital ones with affordable price and design complexity. Certainly, analog receivers can be digitized by practical ADCs. Their performance may improve if advanced digital signal processing (DSP) techniques are incorporated into the design. The advances in device technology will likely lead to convergence of these two trends in a fully digital domain.

7.2 Status of UWB Transceiver Design

Research interest in UWB communication systems has lasted for about half a century, and continues to grow. Beginning with focus on radar applications in military networks [9], the topic has spanned a wide range of applications, including commercial and government sectors. Recent development has shown promise in personal area, home, and sensor networks. In these applications, it is desirable to minimize the power consumption of each device while maintaining low implementation cost. For example, portable electronics are usually battery-driven and have a limited power supply. Meanwhile, with large spectrum allocation, significant differences are incurred to UWB channel characteristics compared with narrowband/wideband systems [10]. The call for low implementation cost and low power consumption will require novel transceiver design techniques.

Mitigation of severe multipath distortion and design of low-complexity UWB transceivers emerge as concurrent objectives to achieve a desirable performance/cost trade-off. Constrained by the current ADC technology, methods in two categories are envisioned as feasible to pursue – digital and mixed analog/digital. In the digital design category, it is expected to apply advanced processing algorithms on digital signals converted from received analog signals [11–14]. Approaches in this direction allow high-speed sampling, but only at finite resolution in ADC such as mono bit. Chip design technology has enabled implementation of GHz band mono-bit receivers [15, 16]. Commercial ADC in CMOS technology that achieves 6-bit resolution at a 1.3 Giga samples per second sampling rate has been reported [7, 17]. Sigma-delta conversion can increase mono-bit resolution to a much higher level (16-bit) by oversampling [14, 18–20], with moderately increased complexity. Through the investigation of various detection techniques in this direction, free choices of statistical signal processing algorithms [21] yield advantages in design flexibility. However, because of finite resolution ADC, performance degradation compared to full resolution counterparts is of particular concern. Thus, comprehensive performance evaluation needs to be conducted both analytically and numerically. Preliminary study has demonstrated optimism with these finite resolution digital processing techniques for UWB communications [11, 14, 17]. In the other feasible design direction, signal detection is primarily based on correlation of received signals in the analog domain before digitization, significantly relaxing the sampling rate requirement [22]. Corresponding approaches include threshold detectors [23, 24], RAKE receivers [1, 25–27], autocorrelation receivers [28, 29], and transmitted-reference (TR) based receivers [30–33]. Those approaches perform full resolution (analogous to extremely high resolution) acquisition of analog signals but at a low sampling rate after correlation such as frame rate. The typical delay-and-add operations might be necessary to process analog signals at different time intervals. Currently, realization of a large delay by an analog delay line may not be feasible at a reasonable implementation cost. Meanwhile, each delay line may cause additional distortion to the signal. To obtain desirable high quality analog signals at an affordable cost is of practical interest. However, along with the maturity of device technology, these two trends are anticipated to converge in a fully digital domain, assisted by high performance and low cost ADCs and powerful DSP microprocessors.

In this chapter, we will study UWB channel estimation and transceiver design techniques along the two aforementioned directions. Differing from most existing approaches, which are essentially proposed for a single-user system, we will explicitly account for not only self-interferences from rich multipath propagation including possible intersymbol interference (ISI), interframe interference (IFI), interchip interference (ICI), and interpulse interference (IPI), but also multiuser interference (MUI) in a multiple access channel. Typical data-modulation formats, namely, binary phase shift keying (BPSK), pulse amplitude modulation (PAM), or pulse position modulation (PPM), will be considered.

Signal processing in the digital domain is very manipulable because many powerful DSP algorithms are available and can be easily tailored for UWB systems. To maintain reasonably low processing complexity, however, designed algorithms must remain simple. For example, one can utilize low order statistics of received signals such as those up to the second-order statistics (SOS) that are easy to estimate in practice. In general, digital channel estimators and receivers have already been developed for existing narrowband or wideband systems using well-known techniques including subspace projection [34–36], covariance matching (CM) [37–39], matched filtering, minimum mean-square-error (MMSE) and zero-forcing (ZF) optimization [40, 41]. Constraints from practical ADCs are not a big concern in those applications because high-resolution digital conversion is possible at a relatively low sampling rate. However, this does not apply to UWB systems. Sampling rate should be sufficiently high to preserve the spectral characteristics of a UWB signal. Accordingly, given a sampling rate, ADC resolution will have a significant impact on the detection performance. Toward this end, full resolution transceiver design techniques are discussed first, followed by performance justification when finite resolution ADC is utilized. Preliminary study shows that a few bits suffice to represent each signal used for detection, with tolerable performance loss compared with full resolution counterparts. Hence, it is feasible to design practically useful digital UWB transceivers. Analog or mixed analog/digital design may also benefit from low cost implementation. Currently, analog domain design relies heavily on simple correlation detection that yields considerable performance degradation by an unknown UWB channel and possibly MUI. The major difficulty in acquiring channels with large delay spreads can be overcome by transmitting some reference signals that are known to a receiver. This task can be achieved by different transmission schemes including TR [32], superimposed training, or differential modulation-based transmissions [28, 42]. Therefore, we will mainly focus on these schemes in the rest of this chapter, with a close look at TR-based transceiver design. In contrast to some existing approaches documented in the extensive literature, those proposed methods are able to mitigate MUI and other interference, and increase data transmission rate from 50 % to full-rate. It will become clear how a pseudorandom (PN) coding idea facilitates design of UWB transceivers for a multiuser communication system.

7.3 Digital UWB Receivers

UWB is recognized as a spread-spectrum technology. Similarities to existing spread spectrum technologies can be identified, especially to the well-studied code-division multiple access (CDMA) technology [43]. Therefore, it is interesting to examine first whether those developed channel estimation or detection techniques can apply to UWB communication systems. Clearly, a direct-sequence (DS) UWB system using BPSK or PAM data modulation closely follows the DS-CDMA system model [44]. Direct application of existing techniques is feasible. Thus, in addition to historical reasons, many research efforts have placed emphasis on time-hopping (TH) impulse radio (IR) UWB systems with PPM data modulation [22]. A typical receiver adopts a bank of correlators at different delays, each with a locally generated template. Their outputs are then properly combined to generate a decision statistic. Both delays and combining weights are pertinent to the multipath delay profile. In a dense multipath environment, channel information is neither known *a priori* nor fixed, because of system dynamics. Channel characteristics can be either measured or estimated. However, field test is sensitive to location and time, and not feasible for an unknown/inaccessible environment in general. Real-time channel estimation helps to enhance robustness of the system. In the digital domain, many estimation algorithms are available. The optimal solution is based on a maximum-likelihood (ML) estimation criterion [10, 45]. Those estimators provide a theoretical guidance for evaluation of other channel estimators, but are computationally expensive and not very practical. Meanwhile, MUI is approximated as a white Gaussian process, leading to performance degradation. Therefore, low-complexity channel estimators with explicit consideration of the MUI structure are desirable. Those channel estimates are also crucial for designing different linear receivers. We primarily focus on PPM-based TH-UWB systems in this section and present channel estimation and receiver design techniques, since it is very straightforward to generalize obtained results to other scenarios. For low complexity and fast convergence, our approaches mainly rely on the SOS of received data. Connections of such systems with DS-UWB systems are not easily observed at the moment, but they will be revealed after overparameterization of the input/output model by viewing PPM-UWB as a DS-UWB system. Afterward, typical existing techniques like moment matching [38, 39], subspace decomposition [36], and relatively robust power of R (POR) [46] are applied for channel estimation. Despite similarities, a PPM signal shows unique statistics compared to a PAM signal. Thus, necessary transformation and adaptation are anticipated. Different complexity reduction ideas are further suggested. After the discussion on full resolution digital methods, study on effects of finite resolution ADC continues in the last subsection.

7.3.1 PPM-Based TH-UWB System Model

Assume K users simultaneously share the spectrum in a PPM-based multiple access TH-UWB system. The transmitted baseband UWB signal from user k can be described by [22]

$$\alpha_k(t) = \sqrt{\mathcal{P}_k} \sum_{i=-\infty}^{\infty} w(t - iT_f - c_k(i)T_c - d\tau_{I_{k,\lfloor i/N_f \rfloor}}), \tag{1}$$

where $\mathcal{P}_k$ is transmission power, $w(t)$ is the baseband monopulse, T_f is the frame duration, N_f is the number of frames in one symbol interval, $c_k(i)$ is the TH sequence that repeats from symbol to symbol, T_c is the chip duration, $I_{k,\lfloor i/N_f \rfloor} \in \{0, 1, \ldots, M-1\}$ is the information symbol during the ith frame, $\lfloor \cdot \rfloor$ is an integer floor operator, and $d\tau_{I_{k,\lfloor i/N_f \rfloor}} = I_{k,\lfloor i/N_f \rfloor}\sigma$ is the corresponding modulation delay with parameter σ. Assume $T_f = N_c T_c$ and $T_c = M\sigma$. Then, this transmitted signal consists of trains of pulses at time instants of multiples of σ. This observation is key to our pulse rate discrete-time modeling. Suppose the signal propagates through an unknown multipath channel to be estimated. After sampling the received signal every σ seconds and stacking samples corresponding to one symbol interval, a virtually linear vector model follows [47–49]

$$\boldsymbol{y}_n = \sum_{k,m,l} \boldsymbol{C}_{k,l,m} \boldsymbol{g}_k s_{k,m}(n+l) + \boldsymbol{v}_n = \sum_{k,l} \boldsymbol{H}_{k,l} \boldsymbol{s}_{k,n,l} + \boldsymbol{v}_n, \tag{2}$$

where $s_{k,m}(n) = \delta(I_{k,n} - m)$, $\delta(\cdot)$ is the Kronecker delta function, and $\boldsymbol{s}_{k,n,l} = [s_{k,0}(n+l), \ldots, s_{k,M-1}(n+l)]^T$. In this model, m takes integers from 0 to $M-1$, symbol index l indicates how many symbols are involved because of multipath, $\boldsymbol{C}_{k,l,m}$ is a code matrix constructed from corresponding TH sequences, $\boldsymbol{H}_{k,l} = [\boldsymbol{C}_{k,l,0}\boldsymbol{g}_k, \ldots, \boldsymbol{C}_{k,l,M-1}\boldsymbol{g}_k]$ can be termed as a *signature waveform matrix*, $\boldsymbol{g}_k$ is an unknown front-end effective channel vector for user k including effects of transmission power, modulated pulse, multipath channel, and pulse matched filter (MF), $\boldsymbol{v}_n$ is a zero-mean white Gaussian noise vector. It is observed that each information symbol $I_{k,n}$ spreads over M virtual inputs in $\boldsymbol{s}_{k,n,l}$ in a unique form. This $M \times 1$ input vector has only one dominant element whose value equals '1' while all other $M-1$ elements are zeros. The position of '1' depends on $I_{k,n}$. Under such modeling, $s_{k,m}(n+l)$ can be viewed as a new input of the system linearly modulating the pulse, although actual input is $I_{k,n}$. Linear relation of new input and output allows easier design of channel estimators and receivers. Meanwhile, a PAM-like modulation prompts some similarities to a DS-CDMA system in a tri-linear form from components such as code matrix, channel, and input [50]. Therefore, it is anticipated that many of the SOS-based existing multiuser detection and channel estimation techniques can be possibly exploited. However, significant difference in inputs causes difference in processing of received UWB signals from CDMA signals. Clearly, $s_{k,m}(n+l)$ has a nonzero mean $1/M$. Then the mean of the received data vector becomes

$$\bar{\boldsymbol{y}} = E\{\boldsymbol{y}_n\} = \frac{1}{M} \sum_{k,m,l} \boldsymbol{C}_{k,l,m} \boldsymbol{g}_k = \sum_k \mathcal{C}_k \boldsymbol{g}_k = \mathcal{C}\boldsymbol{g}, \tag{3}$$

where all channel vectors are stacked in a big vector $\boldsymbol{g}$, definitions of code matrices similarly follow. The nonzero mean immediately suggests an applicable mean-matching (or least squares – LS) technique to estimate all channels as discussed later. To facilitate processing of high order statistics of $\boldsymbol{y}_n$, we obtain a zero-mean vector z_n from $\boldsymbol{y}_n$ by subtracting the mean [47–49]

$$z_n = \boldsymbol{y}_n - \bar{\boldsymbol{y}} = \sum_{k,l} \boldsymbol{H}_{k,l}\boldsymbol{a}_{k,n,l} + \boldsymbol{v}_n, \tag{4}$$

where $\boldsymbol{a}_{k,n,l} = \boldsymbol{s}_{k,n,l} - \frac{1}{M}\mathbf{1}_M$ is a new zero-mean input vector, $\mathbf{1}_M$ is a vector of length M whose M elements all equal one. $\boldsymbol{a}_{k,n,l}$, similar to $\boldsymbol{s}_{k,n,l}$, has only one dominant element. However, different from PAM modulated data, its covariance is structured as $\boldsymbol{A} = E\{\boldsymbol{a}_{k,n,l}\boldsymbol{a}_{k,n,l}^T\} = \frac{1}{M}(\boldsymbol{I}_M - \frac{1}{M}\mathbf{1}_M\mathbf{1}_M^T)$, where $\boldsymbol{I}_M$ is an identity matrix of dimension M. It is rank deficient with rank $M-1$. Its eigenvalue decomposition (EVD) obeys a form $\boldsymbol{A} = \boldsymbol{B}_a\boldsymbol{\Lambda}_a^2\boldsymbol{B}_a^H$, where $\boldsymbol{B}_a$ is of dimension $M \times (M-1)$ and $\boldsymbol{\Lambda}_a$ is a $(M-1) \times (M-1)$ diagonal matrix with all positive diagonal entries.

7.3.2 Channel Estimation Techniques

Our subsequently developed techniques employ the mean or SOS of either $\boldsymbol{y}_n$ or z_n without relying on training. If training symbols are available, the LS channel estimation method detailed in the literature is directly applicable [51] and thus omitted. If data-modulation formats other than PPM such as PAM and BPSK are used, then all following methods are applicable as well, except the LS one.

The mean-matching idea is a reasonable choice for estimation if the signal has a nonzero mean. Assume N_s data vectors are received. The LS criterion to estimate $\boldsymbol{g}$ is to match the estimated mean of $\boldsymbol{y}_n$ with its true mean. According to (3), it can be described as [48, 52]

$$\hat{\boldsymbol{g}} = \arg\min ||\mathcal{C}\boldsymbol{g} - \hat{\bar{\boldsymbol{y}}}||^2, \quad \hat{\bar{\boldsymbol{y}}} = \frac{1}{N_s}\sum_{n=1}^{N_s} \boldsymbol{y}_n. \tag{5}$$

The solution can be found to be $\hat{\boldsymbol{g}} = \boldsymbol{W}\hat{\bar{\boldsymbol{y}}}$, where $\boldsymbol{W} = (\mathcal{C}^H\mathcal{C})^{-1}\mathcal{C}^H$. Then the estimate of $\boldsymbol{g}_k$ corresponds to the k-th subvector in $\hat{\boldsymbol{g}}$. This method is simple and easy to implement. It is suitable for a system with a small number of users since existence of $\boldsymbol{W}$ imposes some conditions on K.

The SOS of a multidimensional signal is in a form of either correlation or covariance. It can be explored in different ways. The SOS methods to be discussed include, but are not limited to, CM, subspace decomposition, and POR. Covariance $\boldsymbol{R}$ of z_n carries channel information in a form $\boldsymbol{G}_k = \boldsymbol{g}_k\boldsymbol{g}_k^H$. The advantage to consider covariance instead of correlation of $\boldsymbol{y}_n$ is to eliminate cross products of $\boldsymbol{g}_{k_1}\boldsymbol{g}_{k_2}^H$ for $k_1 \neq k_2$ and thus simplify estimation. If all unknowns are lumped into a big vector $\boldsymbol{x}$ [39] including vectors $\boldsymbol{x}_k = vec(\boldsymbol{G}_k)$ and noise power σ_v^2 where vec is an operator to stack all columns of a matrix into a big vector, one can find $\boldsymbol{r} = vec(\boldsymbol{R}) = \boldsymbol{S}\boldsymbol{x}$ where $\boldsymbol{S}$ is a matrix determined by all code matrices [48, 52]. Therefore, $\boldsymbol{r}$ can be matched with its estimate $\hat{\boldsymbol{r}}$ from N_s data vectors

$$\hat{\boldsymbol{x}} = \arg\min ||\boldsymbol{S}\boldsymbol{x} - \hat{\boldsymbol{r}}||^2, \quad \hat{\boldsymbol{r}} = \text{vec}\,(\hat{\boldsymbol{R}}), \quad \hat{\boldsymbol{R}} = \frac{1}{N_s}\sum_{n=1}^{N_s}(\boldsymbol{y}_n - \hat{\bar{\boldsymbol{y}}})(\boldsymbol{y}_n - \hat{\bar{\boldsymbol{y}}})^H. \tag{6}$$

Then $\hat{\boldsymbol{x}}_k$ can be extracted from $\hat{\boldsymbol{x}}$, and $\hat{\boldsymbol{G}}_k$ can be reconstructed by the reverse vec operation. After singular value decomposition (SVD) on $\hat{\boldsymbol{G}}_k$, the singular vector corresponding to its maximum singular value becomes an estimate of $\boldsymbol{g}_k$ up to a multiplicative scalar. If hopping sequences are aperiodic, single-user channel estimators can be derived in the presence of unknown quasi-cyclostationary interference, whose mean and correlation are jointly estimated with the desired user's channel based on either the LS or CM criterion [53, 54].

A subspace technique also employs the SOS of a signal. It is built upon the fact that the signal subspace (span of useful signals) is orthogonal to its complementary subspace (the noise subspace) [35] and is also characterized by channel parameters. To identify each subspace, $\boldsymbol{R}$ is decomposed by EVD as $\boldsymbol{R} = \boldsymbol{U}_s\boldsymbol{\Lambda}_s\boldsymbol{U}_s^H + \sigma_v^2\boldsymbol{U}_n\boldsymbol{U}_n^H$, yielding signal subspace $\boldsymbol{U}_s$ and noise subspace $\boldsymbol{U}_n$. Considering the autocorrelation $\boldsymbol{A}$ of $\boldsymbol{a}_{k,n,l}$ described earlier and taking note of (4), the orthogonality principle should read $\boldsymbol{U}_n^H\boldsymbol{H}_{k,l}\boldsymbol{B}_a = \boldsymbol{0}$ for all possible k and l while $\boldsymbol{U}_n^H\boldsymbol{H}_{k,l} \neq \boldsymbol{0}$, unique for PPM-based TH-UWB systems. After expressing $\boldsymbol{H}_{k,l}$ explicitly as a linear function of $\boldsymbol{g}_k$ based on our definition and multiplying out $\boldsymbol{H}_{k,l}\boldsymbol{B}_a$, this orthogonality is readily translated into the following channel estimation criterion for user k that minimizes the total projection error [49, 55]

$$\hat{\boldsymbol{g}}_k = \arg\min \sum_{l,j} ||\boldsymbol{U}_n^H\boldsymbol{D}_{k,l,j}\boldsymbol{g}_k||^2 = \arg\min \boldsymbol{g}_k^H\boldsymbol{O}_k\boldsymbol{g}_k, \tag{7}$$

where $\boldsymbol{O}_k = \sum_{l,j}\boldsymbol{D}_{k,l,j}^H\boldsymbol{U}_n\boldsymbol{U}_n^H\boldsymbol{D}_{k,l,j}$, $\boldsymbol{D}_{k,l,j}$ is a matrix constructed from code matrix of user k and entries of $\boldsymbol{B}_a$, $j = 1, \ldots, M-1$. Channel estimate for user k is then the minimum eigenvector of matrix $\boldsymbol{O}_k$.

Recently, a POR technique has been developed that may show better robustness than the subspace channel estimator/receiver and renders the rank estimation of the noise subspace unnecessary [46]. This is based on the following identity stemming from EVD of $\boldsymbol{R}$: $\lim_{p\to\infty}\sigma_v^{2p}\boldsymbol{R}^{-p} = \boldsymbol{U}_n\boldsymbol{U}_n^H$, where p is a positive integer. Notice that the signal to noise ratio does not affect the identity. It indicates that the subspace component $\boldsymbol{U}_n\boldsymbol{U}_n^H$ in the subspace cost function can be approximated by $\sigma_v^{2p}\boldsymbol{R}^{-p}$ for a sufficiently large p. Ignoring the scalar, the POR criterion is described as follows [47, 56]

$$\hat{\boldsymbol{g}}_k = \arg\min \sum_{l,j} \boldsymbol{g}_k^H\boldsymbol{D}_{k,l,j}^H\boldsymbol{R}^{-p}\boldsymbol{D}_{k,l,j}\boldsymbol{g}_k. \tag{8}$$

Although the subspace method theoretically outperforms the POR method based on any finite p, better and more robust performance of the latter even for small p, such as $p = 2$, is observed in various practical conditions. This is true because subspace decomposition is sensitive to various perturbations (noise subspace and its rank estimation errors caused by system uncertainties, such as noise, finite data sample size, channel order mismatch, load variation, or unknown interference). However, it is unnecessary to estimate $\boldsymbol{U}_n$ in the POR method.

7.3.3 Design of Linear Receivers

Our objective is to detect the desired user's input $I_{k,n}$. It is included in our newly defined data vector $\boldsymbol{a}_{k,n,l}$ $(l = 0)$, which has only one maximum while all others take a smaller value. For low complexity, we design M linear receivers $\boldsymbol{f}_{k,i}$ $(i = 0, \ldots, M-1)$ such that each one is to detect one corresponding element in $\boldsymbol{a}_{k,n,l}$ instead of detecting $I_{k,n}$

directly. Then outputs of M receivers are compared and the index of the maximum element is identified. The symbol detection criterion can be described as follows [47, 49]

$$I_{k,n} = \arg \max_{i \in \{0,\dots,M-1\}} \text{Real } \{\boldsymbol{f}_{k,i}^H z_n\}. \quad (9)$$

Those linear receivers $\boldsymbol{f}_{k,i}$ can be constructed after channel parameters are estimated as before. For convenience, collect all those M receivers in a matrix $\boldsymbol{F}_k$ for user k. They may take different forms such as MF or Rake receiver, ZF receiver, and MMSE receiver. The MF is given by $\boldsymbol{F}_{k,mf} = \boldsymbol{H}_{k,0}$. The ZF receiver has a form $\boldsymbol{F}_{k,zf} = \boldsymbol{H}_{k,0}(\boldsymbol{H}_{k,0}^H \boldsymbol{H}_{k,0})^{-1}$. An MMSE receiver may take two different forms. The direct matrix inversion (DMI) or conventional MMSE receiver can be found as follows after noticing the covariance of $\boldsymbol{a}_{k,n,l}$ is $\boldsymbol{A}$: $\boldsymbol{F}_{k,\text{dmi}} = \boldsymbol{R}^{-1}\boldsymbol{H}_{k,0}\boldsymbol{A}$, while the subspace MMSE receiver can be derived as [35, 49]: $\boldsymbol{F}_{k,\text{sub}} = \boldsymbol{U}_s \boldsymbol{\Lambda}_s^{-1} \boldsymbol{U}_s^H \boldsymbol{H}_{k,0}\boldsymbol{A}$ after applying EVD of $\boldsymbol{R}$ and noticing orthogonality of the signal subspace and the noise subspace. Performance of those receivers based on proposed LS, CM, subspace decomposition, POR channel estimation methods, and full resolution ADCs have been extensively tested [47–49]. They have demonstrated considerable superiority to receivers constructed from existing data-aided and non-data-aided ML channel estimates [45] in a multiuser system. The POR method based on a small p such as $p = 2$ performs almost as good as those based on a large p but at relatively low complexity. It performs better than all other methods in practical conditions. It is more robust than the subspace method even when the dimension of the noise subspace is perfectly known.

All proposed channel estimators and corresponding receivers can be analytically studied as well when finite noisy data samples are received and processed. The results on perturbations of subspace components of a matrix [57] and SOS of an estimated sample covariance matrix [58, 59] will be very helpful. They have been successfully applied to analyze different channel estimators and receivers for CDMA systems [40, 41, 46].

A block diagram for a digital UWB receiver is presented in Figure 7.1. The low noise amplifier (LNA) amplifies the signal. In an ideal situation, full resolution ADC is considered. It may be replaced by any finite resolution ADC to be used in design of finite resolution digital receivers, which is discussed later. The digital receiver block and channel estimator block can represent any previously discussed method.

7.3.4 *Some Thoughts about Complexity Reduction*

Adaptive processing is effective in the further reduction of design complexity for obvious reasons. It also has good channel tracking capabilities. For our LS and CM approaches,

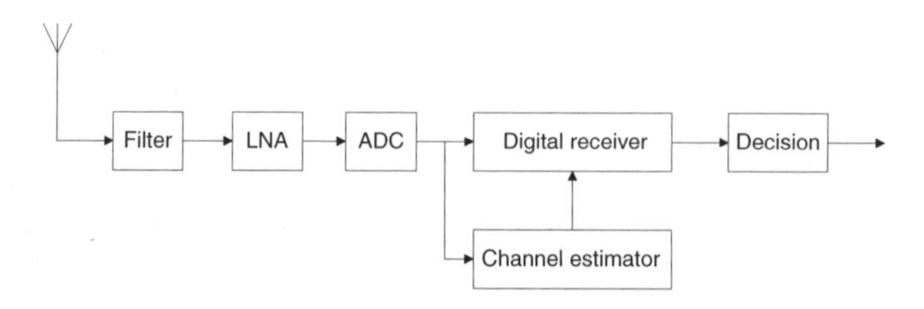

Figure 7.1 Block diagram of a digital UWB receiver

where channel parameters are first estimated and receivers are then built, channel coefficients can be adaptively estimated using stochastic gradient-based techniques and then fed into adaptive MMSE receivers. For the proposed subspace channel estimation, a subspace tracking technique can be applied [60]. The adaptive version of the POR method in [46] can be adopted as well.

The above approaches are implemented in a full-dimensional space that is large in general due to many received samples within one symbol interval. Rank reduction can provide an efficient updating rule and reduce computational cost. It projects the observation onto a small dimensional Krylov subspace [61]. Then, reduced-rank adaptive filtering is possible for substantial computational savings while still yielding close performance to optimal MMSE filtering. Since various structures of reduced-rank adaptive filters have been presented and comprehensively analyzed in the literature, their applicability such as MSWF [62] and CGRRF [63] to the detection of UWB signals can be studied.

It is also observed that previous discussions are based on an overparameterized signal model that mimics a CDMA system. Those methods disregard unique structures of newly defined code matrices that show considerable sparseness. Most entries are zeros except for a limited number of elements that take a value of one to indicate the position of a transmitted pulse. Their structures are believed to be very helpful to further reduce computations invoked by the proposed algorithms.

7.3.5 Finite Resolution Digital Receivers

Previous discussions have taken place under the assumption that analog signals can be converted to digital ones at full resolution. However, a practical ADC poses many constraints, including limited output signal range, finite resolution, and a prespecified sampling rate. Given an acceptable sampling rate, quantization errors from ADC will affect detection performance when those errors are more dominant than errors induced by other imperfections such as noise and finite sample size. The question arises as to whether those errors are tolerable and how to quantify and qualify detection performance under such practical constraints. Joint analysis of channel estimation, detection, and ADC can be performed in a similar manner as in [64, 65], and these methods will be described later. The analytical results are expected to similarly reflect effects of each perturbation factor and their interaction. As an example, Figure 7.2 shows effects of ADC resolutions on digital subspace MMSE receivers constructed from our LS and CM-based channel estimates. There are 4 users in the system. Received data in 500 symbol intervals are used for channel estimation. At low SNR, the performance is relatively insensitive to ADC resolution since noise-induced error may override quantization error. The figure also demonstrates that only a few bits are necessary for the resulting solutions to approach the full resolution solutions in a typical noisy environment. Tests can be similarly performed on other digital receivers. It thus indicates promising applicability of the proposed digital processing techniques under practical ADC constraints.

Remark: As device technology evolves, the ADC and DSP microprocessors will significantly reduce cost while the performance improves. Coupled with low-complexity adaptive algorithms, complexity and performance by digital receiver design will improve as well.

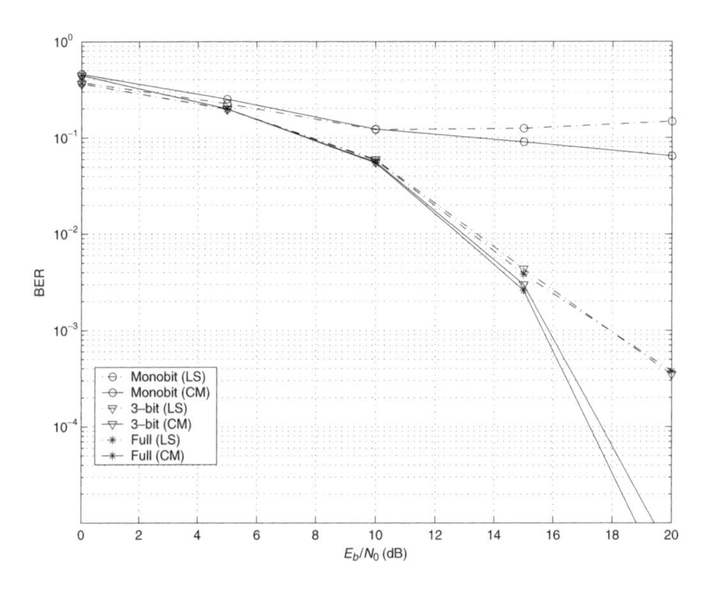

Figure 7.2 Effects of ADC resolution on detection performance of digital receivers

7.4 Analog/Digital UWB Transceivers

Our digital receivers require high rate sampling ADC, usually at a rate around the Nyquist rate. This operative condition incurs some implementation complexity/cost. In this section, we introduce less complex transceiver structures rooted from the transmitted-reference (TR) transmission scheme. They can be implemented in mixed analog/digital circuits based on analog/digital correlators.

7.4.1 Near Full-Rate TR Transceivers

TR modulation was proposed for narrowband systems a few decades ago in order to combat unknown channel distortion [66–68]. It has been applied to transceiver design for a UWB communication system [30–33, 69]. The first pulse of each doublet is information-free, and the second (delayed) pulse carries user's data via its amplitude or delay. The minimum delay of the data pulse is ideally designed to be larger than the channel spread so that the reference pulse does not interfere with the data pulse after multipath propagation (no IPI), although this may be difficult to achieve in an analog delay implementation. The received waveform resulting from the reference pulse can then serve as a template to demodulate the data pulse using a low-complexity correlation receiver [30, 69]. However, minimum spacing of the two pulses inevitably sacrifices data rate, especially when the channel delay spread is large [5]. As the channel is used only half the time for data transmission at best, there is a 50 % rate penalty. In addition, the template may be very noisy, limiting the conventional TR performance. If small spacing between pulses is incorporated in order to increase the data rate, then IPI contaminates the template and may consequently yield poor detection performance.

In order to achieve near full-rate data transmission, we permit arbitrarily small spacing between the reference pulse and data pulse at a price of induced IPI at the receiver due to multipath spreading. Then it is necessary to develop an advanced technique to obtain a clean template from interference-contaminated received signal. We will rely on the first-order statistics of the received signal to restore the signal waveform from noisy data. To further reduce estimation error and improve the template, we introduce a PN sequence to modulate the amplitude of the data pulse. For easy illustration, consider only binary digital PAM or PPM data modulation.

7.4.1.1 PAM Data Modulation

Denote the symbol by $I_n \in \{\pm 1\}$. The transmitted signal with power $\mathcal{P}$ can be described by

$$s(t) = \sqrt{\frac{\mathcal{P}}{2}} \sum_{n=-\infty}^{\infty} \left[w(t - nT_f) + I_{\lfloor n/N_f \rfloor} B_n w(t - nT_f - T_d) \right], \tag{10}$$

where $w(t)$ is the monopulse, B_n is a frame-rate binary PN sequence taking values ± 1, T_f is frame duration, delay T_d of the second pulse is arbitrarily small. Thus, the data rate increases compared to the conventional TR system. After channel propagation, two pulses in each doublet are distorted by multipath in the same way. If we denote the channel response by $g(t)$, then the received signal becomes

$$r(t) = \sum_{n=-\infty}^{\infty} \left[h(t - nT_f) + I_{\lfloor n/N_f \rfloor} B_n h(t - nT_f - T_d) \right] + v(t), \tag{11}$$

where $h(t) = \sqrt{\frac{\mathcal{P}}{2}} w(t) \star g(t)$ is the unknown waveform, $\star$ denotes convolution, and $v(t)$ represents zero-mean Gaussian noise. It should be mentioned that the model reduces to a conventional TR system if we let B_n take value 1 and set the delay T_d to be large. It also represents a special case of superimposed training when T_d is set to be zero. However, introducing T_d brings more design flexibility. Suppose $h(t)$ has maximum support in $(0, T_h)$, where $T_h \gg T_d$. Assume $T_h < T_f$. Discussions can be generalized to other situations. Even so, mutual interference between the reference signal and data signal occurs at the receiver due to small T_d. If the instantaneous received signal is adopted as the template, then the interference to the reference signal introduced by the data signal and noise causes the template to be very noisy, based on which a demodulator yields a large demodulation error [33]. Even when a perfect template is available, the interference to the data signal introduced by the reference signal also degrades demodulation performance.

Our goal is to restore a clean template $h(t)$ from received noisy signal $r(t)$ in N_s symbol intervals first, and then perform the estimation of input sequence I_n. In [69], average of $r(t)$ over N_f frames within one symbol period is performed to reduce noise effect. That effect can be further reduced if we average signals in N_s symbol intervals. Interference from the data signal is also reduced after smoothing because of the pseudorandom property of B_n and zero mean of I_n. Therefore, an improved estimate of the template can be obtained from $N_f N_s$ frame-long segments of $r(t)$ as

$$\hat{h}(t) = \frac{1}{N_f N_s} \sum_{m=1}^{N_f N_s} r(t + mT_f). \tag{12}$$

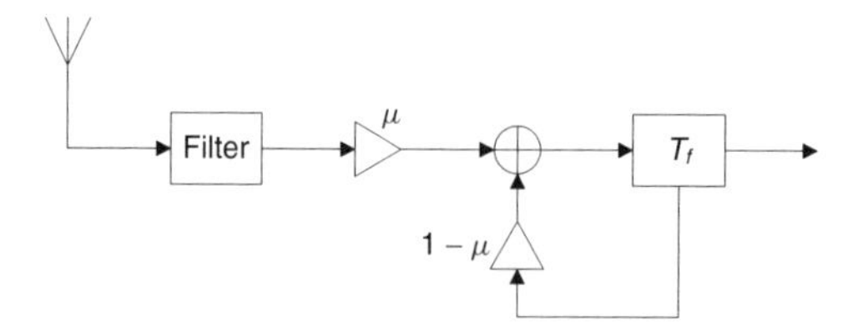

Figure 7.3 Block diagram of an adaptive template estimator

The estimator involves delay elements, multipliers, and adders. It can be implemented in the analog circuit by delay-and-add operations. According to (12), an adaptive algorithm with a small step size μ can be adopted to update the estimated waveform from frame $m-1$ to frame m

$$\hat{h}^{(m)}(t) = (1-\mu)\hat{h}^{(m-1)}(t) + \mu r(t + mT_f). \tag{13}$$

This update significantly relaxes a restrict requirement on the delay element. Figure 7.3 shows a block diagram for the adaptive template estimator.

At the receiver, estimated reference signal (same as template in this case) in each of the N_f frames for $m = nN_f, \ldots, (n+1)N_f - 1$ in the n-th symbol interval is subtracted from $r(t)$. Denote the signal after subtraction by $\tilde{r}(t)$. Then input I_n can be estimated based on N_f correlators' outputs as

$$\hat{I}_n = \text{sign}\left(\sum_{m=nN_f}^{(n+1)N_f-1} \int_0^T \hat{h}(t)\tilde{r}(t + mT_f + T_d) B_m \, dt \right), \tag{14}$$

where 'T' can be chosen according to some *a priori* knowledge about the maximum channel delay spread, division by B_m has been replaced by multiplication since B_m takes ± 1.

7.4.1.2 PPM Data Modulation

A binary information sequence I_n takes $\{0, 1\}$. The transmitted signal is given by

$$s(t) = \sqrt{\frac{\mathcal{P}}{2}} \sum_{n=-\infty}^{\infty} \left[w(t - nT_f) + B_n w(t - nT_f - T_d - d\tau_{I_{\lfloor n/N_f \rfloor}}) \right], \tag{15}$$

and the received signal by

$$r(t) = \sqrt{\frac{\mathcal{P}}{2}} \sum_{n=-\infty}^{\infty} \left[h(t - nT_f) + B_n h(t - nT_f - T_d - d\tau_{I_{\lfloor n/N_f \rfloor}}) \right] + v(t), \tag{16}$$

where the modulation delay is determined by user's data as $d\tau_{I_{\lfloor n/N_f \rfloor}} = I_{\lfloor n/N_f \rfloor}\sigma$. Modulation parameter σ can be designed to optimize the detection performance [22]. Equation (12) can still be used to estimate the waveform because average of PN codes reduces contribution of the data signal. When no PN sequence is introduced as in [70], the waveform estimator has to count the effect of the data signal when applying the smoothing operation. A discrete-time model by sampling at the pulse rate is considered therein.

The data signal is then clearly linked to the unknown waveform. Using a simple matrix operation, the discrete-time counterpart of the waveform is estimated by a modified mean estimator. In the current context, analog operation is mainly involved. Using estimated waveform, a PPM template is constructed as $\hat{h}(t) - \hat{h}_k(t - \sigma)$ to replace $\hat{h}(t)$ in (14) [22], resulting in

$$u_n = \text{sign}\left(\sum_{m=nN_f}^{(n+1)N_f-1} \int_0^T \left[\hat{h}(t) - \hat{h}_k(t-\sigma)\right]\tilde{r}(t + mT_f + T_d)B_m \, dt \right). \qquad (17)$$

Then a simple mapping from $\{\pm 1\}$ to $\{0, 1\}$ is performed

$$\hat{I}_n = \text{int}\left[(1 - u_n)/2\right] \qquad (18)$$

to obtain an estimate of PPM input. An integer operator 'int' has been introduced to ensure that the demodulator's output is consistent with the signal constellation.

Block diagrams for the near full-rate TR transceiver structures are given in the next subsection, which deals with a more general multiuser communication scenario.

Although presented in an analog form, each receiver can be implemented digitally based on practical ADCs and finite received samples. Waveform estimation and detection performance thus depends on estimation error and quantization error. The estimation error due to finite sample size can be analyzed easily from the channel input/output model [70] as in [40, 41]. The quantization error from finite resolution ADC can be analyzed in the light of [71] by deriving distributions of the quantization noise and quantized signal [64, 65]. Either a midriser or midtread quantizer can be considered. A midriser quantizer has an even number of output levels and does not contain a zero in its output while a midtread quantizer has an odd number of output levels including zero [18]. For a b-bit ADC, the quantization step for a midriser is $2V_m/(2^b - 1)$ and that for a midtread is $2V_m/2^b$ where $2V_m$ is the maximum range of the output of a quantizer. Analyses of joint performance due to finite received samples and finite resolution quantization can then be carried out.

Figure 7.4 shows effects of noise and midriser quantization resolution on the near full-rate receiver. Signal waveform is estimated from received data in 200 symbol intervals. Solid curves are obtained from experiment and dashed curves from analysis that takes into account errors from both finite sample size and finite ADC resolution. Convergence of experimental results with analytical ones is observed for all cases except the mono-bit case, where up to the second-order approximation on each perturbation in analysis yields a large error. Meanwhile, a few bits are sufficient for ADC to provide results close to full resolution counterparts. PAM modulation is slightly better than PPM modulation.

The impact of ADC range on detection performance is illustrated in Figure 7.5. A 3-bit midriser ADC is used and the range is normalized by the amplitude of the clean reference signal. Technically, one has to properly choose the range of ADC. Small ADC range causes the quantizer to overload, and large range makes the quantization step too big, resulting in a large quantization error. But this figure shows that performance is not very sensitive to the quantizer range. A factor 2 is a good choice. Experimental results (solid curves) gradually converge to analytical ones (dashed curves) as range increases. Other practical factors worthy of study include sampling rate and power consumption pertaining to an ADC.

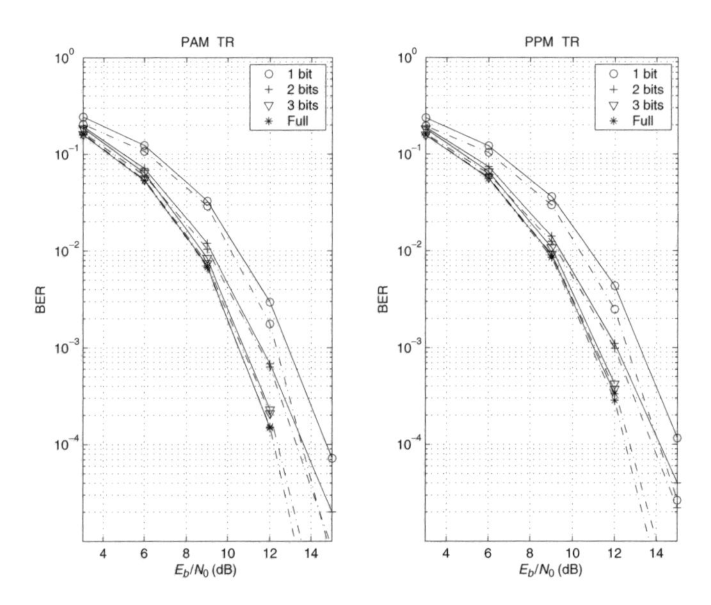

Figure 7.4 Effects of noise and quantization resolution on the near full-rate digital receiver performance

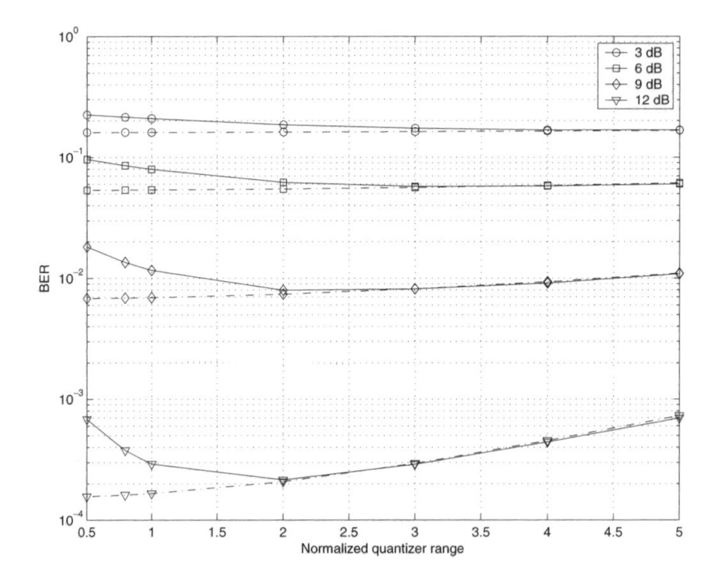

Figure 7.5 Quantization range effect on the near full-rate digital receiver performance

7.4.1.3 Extension to Multiuser Communications

When multiple users access the same channel simultaneously, an effective multiple access method needs to be designed. If a TR transmission scheme is adopted, users' reference signals may interfere with each other, as do data signals. In order to minimize interference during both waveform estimation and data demodulation processes, we modulate the amplitude of the reference pulse by a unique frame-rate PN sequence. That sequence also serves as a signature of the user in order to correctly extract that user's waveform at the receiver end. Meanwhile, the second PN sequence is applied to modulate the amplitude of the data pulse as before to minimize MUI and reduce waveform estimation error as well [72, 73]. In this way, system capacity also increases. This PN coding idea is similar to [74, 75], which is borrowed from that for an overlaying CDMA system [76, 77]. To maximally minimize MUI and offer more design flexibility, the delay of the data pulse is assumed to be user dependent as well. As already observed in previous discussions, two different data-modulation formats require individual discussions in designing such multiuser TR (MTR) transceivers.

For PAM data modulation, denote information sequence of user k by $I_{k,n}$. Then the transmitted signal due to user k in a K-user system can be described by

$$s_k(t) = \sqrt{\frac{\mathcal{P}_k}{2}} \sum_{n=-\infty}^{\infty} \left[A_{k,n} w(t - nT_f) + I_{k,\lfloor n/N_f \rfloor} B_{k,n} w(t - nT_f - \tau_{k,n})\right], \tag{19}$$

where $A_{k,n}$ and $B_{k,n}$ are frame-rate binary PN sequences taking values ± 1, $\tau_{k,n} = C_{k,n} T_c$ that is controlled by a user and frame-dependent hopping code $C_{k,n}$ is the delay of the data pulse in multiples of T_c.

A block diagram for the near full-rate MTR transmitter is presented in Figure 7.6. Clearly, it can represent a single-user case by setting both PN sequences to be ones and fixing delay of the second pulse. When the delay is set to be zero, it reduces to a superimposed training-based single-user transmission scheme [78, 79]. If we denote the channel response of user k by $g_k(t)$, then the received signal becomes

$$r(t) = \sum_{k,n} \left[A_{k,n} h_k(t - nT_f) + I_{k,\lfloor n/N_f \rfloor} B_{k,n} h_k(t - nT_f - \tau_{k,n})\right] + v(t), \tag{20}$$

where $h_k(t) = \sqrt{\frac{\mathcal{P}_k}{2}} w(t) \star g_k(t)$ is the unknown waveform. Accordingly, our waveform estimator and demodulator are modified respectively as

$$\hat{h}_k(t) = \frac{1}{N_f N_s} \sum_{m=1}^{N_f N_s} r(t + mT_f) A_{k,m}, \tag{21}$$

$$\hat{I}_{k,n} = \text{sign}\left(\sum_{m=nN_f}^{(n+1)N_f - 1} \int_0^T \hat{h}_k(t) \tilde{r}_k(t + mT_f + \tau_{k,m}) B_{k,m}\, dt \right). \tag{22}$$

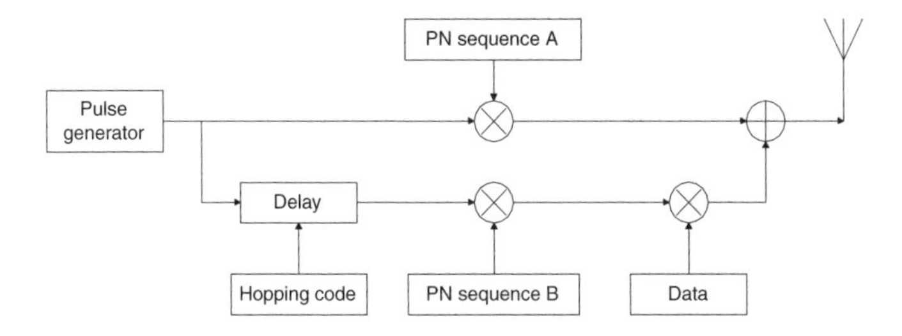

Figure 7.6 Block diagram of a near full-rate MTR-UWB transmitter

Now, $\tilde{r}_k(t)$ is a generic signal after subtraction of reference signal in two different cases. In the first case, only that user's PN sequences are known. This scenario mimics downlink from the access point to a particular user's terminal in a UWB wireless network. Then, only its waveform can be estimated and its reference signal $A_{k,m}h_k(t)$ is reconstructed for subtraction in the n-th symbol interval. In the second case, the receiver knows all users' PN sequences. This represents uplink communication from users to the access point. Then, all their reference signals are reconstructed from estimated waveforms and subtracted from the received signal.

Figure 7.7 plots a block diagram for the near full-rate MTR receiver. The subtraction subblock is omitted for clear presentation of the detection process.

Some representative numerical results are presented to justify performance of proposed near full-rate MTR transceivers. Effect of the sample size used for template estimation on the BER performance is demonstrated in Figure 7.8. The 10 dB noise is added to a four-user system. Severe IPI occurs at the receiver based on our chosen system parameters. Proposed uplink and downlink receivers are compared with conventional receivers that

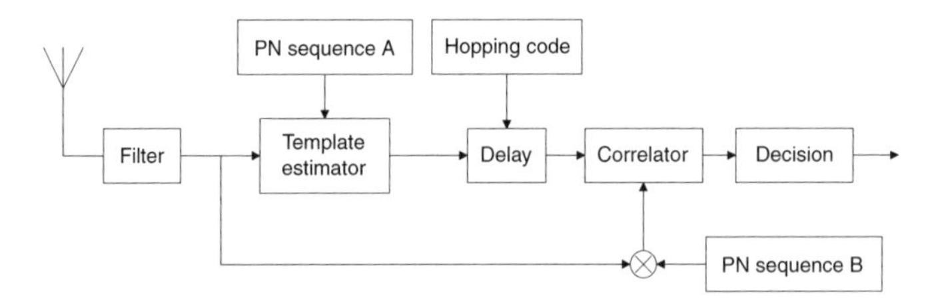

Figure 7.7 Block diagram of a near full-rate MTR-UWB receiver

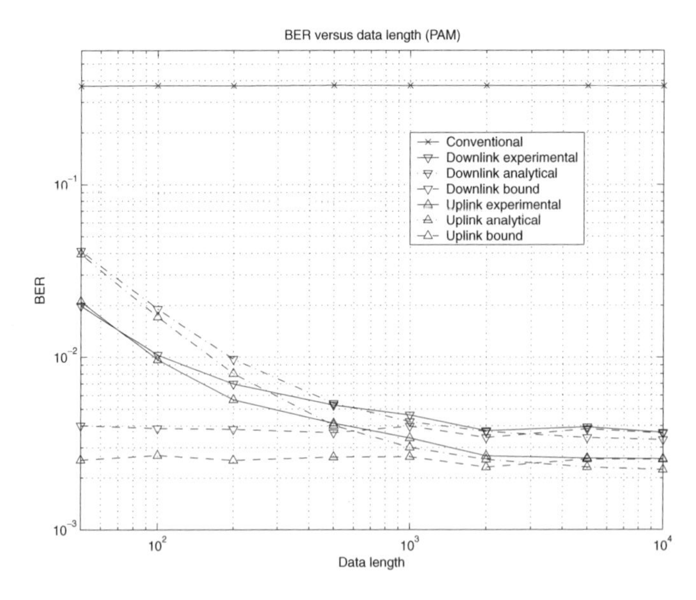

Figure 7.8 Effect of the window size on the near full-rate MTR-UWB receiver performance with PAM modulation

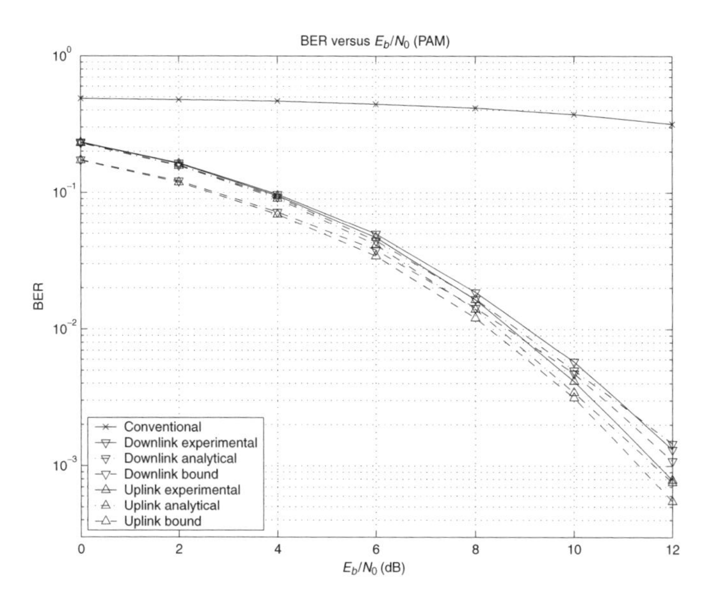

Figure 7.9 Noise effect on the near full-rate MTR-UWB receiver performance with PAM modulation

use instantaneous noisy templates for detection. Bounds correspond to cases when the true waveforms are assumed to be known at the receivers. Experimental results converge with both analytical ones and bounds as sample size increases to about 1000. The uplink detector is better than the downlink one for large sample size. Both detectors significantly outperform the conventional one. Noise is another factor that significantly affects detection performance. Figure 7.9 shows its effect on the BER. Template is estimated based on received samples in 500 symbol intervals. The signal to noise ratio E_b/N_0 ranges from 0 dB to 12 dB. The proposed transceivers substantially outperform the conventional TR transceiver. The uplink detector is better than the downlink detector at high SNR. For each proposed detector, the analytical curve agrees well with the experimental one. Again, small gaps from associated bounds are due to the finite sample effect on the proposed detectors, as already observed from Figure 7.8, for example, at the point 500 samples with $E_b/N_0 = 10$ dB.

When PPM data modulation is used, the second pulse carries information by the pulse position. Transmitted signal due to user k and received signal are given by

$$s_k(t) = \sum_{n=-\infty}^{\infty} \left[A_{k,n} w(t - nT_f) + B_{k,n} w(t - nT_f - \tau_{k,n} - d\tau_{I_{k,\lfloor n/N_f \rfloor}})\right], \tag{23}$$

$$r(t) = \sum_{k,n} \left[A_{k,n} h_k(t - nT_f) + B_{k,n} h_k(t - nT_f - \tau_{k,n} - d\tau_{I_{k,\lfloor n/N_f \rfloor}})\right] + v(t), \tag{24}$$

where $h_k(t)$ is the waveform defined earlier, $d\tau_{I_{k,n}} = I_{k,n}\sigma$ is the delay controlled by a binary information sequence $I_{k,n}$ that takes $\{0, 1\}$ mapped from $\{\pm 1\}$. Subsequently, the waveform $h_k(t)$ is estimated as (21). Then data is demodulated by

$$u_{k,n} = \text{sign}\left(\sum_m \int_0^T \left[\hat{h}_k(t) - \hat{h}_k(t-\sigma)\right]\tilde{r}_k(t + mT_f + \tau_{k,m})B_{k,m}\,dt\right), \tag{25}$$

$$\hat{I}_{k,n} = \text{int}\left[(1 - u_{k,n})/2\right], \tag{26}$$

where m is from nN_f to $(n+1)N_f - 1$ as before.

A block diagram for the transmitter is similar to Figure 7.6. It suffices to replace the data multiplier by a data controlled delay element due to PPM modulation. The receiver block is in the same form as Figure 7.7. Now the template estimator has to be the one designed for the PPM modulation.

7.4.2 Full-Rate TR Transceivers

The conventional single-user TR scheme suffers from a rate penalty besides performance loss. The proposed single-user near full-rate TR scheme can increase data rate by small pulse spacing and improve detection performance by an improved template. Another transmission scheme to achieve full-rate data transmission is assisted by modulation diversity (MD). Two information symbols modulated in PPM and PAM formats are considered simultaneously. Depending on the ways available to distribute those two symbols, two cases exist – unbalanced modulation that allocates both PAM and PPM symbols to the second pulse, and balanced modulation that respectively distributes two symbols to two pulses. The pulse spacing is assumed to be larger than the channel delay spread in this subsection although it is unnecessary in the case of unbalanced modulation.

7.4.2.1 Unbalanced Modulation

The second pulse is modulated by both information symbols in its delay and amplitude respectively

$$s(t) = \sqrt{\frac{\mathcal{P}}{2}} \sum_n \left[w(t - nT_f) + I^{(2)}_{\lfloor n/N_f \rfloor} B_n w(t - nT_f - T_d - d\tau_{I^{(1)}_{\lfloor n/N_f \rfloor}})\right], \tag{27}$$

where the PPM symbol $I^{(1)}_{\lfloor n/N_f \rfloor}$ takes $\{0, 1\}$ and controls the modulation delay as $d\tau_{I^{(1)}_{\lfloor n/N_f \rfloor}} = I^{(1)}_{\lfloor n/N_f \rfloor}\sigma$, the PAM symbol $I^{(2)}_{\lfloor n/N_f \rfloor}$ takes $\{\pm 1\}$ converted from $\{0, 1\}$, T_d is the nominal delay of the second pulse, and B_n is a PN sequence. After propagating through a multipath channel $g(t)$, each pulse is distorted by the transmitted antenna, channel, and receiver antenna as $h(t)$. Then the received signal takes the following form

$$r(t) = \sum_n \left[h(t - nT_f) + I^{(2)}_{\lfloor n/N_f \rfloor} B_n h(t - nT_f - \tau_n - d\tau_{I^{(1)}_{\lfloor n/N_f \rfloor}})\right] + v(t). \tag{28}$$

In order to decode both inputs, the waveform is first estimated by (12). Then in those N_f current frame intervals ($m = nN_f, \ldots, (n+1)N_f - 1$), a generic signal $\tilde{r}(t)$ is obtained as before by subtracting estimated reference signal from the received signal. Suppose the autocorrelation of $h(t)$ at lag zero is $R_{h,0}$ and at lag σ is $R_{h,-1}$. When $h(t)$ is correlated with $\tilde{r}(t + mT_f + T_d)B_n$ to yield an output signal u_n

$$u_n = \frac{1}{N_f} \sum_{m=nN_f}^{(n+1)N_f-1} \int_0^T \hat{h}(t)\tilde{r}(t + mT_f + T_d)B_m \, dt, \tag{29}$$

four possible output signal levels are observed in the absence of noise depending on the input pair $(I_n^{(1)}, I_n^{(2)})$: (i) $u_n = R_{h,0}$ for the input pair $(0, 1)$, (ii) $u_n = -R_{h,0}$ for the input pair $(0, -1)$, (iii) $u_n = R_{h,-1}$ for the input pair $(1, 1)$, (iv) $u_n = -R_{h,-1}$ for the input pair $(1, -1)$. It is possible to choose a proper σ based on the pulse correlation property [22] for joint detection of both PAM and PPM symbols such that (i) those four cases can be well discriminated by a threshold detector, and (ii) the detection performance is balanced between two symbols.

7.4.2.2 Balanced Modulation

The previous scheme is unbalanced because both information symbols modulate the second pulse only. Some advantages can be achieved, such as improved spectral characteristics and potentially improved detection performance, by distributing them to two pulses. PPM and PAM are applied to a transmitted pulse train in an alternating and balanced way [80]

$$s(t) = \sqrt{\frac{\mathcal{P}}{2}} \sum_n \left[w(t - nT_f - d\tau_{I^{(1)}_{\lfloor n/N_f \rfloor}}) + I^{(2)}_{\lfloor n/N_f \rfloor} B_n w(t - nT_f - \tau_n) \right]. \tag{30}$$

It has been observed that its power spectrum density improves compared with (27). The received signal becomes

$$r(t) = \sum_n \left[h(t - nT_f - d\tau_{I^{(1)}_{\lfloor n/N_f \rfloor}}) + I^{(2)}_{\lfloor n/N_f \rfloor} B_n h(t - nT_f - \tau_n) \right] + v(t). \tag{31}$$

Obtaining a clean template at the receiver is difficult since (12) does not yield a good estimator for $h(t)$. However, because of the nonzero mean of the PPM symbol and zero mean of the PAM symbol, we have

$$2E\left\{ h\left(t - d\tau_{I^{(1)}_{\lfloor n/N_f \rfloor}} \right) \right\} = h(t) + h(t - \sigma). \tag{32}$$

Although it is not in an ideal form of either the PAM template or PPM template, it conveys abundant information about the waveform. If one attempts to use it as the template to decode the PPM symbol, clearly the output for each of two possible values of the PPM symbol is indistinguishable. But when it is correlated with the PAM carrying signal, it yields an output of either $R_{h,0} + R_{h,-1}$ or its opposite that can properly reflect the polarity of the PAM symbol. To maximally separate these two values, σ can be chosen to ensure that $R_{h,-1}$ is as large as possible. Therefore, one can use (32) as a nonoptimal template denoted by $z(t)$. It can be estimated as (12)

$$\hat{z}(t) = \frac{1}{N_f N_s} \sum_{m=1}^{N_f N_s} r(t + mT_f), \tag{33}$$

because of (32). Use this estimate to decode the PAM symbol as (14)

$$\hat{I}_n^{(2)} = \text{sign}\left(\sum_{m=nN_f}^{(n+1)N_f-1} \int_0^T \hat{z}(t) r(t + mT_f + T_d) B_m \, dt \right). \tag{34}$$

Once the PAM symbol is decoded, its estimated waveform

$$\hat{h}(t) = \frac{1}{N_f N_s} \sum_{m=1}^{N_f N_s} \hat{I}^{(2)}_{\lfloor m/N_f \rfloor} r(t + mT_f + T_d) B_m \tag{35}$$

can be used to construct a template $h(t) - h(t - \sigma)$ in order to decode the current PPM symbol, as described in (17) and (18). No subtraction is needed under a large pulse spacing assumption because of the absence of mutual interference between the reference signal and data signal.

To balance detection performance for PAM and PPM symbols, σ should be designed properly. The Euclidean distance in detecting a PAM symbol is dependent on $R_{h,0} + R_{h,-1}$, and that in detecting a PPM symbol is dependent on $R_{h,0} - R_{h,-1}$. Thus, it is desirable to choose σ such that $R_{h,-1} = 0$. In practice, this correlation is generally unknown due to unknown channel parameters. The rule of thumb is to consider the correlation of the transmitted pulse $w(t)$ with its offset $w(t - \sigma)$, and select σ such that detection performance is optimized. In [22], that correlation is made as small as possible, favorable to correlation-based detection of PPM symbols therein. In our problem, σ is chosen to make that correlation equal to zero.

7.4.2.3 Extension to Multiuser Communications

Motivated by single-user approaches, we immediately arrive at the multiuser counterparts as in design of near full-rate MTR transceivers. For each user, two PN sequences modulate amplitudes of both pulses and one hopping code controls delay to the second pulse, leading to MD assisted full-rate MTR transceivers.

In unbalanced modulation, both PAM and PPM symbols modulate the second pulse. The transmitted signal from user k has the following form

$$s_k(t) = \sqrt{\frac{\mathcal{P}_k}{2}} \sum_n \left[A_{k,n} w(t - nT_f) + I^{(2)}_{k,\lfloor n/N_f \rfloor} B_{k,n} w(t - nT_f - \tau_{k,n} - d\tau_{I^{(1)}_{k,\lfloor n/N_f \rfloor}}) \right]. \tag{36}$$

The transmitter can be described by Figure 7.10.

Accordingly, received signal takes a form

$$r(t) = \sum_{k,n} \left[A_{k,n} h_k(t - nT_f) + I^{(2)}_{k,\lfloor n/N_f \rfloor} B_{k,n} h_k(t - nT_f - \tau_{k,n} - d\tau_{I^{(1)}_{k,\lfloor n/N_f \rfloor}}) \right] + v(t). \tag{37}$$

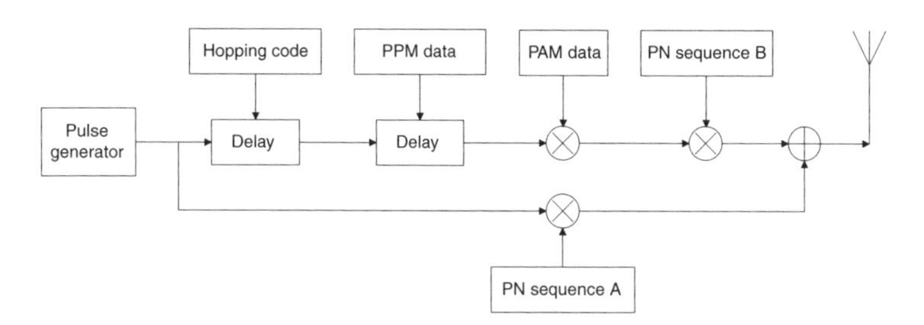

Figure 7.10 Block diagram of a full-rate MTR-UWB transmitter with unbalanced data modulation

Now (21) is directly applicable to estimate the waveform $h_k(t)$. In detection, (29) is modified to indicate the correspondence to user k as

$$u_{k,n} = \frac{1}{N_f} \sum_{m=nN_f}^{(n+1)N_f-1} \int_0^T \hat{h}_k(t)\tilde{r}_k(t + mT_f + \tau_{k,m})B_{k,m}\,dt. \tag{38}$$

As in the single-user case, a threshold detector based on a four level comparison described earlier can determine which pair of PPM and PAM symbols are transmitted. The receiver diagram takes the same form as Figure 7.7. However, the decision block will yield two outputs, one for each of the PAM and PPM symbols.

When two symbols are distributed into two pulses to achieve balanced modulation, transmitted and received signals follow these forms

$$s_k(t) = \sqrt{\frac{\mathcal{P}_k}{2}} \sum_n \left[A_{k,n}w(t - nT_f - d\tau_{I^{(1)}_{k,\lfloor n/N_f \rfloor}}) + I^{(2)}_{k,\lfloor n/N_f \rfloor}B_{k,n}w(t - nT_f - \tau_{k,n})\right], \tag{39}$$

$$r(t) = \sum_{k,n} \left[A_{k,n}h_k(t - nT_f - d\tau_{I^{(1)}_{k,\lfloor n/N_f \rfloor}}) + I^{(2)}_{k,\lfloor n/N_f \rfloor}B_{k,n}h_k(t - nT_f - \tau_{k,n})\right] + v(t). \tag{40}$$

Modifying (33), an estimated template for user k becomes

$$\hat{z}_k(t) = \frac{1}{N_f N_s} \sum_{m=1}^{N_f N_s} r(t + mT_f)A_{k,m}. \tag{41}$$

The PAM symbol is decoded as (34) based on the estimated template

$$\hat{I}^{(2)}_{k,n} = \text{sign}\left(\sum_{m=nN_f}^{(n+1)N_f-1} \int_0^T \hat{z}_k(t)r(t + mT_f + \tau_{k,m})B_{k,m}\,dt \right). \tag{42}$$

Using this decoded input, the waveform of the PAM symbol is estimated by

$$\hat{h}_k(t) = \frac{1}{N_f N_s} \sum_{m=1}^{N_f N_s} \hat{I}^{(2)}_{k,\lfloor m/N_f \rfloor} r(t + mT_f + \tau_{k,m})B_{k,m}, \tag{43}$$

as in (35). Afterward, construct a template based on the estimated waveform and then correlate it with the received signal to decode the current PPM symbol, as (17) and (18).

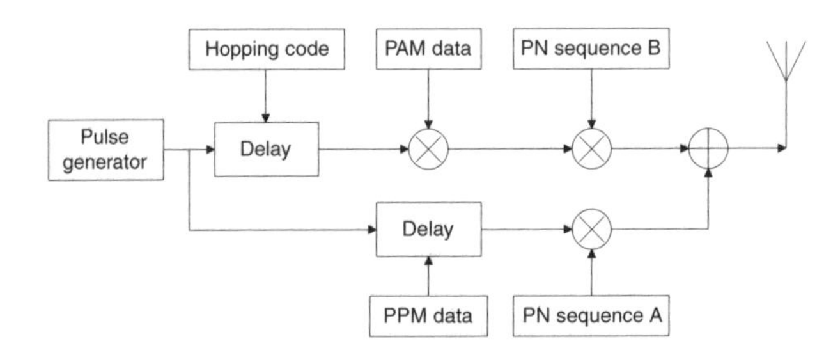

Figure 7.11 Block diagram of a full-rate MTR-UWB transmitter with balanced data modulation

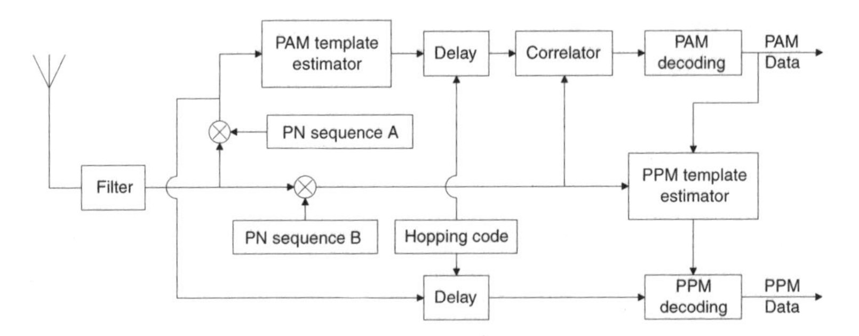

Figure 7.12 Block diagram of a full-rate MTR-UWB receiver with balanced data modulation

Figure 7.11 shows a MD assisted full-rate MTR transmitter structure with balanced modulation, while Figure 7.12 shows its receiver counterpart.

Remark 1: Instead of batch processing, adaptive implementation using a small step size is feasible to update each template estimate online to reduce implementation cost. It is also possible to consider advanced digital detectors instead of analog correlators for all the above receiver structures with practical ADCs, as argued before.

Remark 2: The near full-rate TR transmission scheme will become a superimposed training scheme when the delay of the second pulse in the doublet reduces to zero. Then, previous discussions can be easily adapted to such a scenario.

Remark 3: Besides training-based transmission schemes, differential data modulation is also an effective means to combat multipath channel distortion [28].

Remark 4: In UWB transceiver design, many other factors, such as UWB signal timing and mutual interference between UWB and existing overlaying communication systems, also need to be taken into account. Readers may refer to other chapters of this book for details. It is worth pointing out that unknown timing can be absorbed into channel impulse response with overestimated zeros. The proposed TR variants can tolerate mistiming within a certain range, an inherent property of a conventional TR transceiver. For accurate timing acquisition, however, in addition to applying existing methods [42, 81], other techniques may also be developed that are unique to proposed TR systems [82].

7.5 Conclusions

Novel low-complexity UWB transceiver design techniques are proposed to mitigate both MUI and channel distortion. Solutions are proposed in two complementary directions – digital and analog. For given transmission schemes, design of digital receivers has abundant flexibility in choosing appropriate DSP algorithms. In practice, it has to take into account constraints of ADCs on limited output signal range, reasonably high sampling rate and finite quantization resolution. Such digital design may be assisted by channel estimation. Analytical and experimental study can be carried out to test satisfaction in practical conditions. Simulation results show that a few bits of ADCs at a reasonable sampling rate suffice to yield detection performance close to full resolution digital receivers. Alternatively, analog approaches incur less restrictive requirements upon implementation. Those detailed approaches rely on TR transmission and correlation-type detection mechanisms. Different TR variants are presented to count adverse channel effects and increase data transmission rate. In addition, their extensions to multiuser communications provide effective multiple access capabilities for multiple users and boost the system capacity. Those methods cast away various limitations of a conventional TR transceiver and are highly likely to be applied to practical UWB systems. Meanwhile, the transition from analog to digital domain may occur as ADC devices mature. It is anticipated that fully digital implementations of analog receivers will be realized in the near future to achieve design flexibility and maximally deliver processing capability of digital processors.

Acknowledgments

This work was supported in part by the University of California MICRO Program and the Army Research Laboratory CTA on Communications and Networks under grant DAAD19-01-2-0011. The U.S. Government is authorized to reproduce and distribute reprints for Government purposes notwithstanding any copyright notation thereon. The views and conclusions contained in this document are those of the authors and should not be interpreted as representing the official policies, either expressed or implied, of the Army Research Laboratory or the U.S. Government.

The author wishes to thank Dr. Brian M. Sadler and Dr. Ananthram Swami from the Army Research Laboratory for providing some insightful comments and book editors Profs Xuemin (Sherman) Shen and Robert Qiu for their invitation and encouragement to write this chapter.

References

[1] R. A. Scholtz, "Multiple Access with Time-Hopping Impulse Modulation," *Proc. MILCOM'93*, pp. 447–450, Oct. 1993.

[2] M. Z. Win and R. A. Scholtz, "Impulse Radio: How it Works," *IEEE Commun. Lett.*, vol. 2, no. 2, pp. 36–38, Feb. 1998.

[3] Federal Communications Commission News Release, "Revision of Part 15 of the Commission's Rules Regarding Ultra-Wideband Transmission Systems," ET Docket 98-153, Washington, DC, Feb. 14, 2002.

[4] M.-G. Di Benedetto, *Understanding Ultra Wide Band Radio Fundamentals*, Prentice Hall, Upper Saddle River, NJ, 2004.

[5] IEEE 802.15.3a, http://www.ieee802.org/15/pub/TG3a.html.

[6] IEEE 802.15.4a, http://www.ieee802.org/15/pub/TG4a.html.

[7] M. Choi and A. Abidi, "A 6b 1.3 GSamples/s A/D Converter in 0.35 μm CMOS," *Proc. IEEE International Solid-State Circuits Conference*, Arlington, Virginia, vol. 438, pp. 126–127, 2001.

[8] R. H. Walden, "Analog-to-Digital Converter Survey and Analysis," *IEEE J. Select. Areas Commun.*, vol. 17, no. 4, pp. 539–550, Apr. 1999.

[9] C. L. Bennett and C. F. Ross, "Time-Domain Electromagnetics and its Applications," *Proc. IEEE*, vol. 66, no. 3, pp. 299–318, 1978.

[10] M. Z. Win and R. A. Scholtz, "Characterization of Ultra-Wide Bandwidth Wireless Indoor Channels: A Communication-Theoretic View," *IEEE J. Select. Areas Commun.*, vol. 20, no. 9, pp. 1613–1627, Dec. 2002.

[11] D. Dabeer and U. Madhow, "Detection and Interference Suppression for Ultra-Wideband Signaling with Analog Processing and One bit A/D," *Proc. Asilomar*, Monterey, CA, vol. 2, pp. 1766–1770, 2003.

[12] S. Hoyos, B. M. Sadler, and G. R. Arce, "Analog to Digital Conversion of Ultra-Wideband Signals in Orthogonal Spaces," *Proc. UWBST 2003*, Reston, VA, 2003.

[13] S. Hoyos, B. M. Sadler, and G. R. Arce, "High-Speed A/D Conversion for Ultra-Wideband Signals Based on Signal Projection Over Basis Functions," *Proc. ICASSP*, Montreal, Quebec, Canada, 2004.

[14] S. Hoyos, B. M. Sadler, and G. R. Arce, "Mono-Bit Digital Receivers for Ultra-Wideband Communications," *IEEE Trans. Wireless Commun.*, to appear.

[15] D. S. K. Pok, C.-I. H. Chen, J. J. Schamus, C. T. Montgomery, and J. B. Y. Tsui, "Chip Design for Monobit Receiver," *IEEE Trans. Microw. Theory Tech.*, vol. 45, no. 12, pp. 2283–2295, Dec. 1997.

[16] D. S. K. Pok, C.-I. H. Chen, C. T. Montgomery, J. B. Y. Tsui, and J. J. Schamus, "ASIC for 1-GHz Wide Band Monobit Receiver," *Proc. International Symposium on Circuits and Systems*, Monterey, California, vol. 2, pp. 77–80, May 1998.

[17] W. Namgoong, "A Channelized Digital Ultra-Wideband Receiver," *IEEE Trans. Wireless Commun.*, vol. 3, pp. 502–510, May 2003.

[18] P. M. Aziz, "An Overview of Sigma-Delta Converters," *IEEE Signal process. Magn.*, vol. 13, pp. 61–84, Jan. 1996.

[19] R. M. Gray, "Oversampled Sigma-Delta Modulation," *IEEE Trans. Commun.*, vol. 35, pp. 481–489, May 1987.

[20] S. Hoyos, B. M. Sadler, and G. R. Arce, "Dithering and Sigma-Delta Modulation in Mono-Bit Digital Receivers for Ultra-Wideband Communications," *Proc. UWBST 2003*, Reston, VA, 2003.

[21] L. L. Scharf, *Statistical Signal Processing: Detection, Estimation and Time Series Analysis*, Addison-Wesley Publishing, New York, 1991.

[22] M. Z. Win and R. A. Scholtz, "Ultra-Wide Bandwidth Time-Hopping Spread-Spectrum Impulse Radio for Wireless Multiple-Access Communications," *IEEE Trans. Commun.*, vol. 48, no. 4, pp. 679–689, Apr. 2000.

[23] T. W. Barrett, "History of Ultra Wideband Communications and Radar: Part I, UWB Communications," *Microw. J.*, pp. 22–56, Jan. 2001.

[24] R. J. Fontana, J. F. Larrick, and J. E. Cade, "An Ultra-Wideband Communication Link for Unmanned Vehicle Applications," *Proc. 1997 AUVSI*, Orlando, Florida, 1997.

[25] F. Ramirez-Mireles, "Performance of Ultrawideband SSMA Using Time Hopping and M-ary PPM," *IEEE J. Select. Areas Commun.*, vol. 19, no. 6, pp. 1186–1196, Jun. 2001.

[26] A. Taha and K. Chugg, "Multipath Diversity Reception of Wireless Multiple Access Time-Hopping Impulse Radio," *Proc. 2002 UWBST*, Baltimore, Maryland, pp. 283–288, May 2002.

[27] M. Z. Win and R. A,. Scholtz, "On the Energy Capture of Ultrawide Bandwidth Signals in Dense Multipath Environments," *IEEE Commun. Lett.*, vol. 2, no. 9, pp. 245–247, Sept. 1998.

[28] N. Guo and R. C. Qiu, "Improved Autocorrelation Receivers Based on Multiple Symbol Differential Detection for UWB Communications," *IEEE Trans. Wireless Commun.*, 2005, accepted.

[29] M. Pausini and G. J. M. Janssen, "Analysis and Comparison of Autocorrelation Receivers for IR-UWB Signals Based on Differential Detection," *Proc. ICASSP*, Quebec, Canada, vol. 4, pp. 513–516, May 2004.

[30] J. D. Choi and W. E. Stark, "Performance of Ultra-Wideband Communications with Suboptimal Receivers in Multipath Channels," *IEEE J. Select. Areas Commun.*, vol. 20, no. 9, pp. 1754–1766, Dec. 2002.

[31] S. Franz and U. Mitra, "On Optimal Data Detection for UWB Transmitted Reference Systems," *Proc. IEEE Globecom*, San Francisco, CA, vol. 2, pp. 744–748, Dec. 2003.

[32] R. T. Hoctor and H. W. Tomlinson, "An Overview of Delay-Hopped Transmitted Reference RF Communications," General Electronic Technical Report 2001 CRD198, pp. 1–29, General Electric Global Research, Niskayuna, New York, Jan. 2002.

[33] R. T. Hoctor and H. W. Tomlinson, "Delay-Hopped Transmitted Reference RF Communications," *Proc. 2002 UWBST*, Baltimore, Maryland, pp. 265–270, May 2002.

[34] E. Moulines, P. Duhamel, J.-F. Cardoso, and S. Mayrargue, "Subspace Methods for the Blind Identification of Multichannel FIR Filters," *IEEE Trans. Signal Process.*, vol. 43, no. 2, pp. 516–525, Feb. 1995.

[35] X. Wang and H. Poor, "Blind Equalization and Multiuser Detection in Dispersive CDMA Channels," *IEEE Trans. Commun.*, vol. 46, no. 1, pp. 91–103, Jan. 1998.

[36] Z. Xu, "Asymptotic Performance of Subspace Methods for Synchronous Multirate CDMA Systems," *IEEE Trans. Signal Process.*, vol. 50, no. 8, pp. 2015–2026, Aug. 2002.

[37] G. B. Giannakis and S. D. Halford, "Asymptotically Optimal Blind Fractionally Spaced Channel Estimation and Performance Analysis," *IEEE Trans. Signal Process.*, vol. 45, pp. 1815–1830, Jul. 1997.

[38] Z. Xu and M. Tsatsanis, "Blind Channel Estimation for Long Code Multiuser CDMA Systems," *IEEE Trans. Signal Process.*, vol. 48, no. 4, pp. 988–1001, Apr. 2000.

[39] Z. Xu, "Asymptotically Near-Optimal Blind Estimation of Multipath CDMA Channels," *IEEE Trans. Signal Process.*, vol. 49, no. 9, pp. 2003–2017, Sept. 2001.

[40] Z. Xu, "Effects of Imperfect Blind Channel Estimation on Performance of Linear CDMA Receivers," *IEEE Trans. Signal Process.*, vol. 52, no. 10, pp. 2873–2884, Oct. 2004.

[41] Z. Xu and X. Wang, "Large-Sample Performance of Blind and Group-Blind Multiuser Detectors: A Perturbation Perspective," *IEEE Trans. Inf. Theory*, vol. 50, no. 10, pp. 2389–2401, Oct. 2004.

[42] L. Yang, G. Giannakis, and A. Swami, "Noncoherent Ultra-Wideband Radios," *Proc. Milcom*, Monterey, CA, Nov. 2004.

[43] E. Fishler and H. V. Poor, "Low-Complexity Multiuser Detectors for Time-Hopping Impulse-Radio Systems," *IEEE Trans. Signal Process.*, vol. 52, no. 9, pp. 2561–2571, Sept. 2004.

[44] Q. Li and L. A. Rusch, "Multiuser Detection for DS-CDMA UWB in the Home Environment," *IEEE J. Select. Areas Commun.*, vol. 20, no. 9, pp. 1701–1711, Dec. 2002.

[45] V. Lottici, A. D'Andrea, and U. Mengali, "Channel Estimation for Ultra-Wideband Communications," *IEEE J. Select. Areas Commun.*, vol. 20, no. 9, pp. 1638–1645, Dec. 2002.

[46] Z. Xu, P. Liu, and X. Wang, "Blind Multiuser Detection: from MOE to Subspace Methods," *IEEE Trans. Signal Process.*, vol. 52, no. 2, pp. 510–524, Feb. 2004.

[47] P. Liu and Z. Xu, "POR-Based Channel Estimation for UWB Systems," *IEEE Trans. Wireless Commun.*, Nov. 2005, to appear.

[48] Z. Xu, J. Tang, and P. Liu, "Multiuser Channel Estimation for Ultra-Wideband Systems Using up to the Second Order Statistics," *EURASIP J. Appl. Signal Process.*, vol. 2005, no. 3, pp. 273–286, Mar. 2005.

[49] Z. Xu, P. Liu, and J. Tang, "A Subspace Approach to Blind Multiuser Detection for Ultra-Wideband Communication Systems," *EURASIP J. Appl. Signal Process.*, vol. 2005, no. 3, pp. 413–425, Mar. 2005.

[50] M. Tsatsanis and Z. Xu, "Performance Analysis of Minimum Variance CDMA Receivers," *IEEE Trans. Signal Process.*, vol. 46, no. 11, pp. 3014–3022, Nov. 1998.

[51] L. Tong, B. M. Sadler, and M. Dong, "Pilot-Assisted Wireless Transmissions: General Model, Design Criteria, and Signal Processing," *IEEE Signal Process. Magn.*, vol. 21, no. 6, pp. 12–25, Nov. 2004.

[52] J. Tang, Z. Xu, and P. Liu, "Mean and Covariance Based Estimation of Multiple Access UWB Channels," *Proc. IEEE Conference on UWBST*, Reston, VA, pp. 458–462, Nov. 16–19, 2003.

[53] J. Tang and Z. Xu, "Channel Estimation in Aperiodic Time-Hopping UWB Systems", *IEEE 6th CAS Symposium on Emerging Technologies*, Shanghai, China, May 31-Jun. 2, 2004.

[54] Z. Xu and J. Tang, "Blind Channel Estimation in Aperiodic Time Hopping Ultra-wideband Communications," *IEEE Trans. on Signal Processing*, 2006, in press.

[55] P. Liu, Z. Xu and J. Tang, "Subspace Multiuser Receivers for UWB Communication Systems," *Proc. IEEE Conference on UWBST*, Reston, VA, pp. 116–120, Nov. 16–19, 2003.

[56] P. Liu and Z. Xu, "Performance of POR Multiuser Detection for UWB Communications," *Proc. IEEE International Conference on Acoustics, Speech, and Signal Processing*, Philadelphia, PA, Mar. 19–23, 2005.

[57] Z. Xu, "Perturbation Analysis for Subspace Decomposition with Applications in Subspace-Based Algorithms," *IEEE Trans. Signal Process.*, vol. 50, no. 11, pp. 2820–2830, Nov. 2002.

[58] Z. Xu, "On the Second-Order Statistics of the Weighted Sample Covariance Matrix," *IEEE Trans. Signal Process.*, vol. 51, no. 2, pp. 527–534, Feb. 2003.

[59] Z. Xu, "Statistical Performance of a Data-Based Covariance Estimator," *IEEE Trans. Veh. Technol.*, vol. 53, no. 3, pp. 939–943, May 2004.

[60] X. Wang and H. V. Poor, "Blind Multiuser Detection: A Subspace Approach," *IEEE Trans. Inf. Theory*, vol. 44, no. 2, pp. 677–690, Mar. 1998.

[61] S. Burykh and K. Abed-Meraim, "Reduced-Rank Adaptive Filtering Using Krylov Subspace," *EURASIP J. Appl. Signal Process.*, vol. 2002, no. 12, pp. 1387–1400, Dec. 2002.

[62] J. S. Goldstein, I. S. Reed, and L. L. Scharf, "A Multistage Representation of the Wiener Filter Based on Orthogonal Projections," *IEEE Trans. Inf. Theory*, vol. 44, no. 7, pp. 2943–2959, Nov. 1998.

[63] G. Dietl, M. D. Zoltowski, and M. Joham, "Reduced-Rank Equalization for EDGE Via Conjugate Gradient Implementation of Multi-Stage Nested Wiener Filter," *Proc. 54th VTC*, Atlantic City, New Jersey, pp. 1912–1916, Oct. 2001.

[64] J. Tang, Z. Xu, and B. M. Sadler, "Digital Receiver for TR-UWB Systems with Inter-Pulse Interference," *Proc. SPAWC*, New York, Jun. 2005.

[65] J. Tang, Z. Xu, and B. M. Sadler, "Performance Analysis of B-bit Digital Receivers for TR-UWB Systems with Inter-Pulse Interference," *IEEE Trans. Wireless Commun.*, 2005, Submitted.

[66] R. Gagliardi, "A Geometrical Study of Transmitted Reference Communication Systems," *IEEE Trans. Commun.*, vol. 12, no. 4, pp. 118–123, Dec. 1964.

[67] G. D. Hingorani and J. C. Hancock, "A Transmitted Reference System for Communication in Random or Unknown Channels," *IEEE Trans. Commun.*, vol. 13, no. 3, pp. 293–301, Sept. 1965.

[68] C. K. Rushforth, "Transmitted-Reference Techniques for Random or Unknown Channels," *IEEE Trans. Inf. Theory*, vol. 10, no. 1, pp. 39–42, Jan. 1964.

[69] Y. Chao and R. Scholtz, "Optimal and Suboptimal Receivers for Ultra-Wideband Transmitted Reference Systems," *Proc. IEEE Globecom*, San Francisco, CA, vol. 2, pp. 759–763, Dec. 2003.

[70] Z. Xu, B. M. Sadler, and J. Tang, "Data Detection for UWB Transmitted Reference Systems with Inter-Pulse Interference," *Proc. IEEE International Conference on Acoustics, Speech, and Signal Processing*, Philadelphia, PA, Mar. 19–23, 2005.

[71] A. B. Sripad and D. L. Snyder, "A Necessary and Sufficient Condition for Quantization Errors to be Uniform and White," *IEEE Trans. Acoust. Speech Signal Process.*, vol. ASSP-25, no. 5, pp. 442–448, Oct. 1977.

[72] Z. Xu and B. M. Sadler, "Multiuser Transmitted-Reference UWB Communication Systems," *Proc. SPAWC*, New York, Jun. 2005 (invited).

[73] Z. Xu and B. M. Sadler, "Multiuser Transmitted Reference Ultra Wideband Communication Systems," *IEEE J. Select. Areas Commun.: Special Issue Ultra Wideband Wireless Commun. - Theory Appl.*, 2006, in press.

[74] J. Tang and Z. Xu, "Multidimensional Orthogonal Design for Ultra-Wideband Downlink", *IEEE International Conference on Acoustics, Speech, and Signal Processing*, Quebec, Canada, May 17–21, 2004.

[75] L. Yang and G. B. Giannakis, "Multistage Block-Spreading for Impulse Radio Multiple Access Through ISI Channels," *IEEE J. Select. Areas Commun.*, vol. 20, no. 9, pp. 1767–1777, Dec. 2002.

[76] H. Sari, F. Vanhaverbeke and M. Moeneclaey, "Extending the capacity of multiple access channels," *IEEE Commun. Magn.*, vol. 38, no. 1, pp. 74–82, Jan. 2000.

[77] F. Vanhaverbeke, M. Moeneclaey, and H. Sari, "DS/CDMA with Two Sets of Orthogonal Spreading Sequences and Iterative Detection," *IEEE Commun. Lett.*, vol. 4, no. 9, pp. 289–291, Sept. 2000.

[78] J. K. Tugnait and W. Luo, "On Channel Estimation Using Superimposed Training and First-Order Statistics," *IEEE Commun. Lett.*, vol. 8, pp. 413–415, Sept. 2003.

[79] G. T. Zhou, M. Viberg, and T. McKelvey, "A First-Order Statistical Method for Channel Estimation," *IEEE Signal Proc. Lett.*, vol. 10, pp. 57–60, Mar. 2003.

[80] J. Tang and Z. Xu, "A Novel Modulation Diversity Assisted Ultra Wideband Communication System," *Proc. IEEE International Conference on Acoustics, Speech, and Signal Processing*, Philadelphia, PA, Mar. 19–23, 2005.

[81] L. Yang and G. Giannakis, "Blind UWB Timing with Dirty Template," *Proc. ICASSP*, Monterey, Canada, May 2004.

[82] B. M. Sadler, Z. Xu and J. Tang, "Data Detection Performance of an MTR-UWB Receiver in the Presence of Timing Errors," *Proc. of Asilomar*, Pacific Grove, California, Oct. 2005 (invited).

8

UWB MAC and Ad Hoc Networks

Zihua Guo and Richard Yao

8.1 Introduction

It is known that the ultra-wideband (UWB) developmental efforts are divided into two camps in industry, that is, the Direct-Sequence (DS) UWB camp led by Motorola and Freescale (DS-UWB, http://www.uwbforum.org/), and the Multiband OFDM Alliance (MBOA) led by Intel, TI, and other companies (MBOA, http://www.wimedia.org/index.html). The physical layer and Medium Access Control (MAC) layer of these two camps are different. The readers may refer to [1] and [2] for the details of the PHY specifications of DS-UWB and MBOA, respectively. In this chapter, we would focus on the MAC layer. The DS-UWB camp still utilizes the IEEE 802.15.3 as the MAC support, which was originally designed for the high speed Wireless Personal Area Networks (WPAN) [3]; while the MBOA Special Interest Group (SIG) is defining its own MAC standard [4]. These two MAC specifications are rather different. Because the MBOA MAC has not been finalized, this chapter will mainly study the IEEE 802.15.3 MAC protocol.

8.1.1 Overview of IEEE 802.15.3 MAC

A WPAN defined by the IEEE 802.15.3 is a wireless ad hoc data communication system that allows a number of independent devices (DEVs) to communicate with each other. Such a network is called a *piconet*, as shown in Figure 8.1(a). The DEV is designed to be low power and low cost. One DEV is required to perform the role of PNC (Piconet Coordinator), which provides the basic timing for the piconet as well as other piconet management functions, such as power management, Quality of Service (QoS) scheduling, security, and so on.

Ultra-wideband Wireless Communications and Networks
Edited by Xuemin (Sherman) Shen, Mohsen Guizani, Robert Caiming Qiu and Tho Le-Ngoc

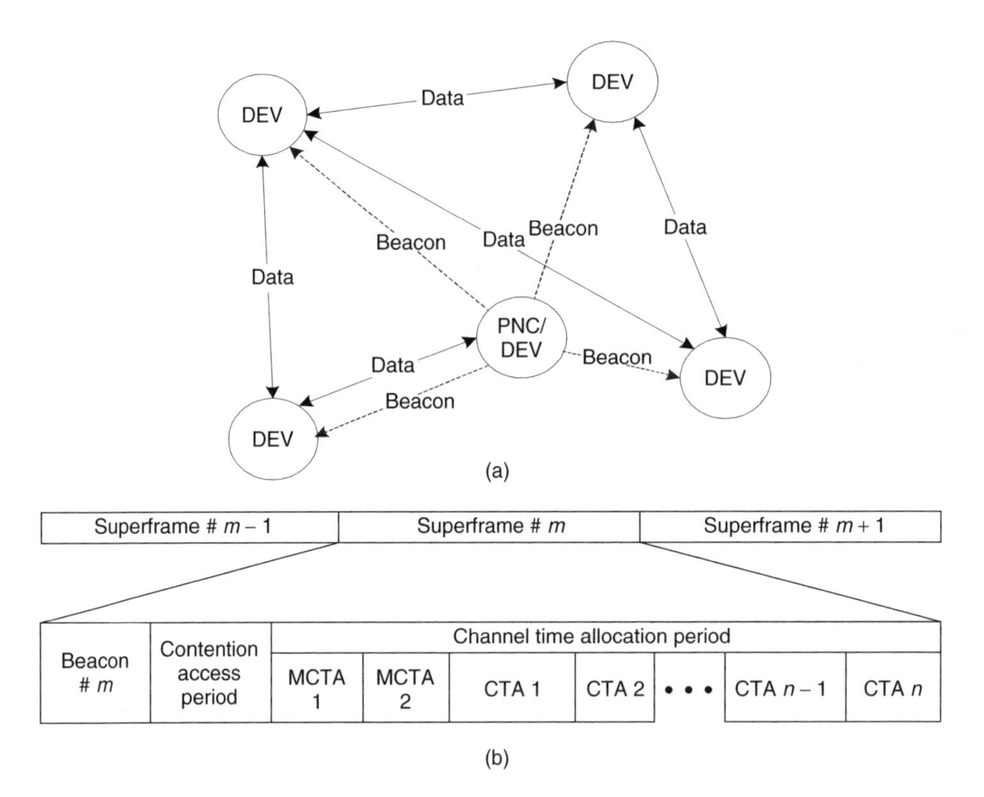

Figure 8.1 (a) Piconet structure in IEEE 802.15.3; (b) Superframe structure of IEEE 802.15.3 MAC. MCTA: Management CTA

In the 802.15.3 MAC protocol, the channel time is divided into superframes which consist of three major components – the beacon, the optional contention access period (CAP), and the channel time allocation period (CTAP), as shown in Figure 8.1(b).

- The beacon frame is sent by the PNC at the beginning of every superframe. It is used to assign the channel time allocation (CTA) and to communicate management information for the piconet.
- The CAP uses CSMA/CA to communicate commands and/or asynchronous data.
- The CTAP, on the other hand, is designed for both asynchronous and isochronous data streams and it uses TDMA protocol where each stream between DEVs is assigned a specified time slot or CTA for medium access. Each CTA is defined by a start time and a duration value. Each DEV knows when and how long it can transmit from CTA information broadcasted in beacon frames. The allocation of the CTA is performed in PNC and the algorithm is out of the scope of the standard being discussed here.

The CSMA/CA scheme used in CAP basically follows the DCF mode in IEEE 802.11 and no QoS is provided, whereas CTAP utilizes the TDMA as the basic access scheme, which is useful for both power saving and QoS. There are two types of CTAs – dynamic CTA and pseudostatic CTA. For the dynamic CTA, the PNC may move the position within the superframe in a superframe by superframe basis. This allows PNC the flexibility to rearrange CTA assignments to optimize the utilization of the assignments. The move of the dynamic CTA by the PNC is announced in the CTA parameters in the beacon. The dynamic CTA may be used for both asynchronous and isochronous streams. The pseudostatic CTA shall be allocated only for isochronous streams and shall not be subrate allocations. If the PNC needs to change the duration or position of a pseudostatic CTA within the superframe, it shall change the corresponding CTA blocks in the beacon.

The specific functionalities of the IEEE 802.15.3 will be described in detail in the following sections. Various piconets can exist simultaneously and form an ad hoc network. They may work in the same or different physical channels. In summary, the design goals of the IEEE 802.15.3 MAC protocol are as follows:

- Fast connection
- Ad hoc networks
- Data transport with QoS
- Security
- Dynamic membership
- Efficient data delivery.

8.1.2 Overview of MBOA MAC[1]

The fundamental difference between the MBOA MAC protocol and 802.15.3 MAC protocol is that the former eliminates the utilization of PNC. Therefore, it is a distributed MAC and every DEV will send a beacon. On the other hand, the local cluster structure is still retained, which is called the *beacon group* instead of a piconet. The structure of the superframe of MBOA MAC is shown in Figure 8.2.

As can be seen from Figure 8.2, in MBOA MAC, the superframe includes the beacon period and the data access period. However, differing from 802.15.3 MAC, MBOA MAC utilizes a very different channel access scheme for data. Basically, the access includes Distributed Reservation Protocol (DRP) and CSMA/CA based Prioritized Channel Access (PCA). In PCA, the access scheme is very similar to the Enhanced Distributed Contention Access (EDCA) in IEEE 802.11e. Therefore, four access categories (ACs) that provide QoS are defined. In DRP, three types of reservations are defined – hard reservation,

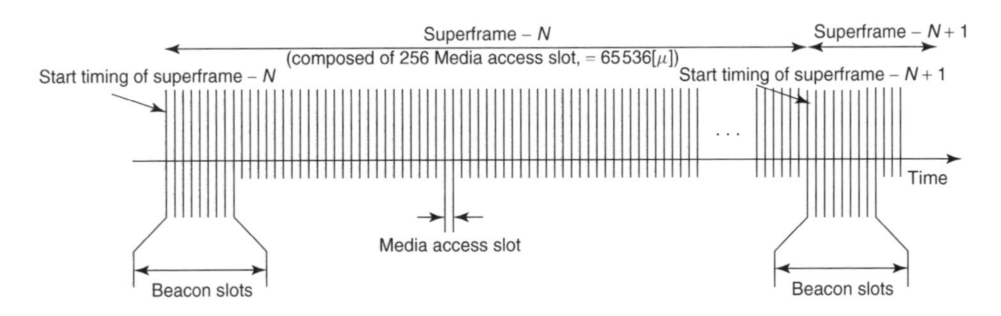

Figure 8.2 Superframe structure of MBOA MAC

[1] The description of MBOA MAC here is based on Draft 0.65 [4] and the specification is subject to change.

Table 8.1 Comparison between MBOA MAC and IEEE 802.15.3 MAC

	IEEE 802.15.3 MAC	MBOA MAC
Local group structure	Piconet	Beacon group
Centralized control device	PNC	None
Send beacon	Only PNC	Every device
CSMA channel access	Optional	EDCA as in 802.11e
TDMA channel access	PNC assigned CTA	Hard reservation
Dependent group	Child/neighbor piconet	Not defined
Superframe duration	1 ms to 65,536 μs	65,536 μs
Power management mode	Active, PSPS, DSPS, APS	Active mode, hibernating mode

soft reservation, and beacon-period reservation. In a hard reservation, only the owner of the reservation can access the medium. Other devices are only allowed to access the medium after the sender and receiver have announced the end of their transmission(s) with an Unused DRP Announcement (UDA) and Unused DRP Response (UDR). In a soft reservation period, other devices can access the medium following the PCA access rules. The owner of the reservation can access the medium with the highest priority without performing backoff.

The major differences between these two MAC specifications are listed in Table 8.1.

In the following text, we will discuss different topics in IEEE 802.15.3 MAC. In particular, the QoS scheduling algorithm in the PNC will be presented in Section 8.2. The power management problem is shown in Section 8.3. In Section 8.4, we will propose an adaptive delayed acknowledgement (Dly-ACK) scheme for UWB-based WPAN. In Section 8.5, the ad hoc network based on 802.15.3 is discussed. Finally, the summary is be given in Section 8.6.

8.2 QoS Scheduling in PNC

In the IEEE 802.15.3, the MAC protocol is based on the standard TDMA technology. The CTA is the resource allocation unit for communication between specified source and destination devices. On the other hand, the CTA is allocated by a PNC in a superframe by superframe manner. However, the 802.15.3 MAC protocol does not define how to implement the scheduler, but leaves this to vendors. The scheduler at PNC makes CTA assignment decisions based on the amount of CTA requested and the total resource available. Although there has been tremendous work regarding the scheduling algorithm in literature [5], the challenges for the problem in the 802.15.3 MAC protocol are that the input information is very limited (only users' requested bandwidth and total available bandwidth) and the required computation complexity should be low since it will be implemented in the portable devices and executed in a frequent manner (needed for each superframe). Therefore, there has been much work regarding the scheduling in the standard body of the 802.15.3 MAC. In [6], Kim *et al.* proposed an application-aware CTA scheduling method. In their algorithm, the CTA is based on the detailed application information; such as the structure of the video frame, in particular; Group of Pictures (GOP) [7]; the sizes of I-frame, B-frame and P-frames; and so on. However, such a scheduling algorithm requires much information from the application layer and is only applicable to MPEG video. In addition, an 11-byte overhead needs to be added to MAC frame header. In [8], after a detailed study of several feasible scheduling algorithms for the 802.15.3 MAC protocol, the authors claimed that the shortest remaining processing time (SRPT) scheduler for the 802.15.3 MAC level scheduling yields the best performance. Their simulation results have demonstrated that the SRPT essentially serves the shortest request first to minimize the total response time [9]. However, SRPT does not take into account the different deadline of the different requested MAC frames. Actually, the accurate deadline information of the frames is not available. In this part, we will propose, on the other hand, a simple but effective way to calculate the relative deadline of the requested frames although the absolute deadline is unknown. With the aid of the relative deadline, a deadline-aware scheduling (DAS) algorithm is proposed. It will be shown that significant QoS performance improvement with regard to job failure rate (JFR) can be achieved as compared with the SRPT scheduler.

8.2.1 Problem Definition

For real-time multimedia streaming, every audio/video frame has a deadline in order to guarantee the smoothness of the output at an end user. If some of the data of a decoded video frame cannot arrive at the decoder on time, a corrupted frame will be displayed or the previous frame will be replayed. This is a deadline miss for the decoder. In wireless system, because of the traffic burstness and the limitation of the channel bandwidth, some of the packets may not be delivered to the receiver timely, which causes the degradation of the audio or video quality. Therefore, the PNC needs to carefully schedule the transmission of the packets from different DEVs to achieve the satisfied QoS.

In contrast to several traditional scheduling algorithms such as SRPT and Weight Round Robin (WRR) that treat all the packets in the DEVs' request identically, the proposed DAS algorithm utilizes a very simple criterion of serving the previously unserved frames first. This is based on a basic observation that for the same class of application, the earlier the frame comes, the more urgent it is. In other words, in each superframe, the PNC first allocates CTAs to the streams whose previous requested transmission time was not satisfied. Only when all the previous frames are served, the DAS scheduler would allocate CTAs to the newly arrived frames. So, it can be seen that although the accurate deadline is not available, we actually utilize the relative deadline in our algorithm. Therefore, the DAS scheduler greatly reduces the complexity and overhead introduced by other sophisticated schedulers requiring accurate timing information, for example, Generalized Processor Sharing (GPS) and Worst-case Fair Weighted Fair Queuing (WF^2Q) [5]. The design of DAS algorithm is targeted to the following features:

- No change of standard. The scheduling algorithm should be compatible with current standard and not need additional information in the MAC header.
- Low-complexity. Since the scheduling algorithm is executed every few milliseconds, a highly sophisticated method is not suitable for MAC level scheduling.

In the following, we will describe the DAS in detail.

8.2.2 Deadline-Aware Scheduling Algorithm

According to the 802.15.3 MAC protocol, a DEV can send the CTA request to PNC in the management CTA (MCTA) that can be at any position within the superframe. If we put the MCTA at the end of the superframe, the CTA requests are just before the PNC performing the scheduling, and then the queue information in the request is the most up-to-date. Of course, the MCTA can be put at the beginning of the superframe. In this case, although the scheduling algorithm will not change much, the overall performance will be a little worse due to the obsolete queue information. So, in this part, we propose arranging the MCTA at the end of the superframe. The DEVs send MCTA requests of each stream indicating their current queue size. Here, we consider only one QoS class; while for the case of multiple QoS classes, we can schedule the frames from high priority to low priority and apply the DAS algorithm to each class. The PNC takes the following steps for CTA allocation:

Step 1: For the CTA from each stream, calculate the number of unserved frames in previous superframe (these are called *Pr-frames*) and how many frames are newly arrived (these are called *Cr-frames*).

Step 2: Assign the CTA to Pr-frame in the ascending order according to their frame number.

Step 3: If current superframe still has extra CTAs, assign the reserved CTAs to the Cr-frames for each stream proportionally.

Step 4: If there are still extra CTAs left in current superframe, assign the CTAs to the overloaded Cr-frame (the Cr-frames in excess of the reservations) for each stream in the ascending order.

Step 5: Combine the CTAs in Steps 2–4 for each stream and calculate the unserved frame number for next round scheduling.

From the above steps, we can see that the DAS is a simplification and combination of the Earliest-Due-Date-First (EDD) [5] and the SRPT scheduling. In contrast to the EDD, we do not require the exact deadline corresponding to each MAC frame. We only divide the requests into two levels, that is, the newly arrived Cr-frames in current superframe and the previously unserved Pr-frames. Moreover, we can divide the previous Pr-frames into more detailed levels for each stream. That is, the Pr-frames arrived in superframe n, the Pr-frames arrived in superframe $n-1$, $n-2$, and so on. However, this will require more complex calculation and more memory in the PNC. Thus, we only consider two levels in the following description.

Here, we give an example in Figure 8.3 to show the requesting/allocation procedure. In superframe n, the DEV requests CTAs for ten frames in which three are previously unserved Pr-frames. The PNC first allocates CTAs for these three Pr-frames and then allocates the reserved CTAs (say, two frames) to its Cr-frames. Then the PNC allocates CTA of two frames to the overloaded Cr-frames if the CTA of current superframe is still not used up. In this example, a total of seven frames are assigned CTA for this stream and there are still three unserved frames. Then, the procedure works in a similar fashion in the superframe $n+1$.

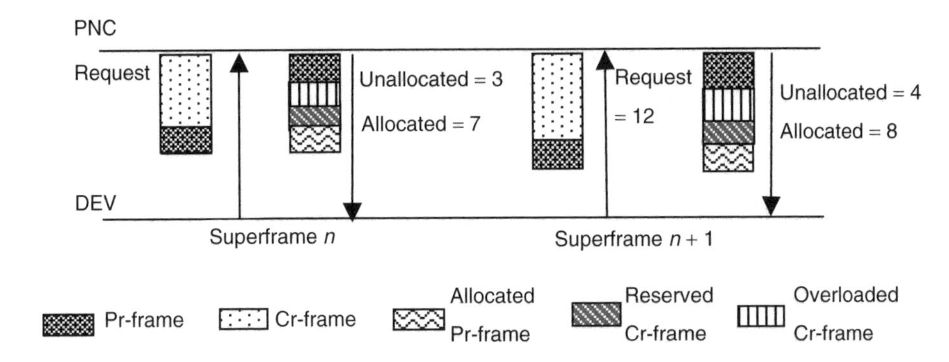

Figure 8.3 Deadline-aware scheduling

8.2.3 Calculation of the Reserved CTA

In each round, the PNC should calculate the number of unserved frames. Assume that the DEV sends the CTA request in superframe n (note that this request is sent at the end of superframe $n-1$) and we denote it as $Requested(n)$. For each stream, the PNC can derive the number of newly arrived frame based on the total request and the previous unserved frame number as follows:

$$NewArrived(n) = Requested(n) - Unserved(n-1),$$
$$Unserved(n-1) = Requested(n-1) - Allocated(n-1), \quad (1)$$

where $Allocated(n)$ is the number of assigned CTA by PNC in superframe n. The reserved CTA that is allocated in Step 3 can be computed as

$$Resered(n) = \alpha \times Reserved(n-1) + (1-\alpha) \times NewArrived(n), \quad (2)$$

where α is the smoothing factor within 0.7–0.9.

From the above discussion, we can see that a simple iterative way is used to calculate the frame number with different deadline requirements, and this relative deadline information is applied to the scheduling algorithm. In the following, with both analysis and simulation, we will show that this algorithm can improve the video QoS performance significantly compared with SRPT, which does not consider the deadline information.

8.2.4 Simulation Results

To show the performance of the proposed DAS scheduling, we present some simulation results that compare DAS with SRPT scheduling. The basic simulation tool is the 802.15.3 ns2 extension [10, 11].

In the simulation, we mainly consider the MPEG video streaming, which has a GOP pattern of IBBPBBPBBPBB. The channel rate is 100 Mbps and all the DEVs send or retrieve the video from PNC. We consider four different scenarios. That is, fixed superframe size (15 ms) and dynamic superframe size (1–30 ms, the duration depends on

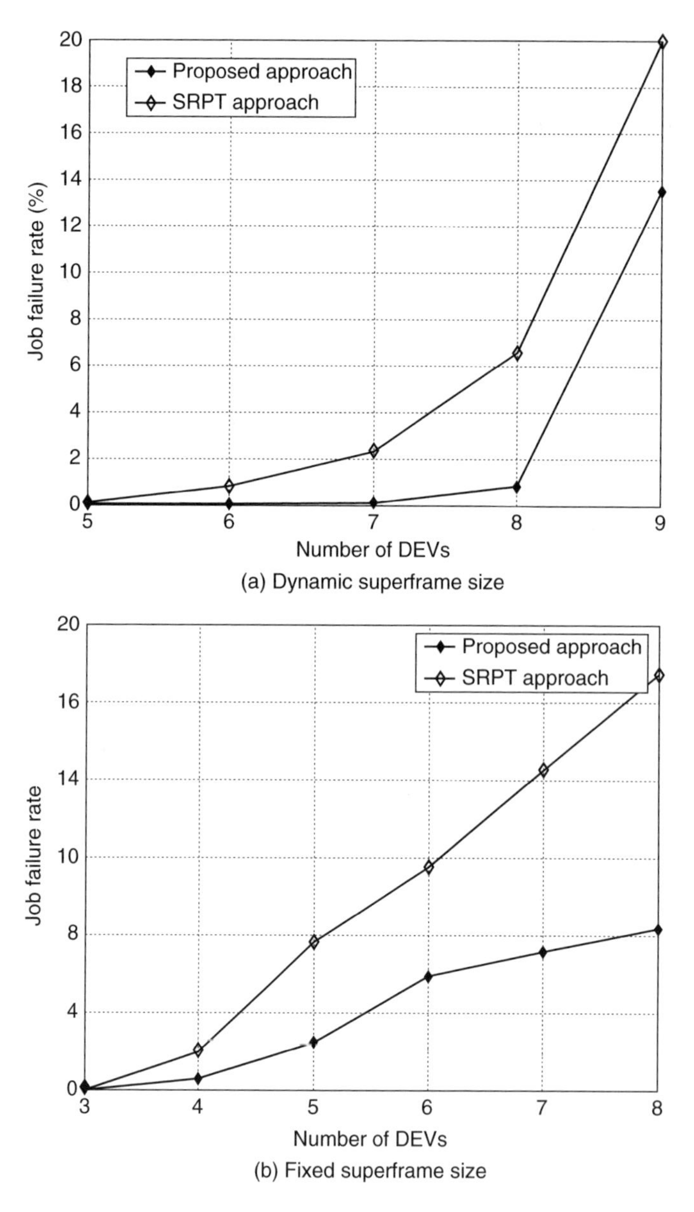

Figure 8.4 Nonsimultaneous starting (FER = 0, frame deadline = 60 ms, bitrate = 8 Mbps, fragmentation size = 2048 kB)

the amount of current total pending traffic); simultaneous flows and nonsimultaneous flows (separated by 100 ms). In Figure 8.4, the nonsimultaneous case with both fixed and dynamic superframe size is shown. In Figure 8.5, the simultaneous case with both fixed and dynamic superframe size is shown. From the figures, it can be observed that the DAS can perform much better than SRPT approach in all cases. This implies that the utilization of relative deadline is critical in the scheduling algorithm.

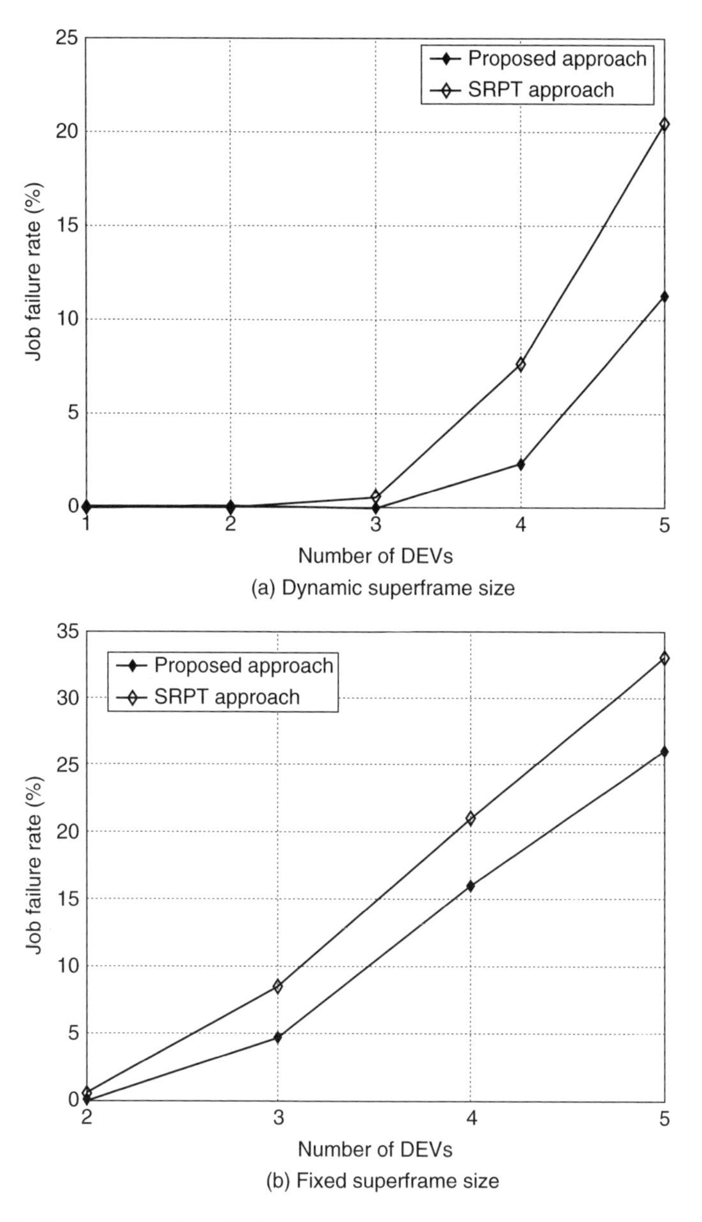

Figure 8.5 Simultaneous starting (FER = 0, frame deadline = 60 ms, bitrate = 10 Mbps, fragmentation size = 2048 kB)

8.3 Power Management in IEEE 802.15.3

Power management is an important issue for wireless portable devices that are usually battery-supplied. The objective of power management is to assist devices to reduce the wakeup times as much as possible. A good power management scheme is a key factor to the success of IEEE 802.15.3 systems. There have been several efforts, in literature, on the MAC design for power management in wireless systems. However, most of them were based on the MAC protocol of IEEE 802.11 Wireless Local Area Networks (WLAN), and there has been no work report on IEEE 802.15.3 systems so far. In [12], the authors studied the effect of the system load and other parameters, such as the user number on the power management, in the ad hoc mode of an 802.11 system. In [13], an Energy Conserving MAC (EC-MAC) was designed and compared with the standardized IEEE 802.11 power management. In [14], the authors defined some directory protocols that may be used by a central node in a wireless environment, such as IEEE 802.11 PCF (Point Coordination Function) mode, to coordinate the data transmission and the sleeping. A transmission scheduling algorithm was proposed to arrange the order of the streams for power saving. Since this algorithm was designed for 802.11 PCF, it can be well applied to the IEEE 802.15.3 WPAN. However, we will show later in this part that the result of this algorithm is far from the optimal one.

In the following, we consider the power management for the TDMA-based 802.15.3 WPAN systems. In the 802.15.3 standard, several power saving modes have been designed, namely, device synchronized power save (DSPS), piconet-synchronized power save (PSPS) and asynchronous power save (APS). All these modes are to enable DEVs to turn off for one or more superframes (superframe will be described in detail in next section) to save power. Different behaviors are defined in the standard for each mode. We call this *intersuperframe power management* problem. On the other hand, within a superframe, the standard defines that a DEV can sleep during the time slots or CTAs not belonging to its streams and only wake up when the CTAs belonging to its streams arrive. In this part, we will show that when a DEV has multiple streams, the ordering, or the relative positioning of the streams from all DEVs within a superframe affects the total number of wakeup and thus has a significant effect on the power saving, because each occurrence of wakeup and sleeping will consume significant power. To distinguish from the previously mentioned power management, we call this as *intrasuperframe power management* problem. In the ensuing part, we abstract this problem to be a graph problem and further show that it is generally a Hamilton path problem. Under some special cases, the Euler path can be used to retrieve the optimal solution. However, the Euler path solution usually is a sufficient, but not a necessary, condition for our studied problem. On the basis of the graph we generate, we propose an efficient suboptimal Min-Degree Searching (MDS) algorithm for this power management in 802.15.3 WPAN systems. Simulation results have demonstrated that, in normal cases, the proposed approach can achieve the optimal solution with a probability of more than 95 % and outperform the existing approach significantly.

8.3.1 Problem Definition

In the 802.15.3 MAC protocol, since the CSMA-based CAP is optional in the standard, we do not consider it in this section. Only the TDMA-based CTAP is considered where multiple users and streams[2] share different time slots within a superframe. The PNC assigns a CTA for each stream and announces the position and the duration of the CTA

[2] A stream is a data flow from one DEV to another DEV.

in the beacon of each superframe. As mentioned before, the standard does not define how to assign the CTAs and leaves this to vendors. In the previous section, we have talked about the CTAs allocation. The next problem is how to assign the positions of these CTAs within the superframe to save power.

Because of the property of TDMA MAC, each DEV can know exactly the CTA position in the superframe from the beacon. Thus, the DEV can go to sleep during the CTAs not assigned for its streams and need only wake up during the CTAs assigned for its streams. As a result, when a DEV has multiple streams, it may wake up and sleep for several times if the CTAs/streams are not well ordered. On the other hand, each occurrence of wakeup and sleeping will consume much power [15]. So, the intrasuperframe power management turns out to be properly arranging the order of the DEVs' streams to minimize the total number of wakeups in a typical multiuser and multistream network. A sample topology given here shows the effect of the stream ordering on power management. Here, DEV B has a stream linking DEV A and DEV C, and DEV D has a stream linking DEV C.

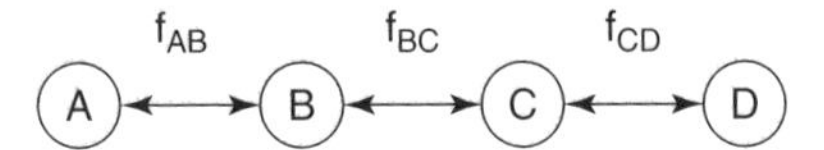

For this topology, we have two possible stream orders:

- Order 1: $f_{AB} \rightarrow f_{BC} \rightarrow f_{CD}$, wakeup times = 4
- Order 2: $f_{AB} \rightarrow f_{CD} \rightarrow f_{BC}$, wakeup times = 5

For order 1, every two contiguous streams share a common DEV. Since each DEV wakes up only in its own CTA and then goes to sleep at the end of the CTA, the total number of wakeups is 4. Note that the first stream f_{AB} requires two DEVs to wake up. For order 2, however, DEV B will go to the sleep state after stream f_{AB}, and then need to wake up again in the third CTA for stream f_{BC}. So, the total number of wakeups is 5, which is worse than order 1.

8.3.2 Proposed Approach

We consider a multiple-DEV, multiple-stream 802.15.3 WPAN. Note that if there is more than one stream between two DEVs, we can combine them as a single stream since it is natural to put them at contiguous CTAs within the superframe. Thus, we can generate a graph with the following rules:

1. Each DEV is denoted as a vertex.
2. Add an edge between two vertexes if there is a stream between the two corresponding DEVs.

The graph so generated is called G1 in the following. An example of G1 is shown in Figure 8.6(a) (Figure 8.6(b) shows graph called G2, which will be explained next). Now, assume that there are M vertexes and N edges in G1. The power management problem is

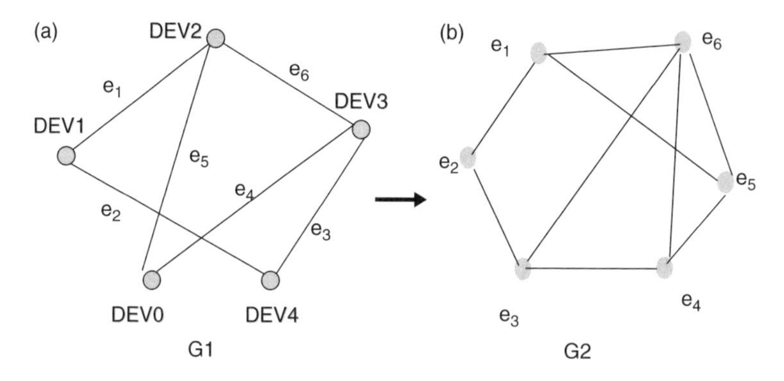

Figure 8.6 Transformation of G1 to G2

translated to finding an order of the edges in G1 so that every two contiguous edges share a common vertex. Since each time we serve a new stream, at least one new DEV needs to wake up, then, we have the following lemma, which is straightforward to prove.

Lemma 1 *The minimum number of the wakeups is lower bounded by $N + D$ with D denoting the number of connected patches in graph G1.*

According to Lemma 1, when G1 is a connected graph, the lower bound of the minimum number of the wakeups is $N + 1$. Obviously, if an Euler Path [16] exists in G1, then the edge order in this path is the optimal solution to achieve the lower bound. An example is shown in G1 of Figure 8.6 where an Euler Path is found to be: $e_1 \rightarrow e_2 \rightarrow e_3 \rightarrow e_4 \rightarrow e_5 \rightarrow e_6$. However, we will point out later that Euler Path is a sufficient, but not necessary, condition for our power management problem here. Nevertheless, on the basis of the property of Euler Path, we have the following lemma

Lemma 2 *Assume that the number of odd vertexes (odd vertex denotes the vertex with odd degree) is L in a connected graph G1. If $L \leq 2$, the minimum number of wakeups for the power management is equal to $N + 1$, which achieves the lower bound; If $L > 2$, the minimum number of wakeups for the power management is upper bounded by $N + L/2$.*

Proof: Note that L is always even in a graph. It is well known that when $L \leq 2$, the Euler Path (not necessary an Euler circuit) exists in G1 [16]. Then, the Euler Path yields the optimal solution and thus, the minimum number of wakeups is $N + 1$. When $L > 2$, we can add $(L - 2)/2$ edges between the $L - 2$ odd vertexes. Then, the Euler Path exists in this modified graph. Next, by removing the $(L - 2)/2$ edges from the Euler Path, we can immediately see that it is a solution with wakeup times being $N + 1 + (L - 2)/2 = N + L/2$. However, it must be noted that this may not be the optimal solution.

It is known that the Euler Path can be found by Fleury Algorithm [16]. Although at the first glance, the Euler Path seems equivalent to the power management problem we defined here, it is actually a sufficient, but not a necessary, condition. This is because in the edges' order obtained from Euler Path, the ending vertex of an edge should be the starting vertex of the successive edge. However, for the power management problem we defined in G1, two contiguous edges in the optimal order can share the common starting vertex. For example, in G1 of Figure 8.6, the edge order e_1-e_6-e_5-e_4-e_3-e_2 is also an optimal order although it is not even a path at all. Therefore, the existence of Euler Path is too strong a condition for our power management problem. In other words, it is sufficient but not necessary. To gain further insight on this problem, we can transform G1 to another equivalent graph G2 by obeying the following rules:

1. Each edge in G1 is denoted as a vertex in G2.
2. Add an edge between two vertexes in G2 if the corresponding two edges in G1 are neighbors.

From G2, we can immediately see that the optimal stream order is equivalent to finding a path in G2 so that G2 can be vertex-traversed; and this path is actually the Hamilton Path. Unfortunately, so far, in academia, unlike the Euler Path, there is no useful method to determine the existence of the Hamilton Path, let alone how to find it [16]. Therefore, in the following, we propose an MDS algorithm to find the suboptimal order of the streams.

Our goal is to find an order of the streams so that the contiguous streams can always share a common DEV. Of course, this is not always achievable. Anyway, for a given graph, there must exist an optimal order although it may not achieve the lower bound. The exhaustive search of the optimal order has a complexity of $N!$, which is unaffordable. Thus, we propose an MDS algorithm [17]. The basic idea of our proposed MDS algorithm is that, each time we select an edge in G1 so that its two associated vertexes have the minimum degree. Its motivation is that the edges with associated small degree of vertexes are the ones most likely to be isolated. So, we serve them first when it is still possible for them to share common vertex with other edges. The diagram of MDS is shown in Figure 8.7 and the details of MDS are described as follows:

1. Let G = G1; $j = 0$.
2. Select an edge as e_j in G so that the two associated vertexes are with the minimum degree. (Here, both two vertexes having the minimum degree means that first, among all candidate vertexes, we find those with the minimum degree (>0). Then, for each of them, among its neighboring vertexes, find one with the minimum degree. Finally, choose the vertexes pair with the minimum degree.)
3. For the current set $e_0 e_1, \ldots, e_j$, in $G = G1 - \{e_0, e_1, \ldots, e_{j-1}\}$, if the neighboring edges of $e_j = \emptyset$, go to step 5; otherwise, among the neighboring edges of e_j, select an edge as e_{j+1} so that the two associated vertexes are with the minimum degree; $j = j + 1$.
4. Continue step 3 until it cannot proceed.
5. If $E - \{e_0, e_1, \ldots, e_j\} \neq \emptyset$, let $G = G1 - \{e_0, e_1, \ldots, e_j\}$ and go back to Step 2; otherwise, stop.

8.3.3 Simulation Results

In this section, to evaluate our MDS, the algorithm in [14] is selected for comparison where the edges belonging to the same DEV are first grouped in a descending order of the vertex degree. That is, first select the vertex with the maximum degree and group all

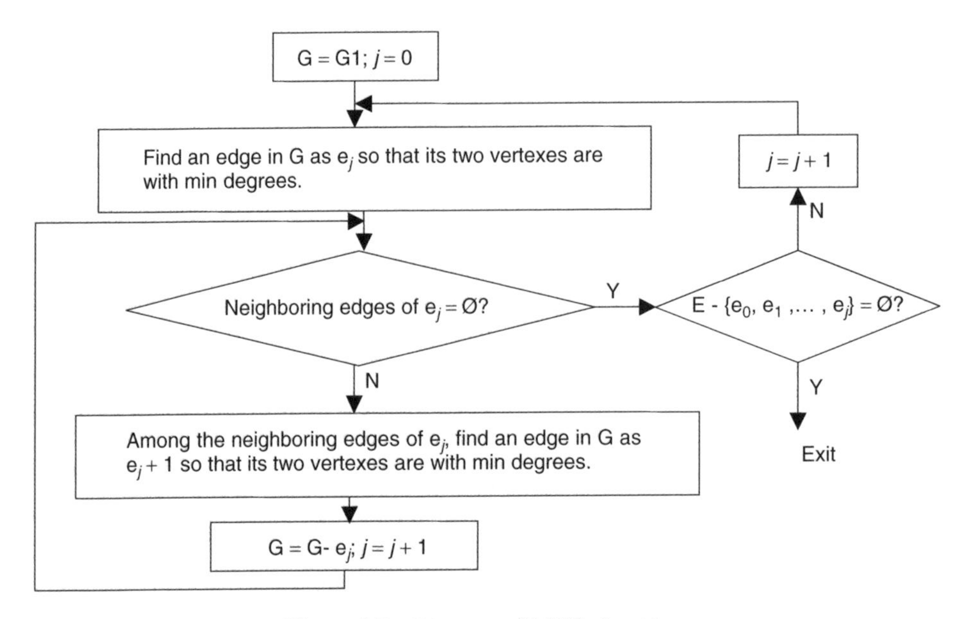

Figure 8.7 Diagram of MDS algorithm

its edges. Then, in the remaining vertexes and edges, select the vertex with the maximum degree, and so on. Obviously, the edges in the same group always share a common vertex. The algorithm then tries to connect different groups as much as possible.

In our simulation, we assume that there is a stream between each pair of the M DEVs with probability p. Obviously, larger the p, the more edges in G1. Here, we mainly evaluate the probability that the algorithm achieves the lower bound for the minimum wakeup times described in Lemma 1. In Figure 8.8, we compare MDS with the algorithm in [14] under $M = 7$ and $M = 9$ DEVs. Totally, 10,000 topologies are simulated for each p. We can see that MDS far outperforms the algorithm in [14] and it can achieve the lower bound in more than 95 % cases. This also indicates that MDS can achieve the optimal solution in most cases.

8.4 Adaptive Dly-ACK

Since the wireless channel is error-prone, the link layer always adopts some error control mechanisms to correct error transmissions. Specifically, the ACK and retransmission mechanism are usually designed to provide a reliable communication for higher layers. Usually, each data frame is followed by an ACK frame from the receiver to indicate its correct reception. In the following, this traditional scheme is referred to as Immediate Acknowledgement (Imm-ACK). However, in high data rate systems, for example, UWB system, the overhead caused by Imm-ACK may consume significant bandwidth, as we will show in Table 8.2 subsequently. To solve this problem, a new ACK policy, Dly-ACK, has been proposed in the 802.15.3 MAC protocol. In the Dly-ACK mechanism, instead of acknowledging each data frame, a burst of frames are received, then these frames are acknowledged by only one ACK frame.

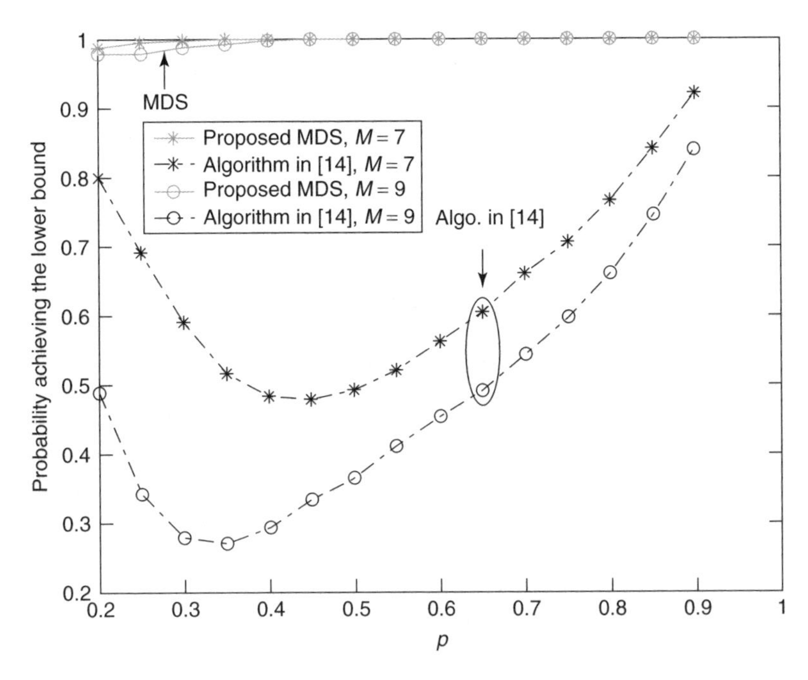

Figure 8.8 Performance comparison between MDS and existing algorithm

Table 8.2 Transmission time under various data rates

Data rate (Mbps)	Imm-ACK		5-Dly-ACK		10-Dly-ACK	
	t_m(μs)	t_p/t_m (%)	t_m(μs)	t_p/t_m (%)	t_m(μs)	t_p/t_m (%)
110	112.7	73.4	90.3	91.6	87.5	94.5
200	80.0	62.5	57.9	86.4	55.0	90.9
480	56.7	47.1	34.5	77.2	31.7	81.6

As a MAC layer technology, it is interesting to evaluate the Dly-ACK performance when working with transport layer protocol in an IP-based network, for example, TCP or UDP. For the 802.15.3 WPAN, the TCP application is usually involved in the multimegabyte file transfer (e.g., for music and image files); while the UDP application is usually involved in high quality audio/video streaming. In this part, we focus on the performances of Dly-ACK interaction with both TCP and UDP. Although the 802.15.3 standard defines the Dly-ACK scheme, how to use the Dly-ACK or determine the parameters of Dly-ACK is open for implementation. Therefore, our objective is to find an appropriate Dly-ACK scheme under different application scenarios. We first tested the Dly-ACK performance with fixed burst-size and found that the fixed Dly-ACK scheme is unable to work well with either TCP or UDP and the performance is rather poor. The reasons behind the poor performance are also analyzed. On the basis of these observations, an adaptive Dly-ACK scheme that dynamically changes the burst-size is proposed. Simulation results show that our proposed adaptive Dly-ACK is applicable to both TCP and UDP and can work perfectly for different applications. Compared with the Imm-ACK and fixed Dly-ACK, our scheme demonstrates a significant performance improvement in terms of, say, TCP throughput.

In the following, we use the term *frame* to denote a MAC layer unit, and *packet* to denote a TCP/IP unit.

8.4.1 Problem Definition

The 802.15.3 MAC protocol defines three types of ACK policies, that is, no-ACK (No-ACK), Imm-ACK, and delayed ACK (Dly-ACK). When using the No-ACK policy, the destination DEV shall not acknowledge the received frame. The two successive frames are separated by a minimum interframe space (MIFS). The No-ACK policy is appropriate for frames that do not require guaranteed delivery. The Imm-ACK policy provides an acknowledgement process in which each data frame is individually acknowledged following the reception of the frame. All frames, including the data frames and the ACK frames, are separated by the short interframe space (SIFS), that is larger than MIFS. The Dly-ACK policy allows the source DEV to send a burst of frames without an ACK frame. To negotiate the use of Dly-ACK, the source DEV adds Dly-ACK request information to an MAC frame's header. Once the destination DEV receives this frame that includes request information, it will send the Dly-ACK frame acknowledging those correctly received frames in current burst. The source shall not start or resume the next burst transmission until a Dly-ACK frame is received. The unacknowledged frames should be retransmitted in the next burst. Moreover, the destination should deliver received data frames to the upper layer according to their frame ID. Because of the possible retransmissions, the destination DEV will not receive the MAC frame in the correct order. So, a receiving buffer (we call it reordering buffer) is needed in each DEV to temporarily accommodate data frames received. The three ACK policies are shown in Figure 8.9. While Figure 8.9 gives the Dly-ACK frame format, the Max Burst filed indicates the number of frames with the maximum size that may be sent in one burst. The Max Frames field denotes the maximum number of frames regardless of size that may be sent in one burst. In Figure 8.9(d), the MAC Protocol Data Unit (MPDU) is the MAC data frame that will be sent over PHY. It is fragmented from the MAC Service Data Unit (MSDU) that is passed from the upper layer. All the MPDUs belonging to the same MSDU have the same size except the last MPDU. The maximum MPDU fragmentation number of an MSDU is 128. Thus, the MPDU-ID in Figure 8.9(d) is composed of the 9-bit MSDU ID plus the 7-bit fragmentation number.

In the following, we call the number of MPDUs that are transmitted in a Dly-ACK burst as burst-size n. For convenience, we use *n-Dly-ACK* to represent the Dly-ACK scheme with burst-size of n. Obviously, the Imm-ACK and No-ACK are special cases of Dly-ACK with n being 1 and infinite, respectively. The No-ACK policy has the best channel utilization, but it is not suitable for reliable transmissions.

Note that for each MAC frame (no matter data or ACK frame), a PHY preamble and header are always attached. Such an overhead is usually transmitted with the lowest rate for robustness and thus fixed in duration regardless of the size of the MAC payload. Therefore, although the ACK frame overhead and the associated interframe space (IFS) may be trivial for low data rate wireless systems, they may occupy relatively significant

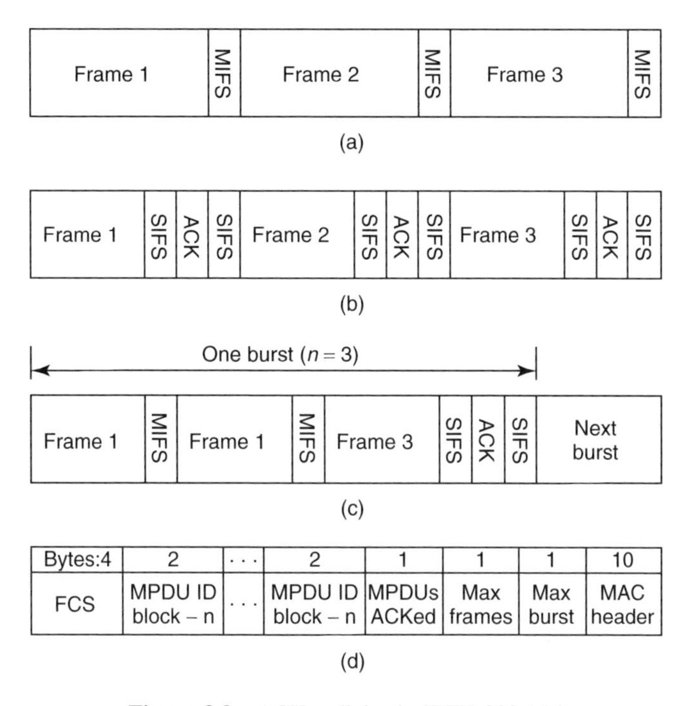

Figure 8.9 ACK policies in IEEE 802.15.3

channel bandwidth in high data rate system. As a result, compared with Imm-ACK, the Dly-ACK improves channel utilization through two ways. One is to reduce the number of ACK frame. The other is to reduce the associated IFS. When the Imm-ACK is used, the average total transmission time t_m of one frame is $t_p + t_{\text{ACK}} + 2t_{\text{SIFS}}$ with t_p, t_{ACK}, and t_{SIFS} denoting the length of the data frame, the ACK frame, and the SIFS, respectively. When the n-Dly-ACK is used, t_m is $(nt_p + t_{\text{ACK}} + (n-1)t_{\text{MIFS}} + 2t_{\text{SIFS}})/n$. Now, we assume the basic data rate is 100 Mbps, the MPDU payload size is 1000 bytes. The other parameters, for example, MAC header, PHY preamble and header, and so on, are according to Table 8.3. In Table 8.2, we list the values of t_m under various data rates and the ratio of t_p/t_m without considering channel error. It can be seen that when the data rate is high, for Imm-ACK, among the total time required to successfully transmit an MAC frame, only a small portion of the time is really used to send the payload. Therefore, we can see that the Dly-ACK policy gives much room to improve the channel utilization. This is also the motivation of adopting the Dly-ACK for high data rate.

Before concluding this section, we present the assumptions that will be used in our following study:

1. Each MPDU encounters an error with an error probability p while the MAC header is always correctly decoded because of the more robust transmission. In addition, we further assume that the ACK frame is error free. This common assumption is supported by the fact that it has a much shorter length compared with MPDU.
2. For UDP-based video streaming, an I, B, or P video frame plus some protocol (UDP+IP) headers usually corresponds to one MSDU (max size is 64 kB). Similarly,

Table 8.3 Simulation parameters

MAC header	10 bytes
PHY header	9.6 μs
Basic data rate	100 Mbps
Channel rate	200 Mbps
Superframe size	15 ms
Retransmission limit r_l	5
MIFS	2 μs
SIFS	10 μs

a TCP/IP packet also corresponds to one MSDU. As described in standard, an MSDU can be fragmented into one or multiple MPDUs.

3. The destination DEV has a finite reordering buffer of B_r bytes. As mentioned in Figure 8.8, when responding to a Dly-ACK frame, it informs the source DEV how large its remaining buffer is in the filed of Max Burst.

Some important parameters are listed in Table 8.3 according to the standard.

8.4.2 Adaptive Dly-ACK

In this section, we first investigate some problems when using a fixed Dly-ACK scheme. On the basis of these observations, we propose an adaptive Dly-ACK scheme to address these problems. In our study, the throughput is used as the performance metric for TCP applications; while the JFR is used for UDP applications. The index, JFR, represents the ratio of those video frames that are not correctly received within a deadline d_l among all transmitted video frames.

8.4.2.1 Problems of Fixed Dly-ACK

First, we use simulations to demonstrate the performance of n-Dly-ACK as indicated in Figure 8.10. In Figure 8.10(a), we present the TCP throughput with fixed n-Dly-ACK for $n = 1$, 3, 5, and 7 over a channel with $p = 20\,\%$, where TCP Reno is used to obtain these results. Figure 8.10(b) shows the JFR of UDP video streaming, where the number of UDP connections varies from 2 to 6, $p = 10\,\%$ and $d_l = 50$ ms. From Figure 8.10, we can see that only Imm-ACK ($n = 1$) can work.

1. *The local information problem*: According to 802.15.3 specification, the MSDUs should be delivered to the upper layer in order. Since MSDU is composed of single or multiple MPDU(s), some MPDUs with higher ID correctly received must wait for the missing MPDUs that have lower ID and are incorrectly received because of transmission errors. In general, each MPDU has the retransmission limit r_l. If the retransmission limit has been reached, the source DEV will discard this MPDU. Then, it is unnecessary for those MPDUs in the reordering buffer to wait for this dropped MPDU any more. However, the destination DEV cannot know that the missing MPDU has been dropped by the source DEV. Thus, those buffered MPDUs may still wait and

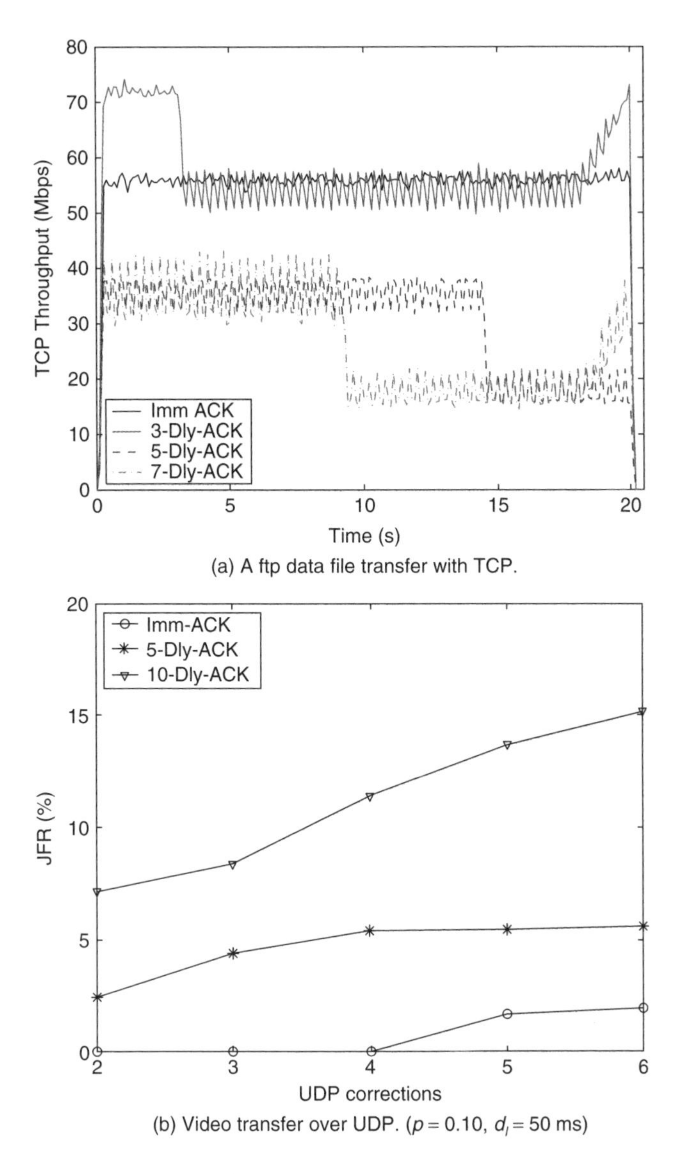

(a) A ftp data file transfer with TCP.

(b) Video transfer over UDP. ($p = 0.10$, $d_l = 50$ ms)

Figure 8.10 Problems with fixed Dly-ACK

cannot be delivered to upper layer timely. For UDP video streaming, this will finally result in an additional delay of the corresponding video frame, which may exceed the deadline miss and so the JFR performance will degrade. For TCP connections, one time-out event may occur at TCP layer of the source DEV and the throughput will be degraded significantly. For one MPDU, the probability that it can be dropped by the source DEV is p_l^r. Although the probability is very small, the effects of local information problem on performance are very serious. Therefore, a mechanism is needed to address this problem. This problem may be solved by adding a delay timer. But, to determine an appropriate size of the timer is not easy. For example, if the timer is too short for UDP video streaming, it may result in unnecessary discarding the video frames and increasing the JFR; on the other hand, if the timer is too long, the waiting MPDUs may exceed their deadlines. Similarly, for TCP, if the timer is set too short, it may result in unnecessary retransmission at TCP layer, but the missing MPDU may be received correctly in the next retransmission. On the other hand, if the timer is set too long, it may cause a time-out at TCP layer. Here, we implement a retransmission counter (Re-Counter) in the destination DEV to overcome this problem; that is, once an MPDU is received in error, the destination DEV will begin to count the retransmission time, which is actually the number of Dly-ACK frame used to acknowledge the incorrect reception of this MPDU. Once the Re-Counter exceeds the limit r_l, the waiting MPDUs in the reordering buffer no longer wait for the missing MPDU and should be delivered to upper layer immediately.

2. *The fixed burst-size problem*: Because of the burstness of traffic, the source DEV's MAC queue may be empty sometimes. When there are no MPDUs in the MAC queue, as compared to Imm-ACK, the fixed Dly-ACK scheme results in some serious problems. Following is the descriptions of these problems for UDP and TCP, respectively.

UDP: We use an example to illustrate these problems for UDP. At time t, there are 4 MPDUs belonging to one MSDU transmitted in one burst. Assuming that the last MPDU in this burst is with MPDU-ID m and the burst-size $n = 5$. After transmitting MPDU m, the source DEV queue is empty. If the MPDU m is incorrectly received because of error, it is unable to be retransmitted timely since the current Dly-ACK burst is not full, the source DEV has to wait for MPDU $m + 1$ before it requests a Dly-ACK from the destination DEV. However, the waiting time may be long and depends on the arrival of another MSDU or video frame. As a result, this may cause MPDU m and thus the whole MSDU to exceed its deadline limit.

TCP: For TCP, its packet transmission depends on the congestion window, and the MAC queue will be empty occasionally. In this case, when using Dly-ACK policy with fixed burst-size, there are two negative effects on the performance. First, it may introduce additional delay for the source MAC to wait for the ACK frame if error occurs. Second, the time-out event will occur at TCP layer. We use an example to explain these two problems. Assuming that the burst-size is fixed to be $n = 10$. For simplicity, in this example, we assume that one MSDU or a TCP/IP packet corresponds to only one MPDU, but the extension to the case of multiple MPDUs is straightforward. At time t, the congestion window at TCP is 16. Thus, there are 16 outgoing MPDUs in source DEV's MAC queue. Afterward, TCP will pass nothing to the MAC layer until it receives some TCP layer acknowledgements for the 16 TCP packets. Let us see what happens at the MAC layer. The 16 MPDUs will be transmitted in two bursts. In the first burst, MPDUs 1-10 are transmitted. Then, there are two possibilities. The first possibility is if there is no error for MPDUs 1–10. The MPDUs 11–20 will be passed in the second burst. Note that there are no MPDUs in source DEV's MAC buffer after MPDU 16 is transmitted. Moreover, after MPDUs 11–16 have been transmitted, the source TCP just receives ACK frames of MSDUs 1–10 and injects MSDUs 17–20 into MAC layer. Only when MPDUs 11–20 are received at destination, the Dly-ACK frame will be sent to source MAC. Therefore, for MPDUs 11–16, they need longer time to obtain Dly-ACK frame than

MPDUs 1–10. This will cause additional delay if retransmission is needed in MPDUs 11–16. The second possibility is if there are some errors in the first burst. For example, MPDU 1 is in error. Then, the destination MAC does not pass any frames to upper layer due to missing MPDU 1, even if MPDUs 2–10 are correctly received. Now, in the source side, MSDUs 17–20 cannot be injected to MAC layer because no TCP acknowledge packets are received at source TCP. Then, the source MAC will send MPDUs 1, 11–16 at the second burst. Unfortunately, MPDU 1 is in error again. Then, after MPDU 16 is transmitted, the source MAC will wait for MPDU 17–19 to fill this burst before requesting the Dly-ACK. However, the destination MAC does not generate the Dly-ACK frame because it has only received seven MPDUs for the second burst. Further, since MPDU 1 is in error, no MSDU is delivered to TCP layer and thus, no TCP acknowledge packets will be sent to the source. Then, there is a deadlock between the source and the destination. For the source, it needs some TCP acknowledge packets to generate more MSDUs filling the second burst so that the destination can generate a Dly-ACK frame. For the destination, MSDUs cannot be delivered to upper layers. This will not be solved until a time-out is triggered at source TCP.

These serious problems can be solved if the source MAC requests the Dly-ACK frame whenever the source MAC queue is empty. We use the same example to explain this scheme. For UDP, if the source DEV requests the Dly-ACK frame when transmitting MPDU m, MPDU m could be retransmitted timely. For TCP, if the source DEV requests the Dly-ACK frame when it transmits MPDU 16, the two problems can be solved efficiently. First, it reduces the additional waiting time corresponding to the MPDUs 17–20. Second, the deadlock will not occur between the source and destination. We propose that the source should request the Dly-ACK frame once it detects that its buffer is empty.

8.4.2.2 Proposed Adaptive Dly-ACK

According to the previous discussion regarding the interaction between Dly-ACK and upper layer protocols (TCP, UDP), we summarize our adaptive Dly-ACK as follows [18]:

At source DEV's MAC:

1. Before sending an MPDU, it checks the traffic buffer. If the buffer is empty, it should set the Dly-ACK frame-requesting bit in the MAC header of the MPDU which is going to be transmitted. (Else, go to 2.)
2. It sends MPDUs by means of k-Dly-ACK policy. The parameter k is determined by $\min(n, b_l)$ where n is an integer and is within $[6, 11]$ and b_l is the destination's remaining reordering buffer size and is obtained from the Max Burst filed of the latest Dly-ACK frame sent from the destination DEV. If the destination DEV's MAC buffer is large enough, the source DEV will request a Dly-ACK regularly for every n frames. We will show the effect of n on both the UDP and TCP performance in Section 8.4.

At destination DEV's MAC:

1. Deliver correctly received and concatenated MSDUs to the upper layer in order.
2. If one MPDU is missing due to channel error, those correctly received MPDUs with higher ID should wait for the missing frame before its Re-Counter exceeds the retransmission limit r_l.

3. Once the missing MPDU's Re-Counter exceeds r_l, the other MPDUs and the corresponding MSDUs in the reordering buffer waiting for this MPDU should be delivered to the upper layer if possible according to their ID order.
4. When receiving the Dly-ACK requesting information, the destination responds a Dly-ACK frame by setting the 'Max Burst' field in the MAC header to be the value of the remaining buffer size.

8.4.2.3 Analysis of Average Burst-Size

In this subsection, we give an approximate estimate of the average burst-size when using our proposed adaptive Dly-ACK scheme. For simplicity, only the case of UDP video streaming is studied here.

Assume that the maximum burst-size is n. Then, the total number of MPDUs, S, for a video frame including those incorrectly received and thus retransmitted MPDUs in the regular n-Dly-ACK is given by

$$S = nT + m, \tag{3}$$

where $\mathrm{T} = \left\lfloor \frac{S}{n} \right\rfloor$ with $\lfloor .x \rfloor$ denoting the largest integer below x and m is the remainder of S. For simplicity, it is reasonable to assume that m is uniformly distributed in $[1, n-1]$. Therefore, before sending the last Dly-ACK burst for the last m MPDUs, the previous Dly-ACK burst-size is always n.

In our adaptive Dly-ACK scheme, the source requests a Dly-ACK frame when the queue is empty. Thus, the subsequent transmissions after the final n-Dly-ACK burst will be an m-Dly-ACK burst. Since for the m MPDUs' transmissions, there are possibly m_1 MPDUs in error, thus, m_1 MPDUs will be retransmitted after the m-Dly-ACK. Again, it is possible that for this m_1-Dly-ACK transmission burst, there are m_2 errors MPDUs, and so on. The burst-size of the Dly-ACK is shown in the following figure for current video frame.

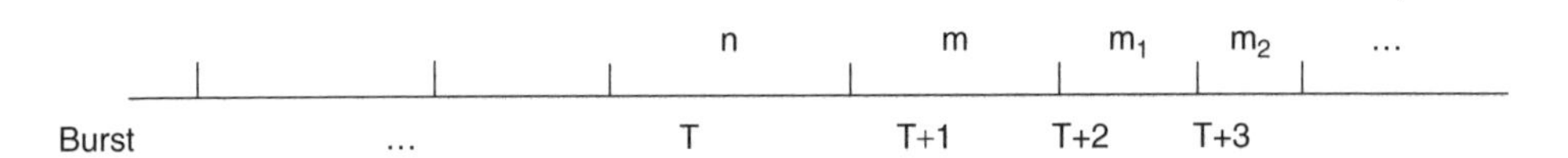

For simplicity of our following derivation, we assume that after at most 2 retransmissions, the MPDUs will be all correctly received. However, the derivations can be extended to more retransmissions in a straightforward manner. Note that the probability of there being m_1 error MPDUs out of m MPDUs is given by:

$$q_{m_1} = \binom{m}{m1} p^{m_1}(1-p)^{m-m_1}. \tag{4}$$

Then, in the next Dly-ACK burst-size, under the condition that there are m_1 MPDUs to be retransmitted, the probability that m_2 MPDUs are in error again is given by

$$q_{m_2|m_1} = \binom{m1}{m2} p^{m_1}(1-p)^{m_1-m_2}. \tag{5}$$

Now, the total number of the burst is given by

$$B(m) = (1-p)^m(\bar{T}+1) + \sum_{m_1=1}^{m} q_{m_1}(1-p)^{m_1}(\bar{T}+2) + \left(\sum_{m_1=1}^{m} q_{m_1}(1-(1-p)^{m_1})\right)(\bar{T}+3), \quad (6)$$

where $B(m)$ is already averaged over the video frame size and $\bar{T} = E[T]$; also note that the second term is the probability of error-free transmission for the m_1-Dly-ACK MPDUs and the final term is the probability that at least one error occurs during the m_1-Dly-ACK and thus another round of m_2-Dly-ACK is needed to finish the current video frame.

On the other hand, the total number of transmitted MPDUs for current video frame is given by

$$M(m) = n\bar{T} + m + \sum_{m_1=0}^{m} \sum_{m_2=0}^{m_1} q_{m_1} q_{m_2|m_1}(m_1+m_2), \quad (7)$$

where the final term is the weighted summary of the m_1-Dly-ACK and m_2-Dly-ACK and it is also averaged over the video frame size. Finally, the average burst-size k is given by

$$E(k) = E\left[\frac{M(m)}{B(m)}\right]. \quad (8)$$

In the above equation, the expectation is with respect to m.

8.4.3 Simulation Results

First, we compare the performance of our proposed adaptive Dly-ACK with the fixed Dly-ACK for UDP video streaming. In Figure 8.11, we show the results of JFR when multiple video streamings are presented in the system. The destination's MAC reordering buffer size is 400 kb, $n = 10$ and $p = 10\,\%$. The deadline is set to be 45 ms. As shown in Figure 8.11, our scheme can improve the performance significantly. In all cases, the JFR is reduced more than 10 times.

Figure 8.12(a) illustrates the aggregate throughput at TCP layer under various loss probability p. In these simulations, TCP Reno is used and the destination's MAC reordering buffer size is 400 kb. Further, each TCP frame size is set to 1000 bytes, that is, one MSDU/MPDU corresponds to one TCP packet. The parameter n is set to 10 and there are five TCP flows. We can see that in all cases, the throughput of our proposed Dly-ACK is much larger than that of the Imm-ACK. In Figure 8.12(a), it is observed that the throughput gain is more than 40 %. For the fixed 5-Dly-ACK and 10-Dly-ACK, their performances are good at low FER, however, they degrade quickly when the FER increases.

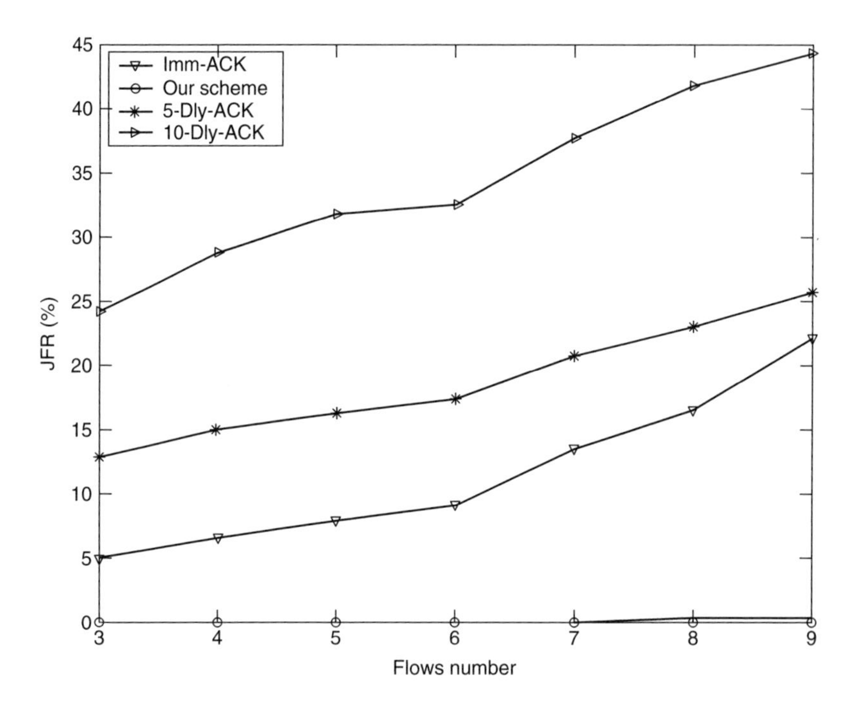

Figure 8.11 Performance enhancement for UDP flow, $d_l = 45$ ms, $p = 0.1$

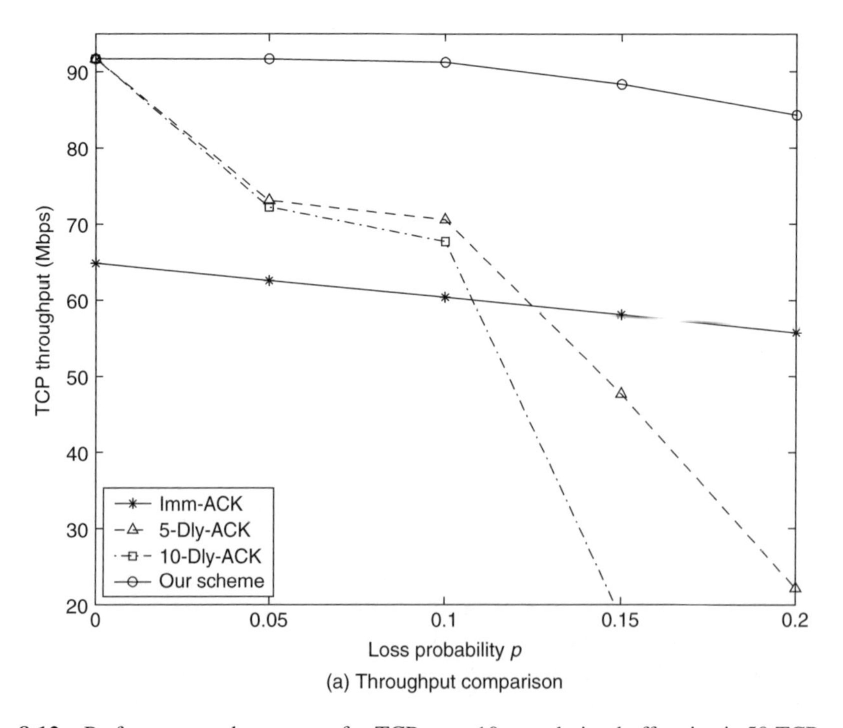

Figure 8.12 Performance enhancement for TCP, $n = 10$, reordering buffer size is 50 TCP packets

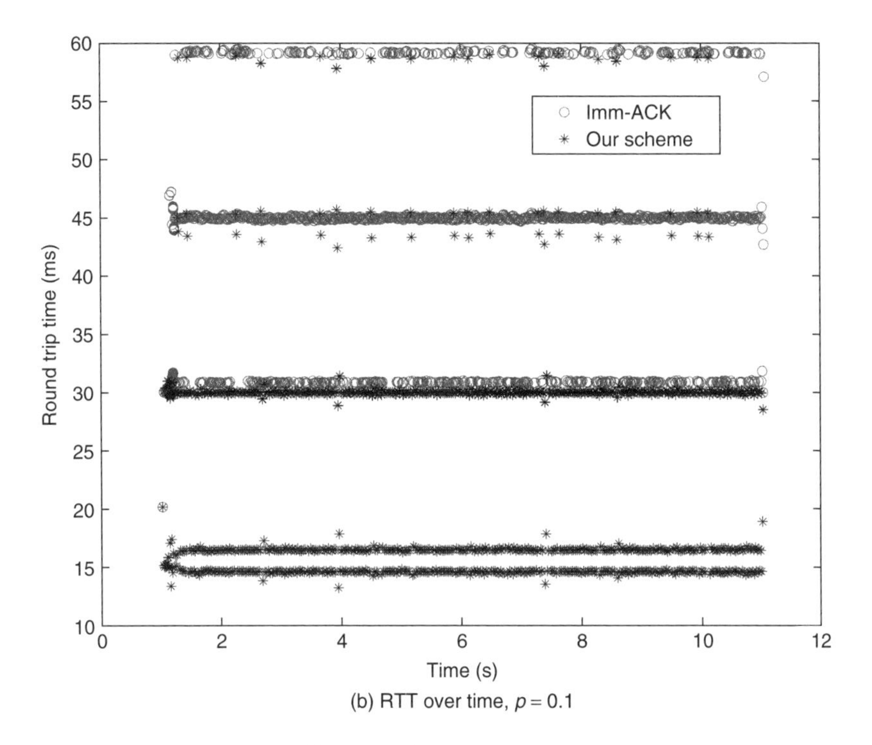

(b) RTT over time, $p = 0.1$

Figure 8.12 (*Continued*)

In general, the throughput of a single TCP flow is related to both the round-trip time (RTT) and the end-to-end loss probability r [19, 20]. That is, smaller the RTT, larger the throughput. Compared with Imm-ACK, the throughput gain of Dly-ACK scheme is because of the decrease of RTT and the increase of the channel utilization. In Figure 8.12(b), the RTT of one TCP flow is shown with $p = 0.1$. From this figure, it is clear that the RTT can be greatly reduced by our scheme. In fact, compared to Imm-ACK, because the number of ACK frames and the amount of IFSs between frames are reduced when using the Dly-ACK scheme, each MPDU can be delivered to the destination more quickly. Thus, the TCP acknowledgement from the destination can be delivered more quickly to the source's TCP layer. Finally, the TCP RTT can be reduced significantly.

We also investigate the effects of fragmentation on TCP performance when using our Dly-ACK scheme. These results are illustrated in Figure 8.13, where $p = 0.1$, $n = 10$ and the number of MPDUs for each TCP frame, t_f, varies from 1 to 8. Again, we can see that the proposed scheme can outperform the Imm-ACK significantly. In this simulation, the maximum congestion window value of TCP is set to 64. Thus, the channel is actually underutilized. The larger the TCP packet, the larger the input traffic load. Thus, the throughput becomes larger with the increase of TCP packet size in both Imm-ACK and Dly-ACK.

In the following, let us discuss the impacts of some parameters in our Dly-ACK scheme on the system performance. As mentioned before, the source sends a Dly-ACK request after k frames with $k = \min(n, b_l)$. Obviously, the parameter n plays an important role

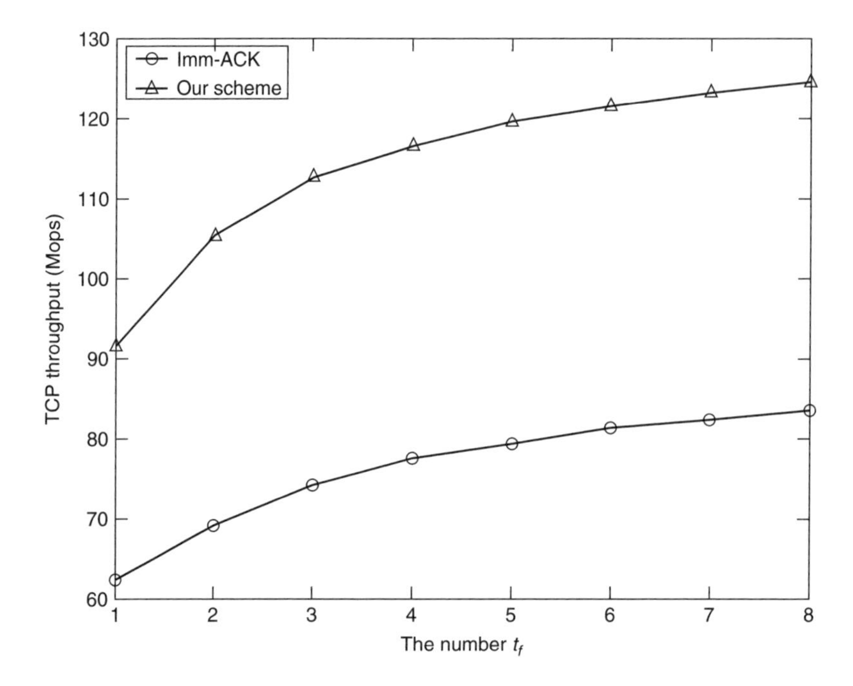

Figure 8.13 Throughput enhancement under large TCP frame

on the performance. The results of throughput versus n are shown in Figure 8.14(a). In this figure, to show the effect of n more clearly, the destination's buffer is set to be large enough to accommodate all correctly received frames. We list the results when $p = 0.05$ and 0.1. We can see that in the two scenarios, the throughput becomes larger with the increase of n when n is lower than 8. However, when n is larger than 8, the throughput is almost fixed. Although the performance may become a little better when n is larger than 10 with $p = 0.05$ or 0.1, a larger reordering buffer has to be used in the destination MAC. In Figure 8.14(b), we present the number of MPDUs in the destination MAC's reordering buffer with $n = 10$ and 20 under $p = 0.1$. This figure is zoomed-out and is entirely a piece of simulation. From this figure, we can see that the queue length is seldom larger than 20 when $n = 10$. However, when $n = 20$, the peak is as large as 50. Since the performance is not improved significantly and there is need for a larger reordering buffer when $n > 10$, our suggestion is that n may be approximately 5–10 and a larger value is unnecessary. A similar conclusion can be drawn from Figure 8.14(c), which illustrates the effects of n on UDP video streaming. In this figure, there are 10 UDP connections and the deadline of each video frame is set to 35 ms. When $n > 10$, it has little effect on JFR.

Finally, we are to investigate the effect of the destination MAC's reordering of buffer size on the TCP performance. As mentioned earlier, larger the n, larger the buffer size that is required to fully accommodate the received MPDU before they are delivered to the upper layer. However, it is interesting to see, for a given n, the effect of the buffer size to the performance. Figure 8.15 shows the TCP throughput as a function of the buffer size under various loss probability p with $n = 10$. When the loss probability is low ($p = 0.05$), the system throughput is almost constant regardless of the buffer size.

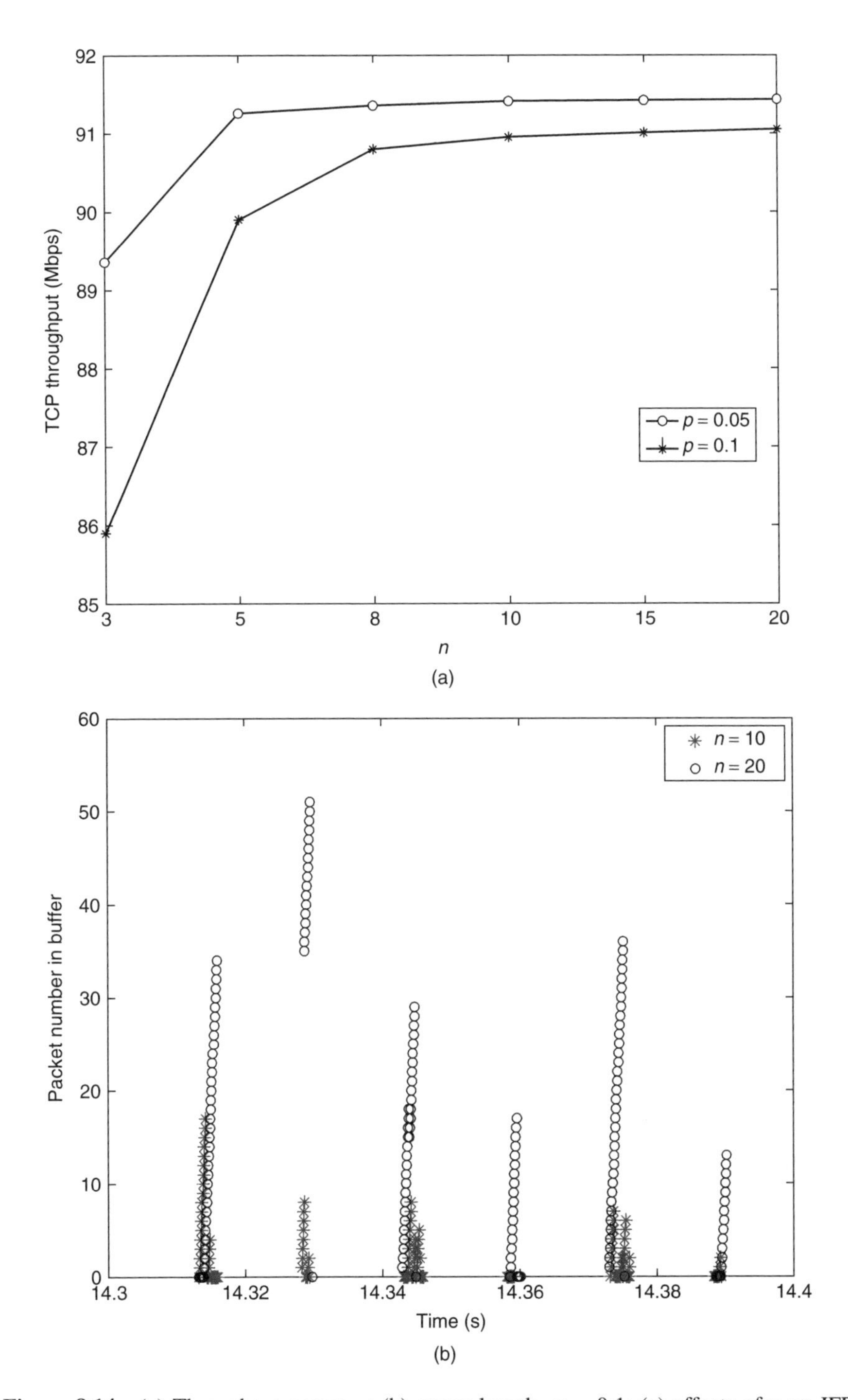

Figure 8.14 (a) Throughput versus n; (b) queue length, $p = 0.1$; (c) effects of n on JFR

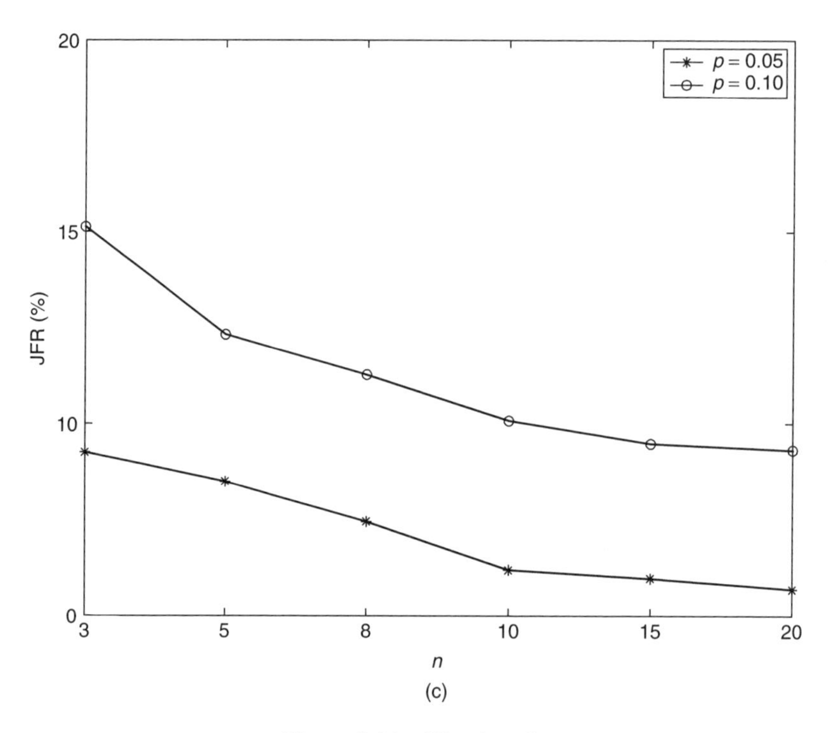

Figure 8.14 *(Continued)*

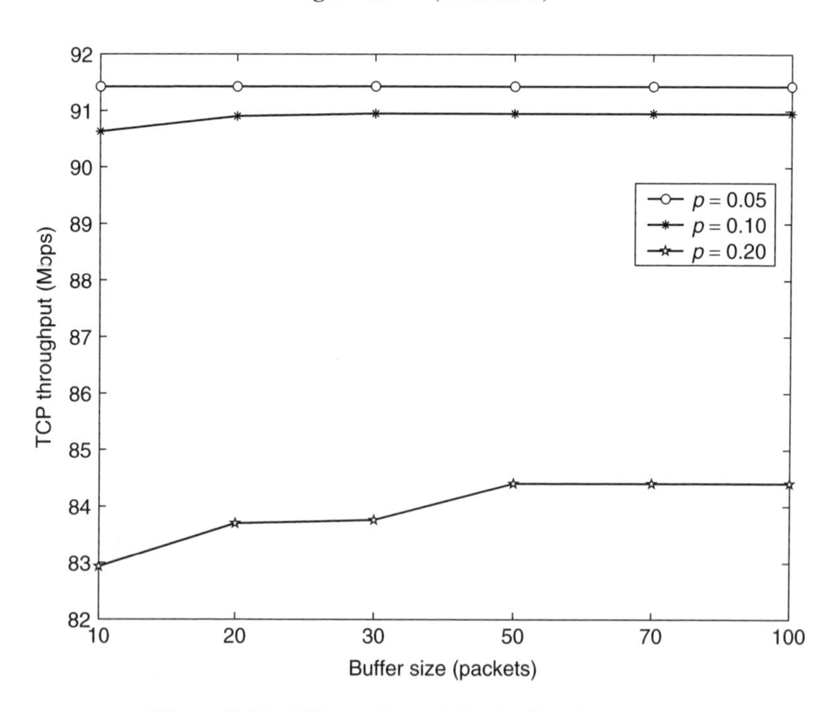

Figure 8.15 Effects of reordering buffer size, $n = 10$

Table 8.4 The distribution of burst-size $k(p = 0.2, n = 10)$

Buffer size (frames)	$k \leq 5$	6	7	8	9	10	Total
10	3128	504	470	451	454	2497	7504
20	925	162	182	186	200	4651	6306
100	655	65	98	93	72	5098	6081

This can be easily explained because there seldom are MPDUs in error and the destination buffer is usually empty since each MPDU can be delivered from MAC layer to the upper layer immediately. However, when there is a high loss probability, for example, $p = 0.2$, the buffer size has a visible impact on the performance. This is because we set the burst-size $k = \min(n, b_l)$. As loss probability is high, there may be a lot of frames in the buffer that cannot be delivered to upper layer timely. Thus, there is little remaining space to accommodate more upcoming frames. Finally, this results in the reduction of k. Therefore, the performance degrades. To gain further insight about this problem, we list the distribution of burst-size k in Table 8.4. The element in the table indicates the number of corresponding burst-size under different buffer size. These results are collected from a simulation that runs within a duration of 20 s under high loss probability. Obviously, k has a higher average value when the buffer size is larger, which indicates a better channel utilization. Further, the total number of transmission burst (i.e., the count of Dly-ACK frames) in the entire simulation duration becomes smaller with larger buffer size.

8.5 Ad Hoc Networks

There are two possible approaches in the IEEE 802.15.3 to support ad hoc networks. One is the dependent piconet, for example, child piconet and the other is the independent piconet or simultaneous operating piconet (SOP), although the latter is not clearly defined in the standard.

8.5.1 Child Piconet

The child piconet is created when one of the members, PNC capable DEV in the current piconet, wants to do so. It will become a child PNC and request a pseudostatic private CTA from the parent PNC. This duration of CTA is used by all the members of the formed child piconet and the assignment is left for the child PNC. Therefore, from the point of view of both the child piconet and the parent piconet, their superframe sizes are the same. However, both of them only have a portion of available time slot within the superframe, that can be used by their DEVs. Figure 8.16 shows the superframe relationship of the parent piconet and child piconet.

Note that the child PNC is a member of the parent PNC. Thus, it can exchange data with any of the devices in the parent piconet. On the other hand, the standard does not define the direct communication between child piconet member and the parent piconet member. Thus, the routing in the ad hoc network formed with child piconet is quite straightforward. That is, the child DEV can only communicate with the child PNC and

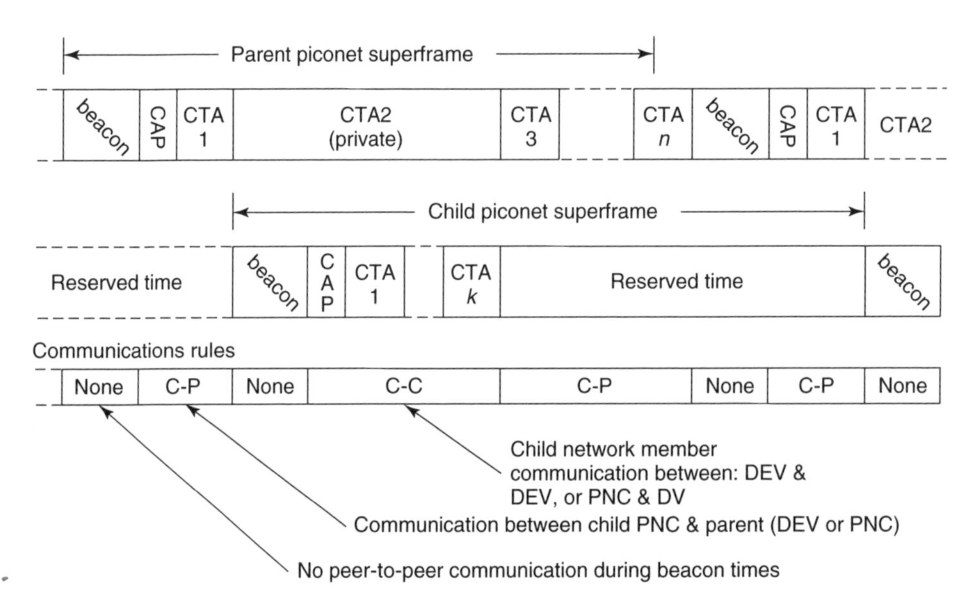

Figure 8.16 Superframe structure of parent piconet and child piconet

the child PNC will forward the packet to parent DEV/PNC if data needs to be exchanged between child DEV and outside.

The 802.15.3 standard also defines another dependent piconet called *neighbor* piconet. However, in contrast to the child PNC, the neighbor PNC is not a member of the parent piconet (and the parent PNC is not a member of the neighbor piconet). Therefore, the neighbor PNC cannot forward the packet for its member to the parent piconet. Therefore, these two piconets cannot talk to each other if no other approach is taken. Figure 8.17 shows the superframe relation between parent piconet and neighbor piconet. The difference between Figure 8.16 and Figure 8.17 is obvious.

8.5.2 Independent Piconets

In child piconet-based ad hoc network, since the child piconet needs to share the superframe resource with the parent piconet, the spectrum efficiency is rather low. As a result, the independent piconets with some bridging DEVs between piconets will form a more efficient ad hoc network. The bridging node is the device that belongs to two or more piconets simultaneously. Note that the major difference between 802.15.3 based ad hoc network and the 802.11 based ad hoc network is that the latter is a flat topology where all nodes are equal while the former is a locally structured topology. Also, note that the devices in an 802.15.3 piconet can communicate in a peer-to-peer mode. Taking into account the difference between these two networks, the research to be conducted in the 802.15.3 based ad hoc networks should include the following issues:

- Piconet formation
- Interpiconet scheduling
- Joint scheduling and routing

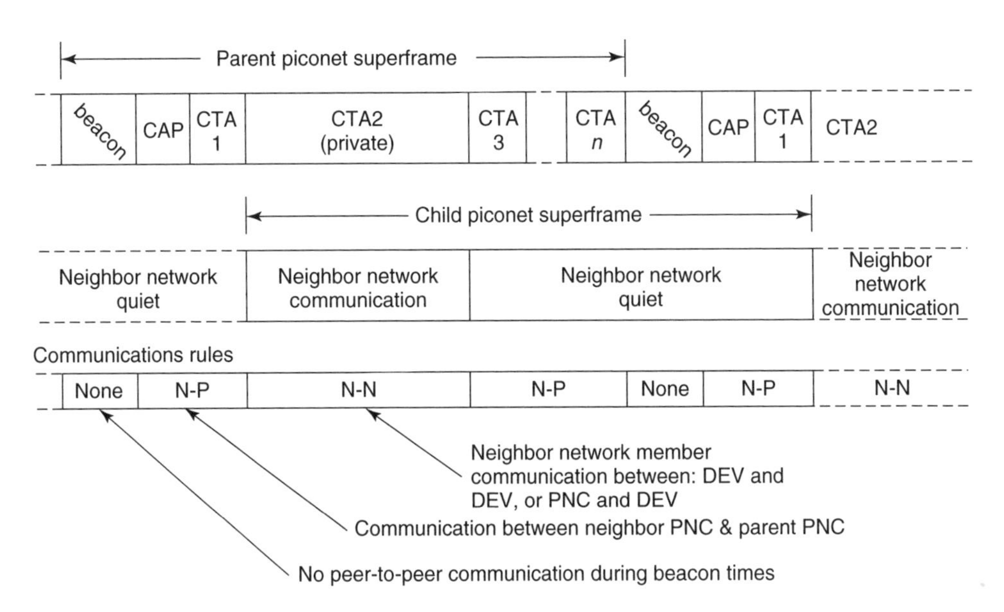

Figure 8.17 Superframe structure of parent piconet and neighbor piconet

- Latency reduction in route discovery and rediscovery
- Piconet-aware routing scheme.

The piconet formation is to generate the piconet topology for a group of devices. Different objectives may be considered during the formation, for example, power, throughput, latency, and so on. The ad hoc network with piconet structure in 802.15.3 bears some similarity to Bluetooth network [21]. There has been much work regarding the Bluetooth piconet formation algorithm, such as BlueStar and BlueTrees [22–25]. However, these algorithms may not be applied to 802.15.3 directly because of the following reasons:

- The Bluetooth piconet formation usually has a restriction on slave/member numbers.
- The Bluetooth piconet usually has a tree structure.
- They usually require the in-range of all devices.
- They seldom consider the multirate feature.
- They seldom consider the multichannel feature.

However, the piconet formation work for 802.15.3 in literature is currently very limited. In [26], Gong *et al.* propose a formation algorithm that mainly considers the power and interference. The basic idea is that the transmission power of the PNC should be minimized to reduce the interference generated by the piconet and an interference factor is defined for each device. The procedure is as follows:

1. Device discovers each other via broadcasting with a predefined power level.
2. Device broadcasts its information to all other devices, including interference factor, ID, neighbor lists, and so on. Therefore, each device holds a table that contains the global information.

3. With the table, each device does the following:
   (a) Find the node with the minimum interference factor and mark it as a PNC.
   (b) Delete the PNC's neighbors from the table.
   (c) Stop the iteration if no PNC is left.
4. Nodes marked as PNC are now the self-claimed PNCs and send beacons.
5. The remaining devices select a PNC closest to it and join the piconet.
6. A device that cannot find any neighbors at the predefined power level increases its transmission power until it can find at least one neighbor. Then, it picks a neighbor that is closest to it to be its PNC.

A more interesting problem is that when we know the traffic pattern, how can we form the piconet so that the resultant end-to-end throughput will be maximized? This is the joint optimization of the piconet formation and routing. Because of the property of 802.15.3 network, there are several new challenges in the optimization. For example, when a device S wants to communicate with device V, the route may involve devices A, B, C, D, E, and so on, as shown in Figure 8.18. Then, the problem is how much time the bridging DEV A should stay in piconet 1 and how much time in piconet 2. On the other hand, for the bridging link DE, how much time should it be in piconet 2 and how much time in piconet 4, because within piconet 2, link DE will conflict with link BC.

In an ad hoc network, the beacon alignment between piconets will be important because the beacons from different PNCs may interfere with each other in the bridging device. MeshDynamics (http://www.meshdynamics.com) has proposed the 'heartbeats' technique to handle this [27]. The heartbeats are essentially retransmissions sent by devices while PNCs send beacons. In addition, the heartbeats-based mesh routing technique addresses the complexity of dynamic environments by modeling them as free market enterprises. Thus, the QoS in the mesh is guaranteed by the concept of 'toll cost' and hop cost [27]. The basic idea of the 'toll cost' is that when a device has too much traffic to forward, and in some cases, for example, when it wants to reduce the traffic or it has its own traffic to send, it may raise the 'price' to forward data. With this approach, the system automatically adjusts its routing paths and maintains QoS despite local congestion. The high priority traffic gets preferred treatment because they are willing to pay for a higher hop cost. While the low priority traffic will go through less direct routes, thus containing potential congestion at popular nodes. These schemes are implemented in the distributed

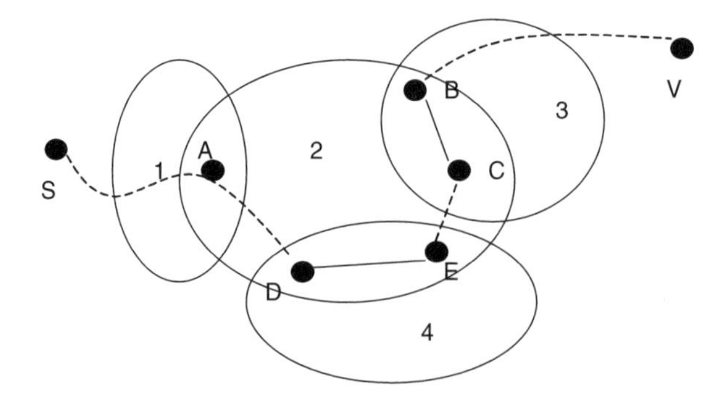

Figure 8.18 Illustration of joint piconet formation and routing

control layer above MAC, which includes the mesh control layer and the MAC-mesh interface.

8.6 Summary

The MAC of UWB system affects the UWB system performance a lot. The DS-UWB utilizes the 802.15.3 as the MAC support; while MBOA UWB camp is defining a new MAC. In this chapter, we mainly focus on the 802.15.3 MAC. In particular, the QoS scheduling algorithm in PNC is presented in Section 8.2. It is shown that by adding the relative deadline information into the scheduler, the JFR performance can be improved significantly. In Section 8.3, the intrasuperframe power management problem is studied. It is transformed to a graph theory problem and a suboptimal algorithm is proposed to find the flows' order. It is shown that this algorithm can achieve the optimal solution in most of the cases. The adaptive Dly-ACK scheme is investigated in Section 8.4. The problem of the fixed Dly-ACK is first presented and then an adaptive scheme is proposed to achieve a much better performance for both TCP and UDP. Finally, the ad hoc network based on 802.15.3 MAC is studied in Section 8.5. Both the dependent and independent piconets structures are discussed. On the basis of the unique features of the piconets, some research topics in the ad hoc networking for UWB are presented.

References

[1] DS-UWB Physical Layer Submission to 802.15 Task Group 3a, IEEE P802.15-04/0137r1, Mar. 2004.

[2] Multi-band OFDM Physical Layer Proposal Layer Proposal for IEEE 802.15 Task Group 3a, IEEE P802.15-03/268r4, Mar. 2004.

[3] IEEE Standard for Information Technology-Telecommunications and Information Exchange between Systems-Local and Metropolitan Area Networks-Specific Requirements-Part 15.3: Wireless Medium Access Control (MAC) and Physical Layer (PHY) Specifications for High Rate Wireless Personal Area Networks (WPANs), Sept. 2003.

[4] MBOA Wireless MAC Specification for High Rate Wireless Personal Area Networks (WPANs), Version 0.65, Apr. 2004.

[5] H. Zhang, Service Disciplines for Guaranteed Performance Service in Packet-Switching Networks, *Proce. IEEE*, vol. 83, no. 10, pp. 1374–1396, Oct. 1995.

[6] Y. Kim, W. Yoon, S. Hyong Rhee, D. Won Kwak, and B. Joo Lee, "Samsung MAC enhancement contribution for IEEE 802.15 Task Group 3a," IEEE 802.15.3 WG, doc.: IEEE 802.15-03/212r1, May 2003.

[7] F. H. P. Fitzek and M. Reisslein, "MPEG-4 and H.263 Video Traces for Network Performance Evaluation," *IEEE Netw.*, vol. 15, no. 6, pp. 40–54, Dec. 2001. http://www-tkn.ee.tu-berlin.de/~fitzek/TRACE/trace.html.

[8] R. Mangharam and M. Demirhan, "Performance and simulation analysis of 802.15.3 QoS," IEEE 802.15.3 WG, doc.: IEEE 802.15-02/297r1, Jul. 2002.

[9] L. E. Schrage and L. W. Miller, "The Queue M/G/1 with the Shortest Remaining Processing Time Discipline," *Oper. Res.*, vol. 14, pp. 670–684, 1966.

[10] Network Simulator 2 (NS2), http://www.isi.edu/nsnam/ns, accessed on 2003.

[11] NS2 with 802.15.3 Module, Originally at http://www.winlab.rutgers.edu/~demirhan/research/.

[12] J. Chen, K. Sivalingam, P. Agarwal, and S. Kishore, "A Comparison of MAC Protocol for Wireless Local Networks Based on Battery Power Consumption," *Proc. IEEE Infocom'98*, San Francisco, CA, pp. 150–157, Mar. 1998.

[13] H. Woesner, J. Ebert, M. Schlager, and A. Wolisz, "Power-Saving Mechanism in Emerging Standards for Wireless LANs: The MAC Level Perspective," *IEEE Personal Communications*, ICC'01: Helsinki, Finland, pp. 40–48, Jun. 1998.

[14] J. Stine and G. Veciana, Improving Energy Efficiency of Centrally Controlled Wireless Data Networks, *Wireless Netw.*, vol. 8, no. 6, pp. 681–700, Nov. 2002.

[15] D. Li, P. Chou, and N. Bagherzadeh, "Mode Selection and Mode-Dependency Modeling for Power-Aware Embedded Systems," *Proc. IEEE International Conference on VLSI Design*, Bangalore, India, 2002.

[16] R. A. Wilson, *Graphs, Colourings, and the Four-colour Theorem*, Oxford University Press, 2002.

[17] Z. Guo, R. Yao, W. Zhu, X. Wang, and Y. Ren, "Intra-Superframe Power Management for IEEE 802.15.3," *IEEE Communications Letters*, vol. 9, no. 3, 2005.

[18] H. Chen, Z. Guo, R. Yao, and Y. Li, "Adaptive Dly-ACK for TCP Over 802.15.3 WPAN", Proc. IEEE Globecom'04, Dallas, TX, 2004.

[19] X. George, C. P. George, M. Petri, and S. Mika, "TCP Performance Issues Over Wireless Links," *IEEE Commun. Magn.*, vol. 39, no. 4, pp. 52–58, Apr. 2001.

[20] M. Mathis, J. Semke, J. Mahdavi, and T. Ott, "The Macroscopic Behavior of the TCP Congestion Avoidance Algorithm," *ACM Comput. Commu. Rev.*, vol. 27, no. 3, pp. 67–82, Jul. 1997.

[21] IEEE Std 802.15.1–2002, IEEE Std 802.15.1, IEEE Standard for Information technology- Telecommunications and information exchange between systems- Local and metropolitan area networks- Specific requirements Part 15.1: Wireless Medium Access Control (MAC) and Physical Layer (PHY) Specifications for Wireless Personal Area Networks (WPANs), 2002.

[22] C. Petrioli, S. Basagni, and I. Chlamtac, "Configuring BlueStars: Multihop Scatternet Formation for Bluetooth Networks," *IEEE Trans. Comput.*, vol. 52, no. 6, pp. 779–790, Jun. 2003.

[23] G. Zaruba, S. Basagni, and I. Chlamtac, "Bluetrees- Scatternet Formation to Enable Bluetooth-Based Personal Area Networks," *IEEE Conference on Ultra Wideband Systems and Technologies*, Reston, VA, 2003.

[24] T. Salonidis, P. Bhagwat, L. Tassiulas, and R. Lamaire, "Distributed Topology Construction of Bluetooth Personal Area Networks," *Proc. IEEE Infocom'01*, Alaska, USA.

[25] S. Basagni, M. Conti, S. Giordano, and I. Stojmenovic, *Mobile Ad Hoc Networking*, IEEE Press, 2004.

[26] M. Gong, S. Midkiff, and R. Buchrer, "A New Piconet Formation Algorithm for UWB Ad Hoc Networks," *Proc. IEEE Conference on Ultra Wideband Systems and Technologies*, 2003, pp. 180–184.

[27] White paper, Meshdynamics, http://www.meshdynamics.com

9

Radio Resource Management for Ultra-wideband Communications

Xuemin (Sherman) Shen, Weihua Zhuang, Hai Jiang and Jun Cai

9.1 Introduction

Ultra-wideband (UWB) transmission is an emerging technology for future wireless communications, although the basic idea can be tracked back to the first wireless communication system in the late 1890s [1, 2]. Similar to spread-spectrum or code division multiple access (CDMA), UWB technology was first used in a military environment and then introduced in the commercial market recently. Its applications have been stimulated by the Federal Communications Commission (FCC) *Notice and Inquiry* in 1998 and the FCC *Report and Order* in 2002. Today, UWB has been considered as one of the most promising candidates for both indoor and outdoor wireless communications within a short range and has been attracting more and more attentions from the research community.

Currently, a UWB system is defined as one having a -10 dB fractional bandwidth of at least 0.20 or a -10 dB bandwidth of at least 500 MHz. The FCC has allowed unlicensed use of UWB devices in the 3.1–10.6 GHz frequency band [3]. At the physical layer, the implementation of a UWB system can be achieved by using a pulse-based approach or a multiband orthogonal frequency division multiplexing (MB-OFDM) based approach. In a pulsed UWB system, pulses of an extremely short duration, typically in the order of a nanosecond, are used for information transmission; whereas in the MB-OFDM, hybrid frequency hopping and OFDM are applied. Each of the two leading UWB technologies has its pros and cons for communications in a multipath propagation environment. For pulse-based UWB, benefiting from a simple transmitter and rich resolvable multipath components, the receiver can exploit multipath diversity effectively, while MB-OFDM offers robustness to narrowband interference, spectral flexibility, and efficiency.

Ultra-wideband Wireless Communications and Networks
Edited by Xuemin (Sherman) Shen, Mohsen Guizani, Robert Caiming Qiu and Tho Le-Ngoc

However, pulse-based UWB needs a long channel acquisition time and requires high-speed analog-to-digital converters (ADCs) for signal processing, while MB-OFDM requires a slightly complex transmitter. The large bandwidth and low transmission power density (−41.25 dBm/MHz for indoor applications) make the UWB technology attractive for high-rate (>100 Mbps) short-range (<10 m) or low-rate (<a few Mbps) moderate/long-range (100–300 m) wireless communications [1, 4]. In addition to the traditional multimedia services such as voice/video conversations, video streaming, and high-rate data, UWB applications include industrial automation and control, medical monitoring, home networking, gaming, imaging, vehicular radar systems, Department of Defense (DoD) systems, and so on. As shown in Figure 9.1, a typical UWB network can be constructed to provide peer-to-peer connections among mobile nodes; via an access point (AP) and a gateway, the mobile nodes can also be connected to the Internet backbone to set up a connection with a correspondence node. Each mobile node assumes a double role of terminal and router, and a mobile node is connected to the Internet probably via a multiple-hop link.

For UWB wireless networks, an efficient radio resource management mechanism is essential for bandwidth utilization and quality of service (QoS) provisioning. The unique UWB characteristics, such as large bandwidth, low transmission power, pulse transmission, precise positioning capacity, and long acquisition time, are all important factors and should be taken into account in the development of radio resource management. In this chapter, our focus is to provide a comprehensive overview on the state-of-the-art in pulse-based UWB radio resource management, and identify the challenges and further research issues in this area. The rest of the chapter is organized as follows. Section 9.2

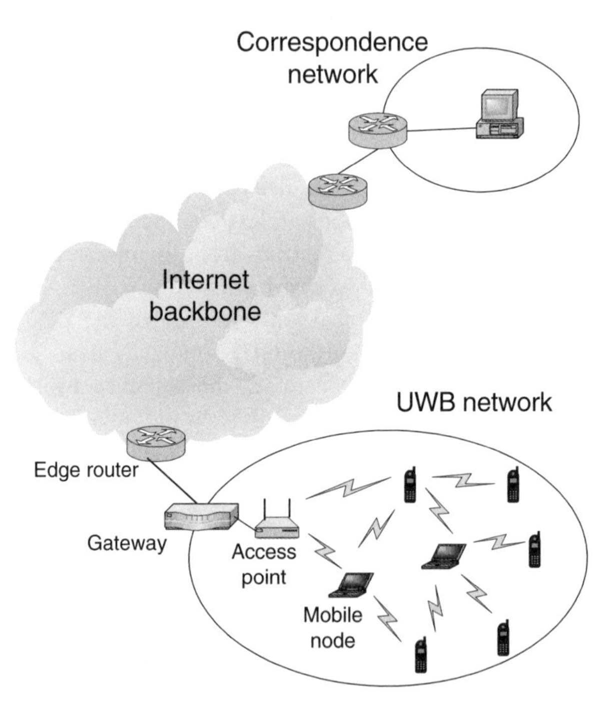

Figure 9.1 UWB network architecture

presents the challenges and opportunities of radio resource management in UWB wireless networks. The subsequent three sections are devoted to the three essential research topics of UWB radio resource management, multiple access, overhead reduction, and power/rate allocation. Concluding remarks are given in Section 9.6.

9.2 Radio Resource Management

In the past few years, first-generation multimedia capabilities have become available on portable PCs, reflecting the increasing mainstream role of multimedia in computer applications. As multimedia features continue their inevitable migration to portable devices such as laptop PCs, personal digital assistants (PDAs), and personal information assistants (PIAs), wireless extensions to wireline broadband networks will have to support an integrated mixture of multimedia traffic (such as voice, high-rate data, and streaming video) with guaranteed QoS. With the potential for high transmission rates, UWB systems are expected to provide multimedia services in a wide set of applications, from wireless personal area networks (WPANs) to wireless ad hoc networks.

Multimedia services can be of any nature, including the constant-rate traffic for uncompressed voice and video, variable-rate traffic for compressed voice and video, and available-rate traffic for data. As a result, effective and efficient resource management is essential to provide UWB multimedia services with QoS provisioning.

Differing from traditional wireless networks, UWB systems exhibit unique physical layer characteristics such as low power condition and precise positioning capability. Hence, novel radio resource management mechanisms should be explored for UWB communications. Because traditional OFDM-based resource management mechanisms can be smoothly applied to MB-OFDM UWB networks, our focus would be on pulse-based UWB networks.

9.2.1 Pulse-Based UWB Physical Layer Characteristics

Pulse-based UWB can be classified as pulse-based time-hopping (TH) UWB and pulse-based direct-sequence (DS) UWB. In both systems, information is transmitted by sending narrow time-domain pulses. Widely used modulation schemes include pulse amplitude modulation (PAM), on-off keying (OOK), and pulse position modulation (PPM). For a single-user system with binary signaling, if one pulse is used to represent one bit, the transmitted signal for these modulation schemes can be written in a general form as

$$s(t) = \sum_{n=-\infty}^{\infty} \sqrt{E_b} b_n^0 p\left(t - nT_b - \frac{\tau}{2}(1 - b_n^1)\right), \tag{1}$$

where E_b is the transmitted energy per bit, $p(t)$ is the UWB pulse, T_b denotes the bit interval, and b_n^0 and b_n^1 are related to information bits. For binary PPM signals, b_n^0 is set to 1, $b_n^1 \in \{-1, 1\}$, and τ is the time-shift relative to the time reference when a '-1' is transmitted. For binary PAM signaling, b_n^1 is set to 1 and $b_n^0 \in \{-1, 1\}$ carries information. For OOK signaling, $b_n^1 = 1$ and $b_n^0 \in \{0, 1\}$ carries information. To support multiuser communications, TH and DS spread-spectrum schemes are normally applied.

From (1), the transmitted TH signal of the ith user can be written in a general form as

$$s_i(t) = \sum_{n=-\infty}^{\infty} \sqrt{E_b} b_{i,n}^0 \sum_{j=0}^{N_s-1} p\left(t - nT_b - jT_f - h_{i,nN_s+j}T_c - \frac{\tau}{2}(1 - b_{i,n}^1)\right), \quad (2)$$

where N_s is the number of pulses used to represent one bit, T_f is the nominal pulse repetition interval, T_c is the chip (or pulse) duration, and $\{h_{i,n}\}$ is the pseudorandom hopping sequence of the ith user. For DS-UWB, the transmitted signal for the ith user is

$$s_i(t) = \sum_{n=-\infty}^{\infty} \sqrt{E_b} b_{i,n}^0 \sum_{k=0}^{N_c-1} a_{i,k} p\left(t - nT_b - kT_c - \frac{\tau}{2}(1 - b_{i,n}^1)\right), \quad (3)$$

where N_c is the number of chips used to represent one bit, $a_{i,k} \in \{-1, 1\}$ is the kth chip of the ith user's pseudorandom sequence [1].

The two main merits of pulse-based UWB are its robustness to multipath propagation and capability in user ranging/positioning. However, the very short pulse duration poses a significant challenge on synchronization, thus requiring a very long acquisition time. In addition, to suppress narrowband interference, notch filters should be employed. It is not beneficial because additional complexity is needed to compensate the distortion of the pulses caused by the notch filters [5].

The unique characteristics at the UWB physical layer provide challenges but they also provide opportunities for designing efficient radio resource management mechanisms.

9.2.2 Challenges and Opportunities

First of all, UWB networks have a very stringent transmission power constraint for coexistence with other narrowband networks. Very low transmission power is also important for non-cooperative UWB networks, which may operate simultaneously at a close range. The low power requirement puts significance on power control, while providing opportunities for supporting simultaneous transmissions as long as the communication pairs are adequately separated spatially.

UWB indoor networks can be designed to support very high transmission rate, for example, more than 100 Mbps. For such high transmission rates, any overhead time introduced by the physical/link layer may cost a large portion of system resources and significantly degrade the system performance in terms of throughput and efficiency.

Another critical issue is the acquisition in UWB transmissions [4], a process to synchronize the receiver's clock with the transmitter's clock to achieve bit synchronization. A long acquisition time is needed because of the highly precise synchronization requirement. To obtain acquisition in a UWB system, at the beginning of each transmission, the sender may send a preamble with duration varying from tens of microseconds to tens of milliseconds (depending on the receiver design) [6]. Apparently, in each UWB transmission, a large portion of the time will be used to perform the acquisition, thereby significantly degrading the efficiency of UWB transmissions, particularly for a very high-rate UWB system.

One of the major applications of UWB technology is in an ad hoc networking environment, which is characterized by distributed control functions and nonfixed infrastructure.

Since no fixed central controller exists in an ad hoc network, only local information is available for each node in the system, and some control mechanisms (such as power allocation) become more complicated. This should be taken into account in UWB radio resource management design.

On the other hand, UWB physical layer characteristics also provide new opportunities for designing effective and efficient radio resource management. For example, its large bandwidth and low transmission power allow the feasibility of exclusion region concept [7], its unique pulse transmission provides more flexibility in resource allocation, and its inherent capability in positioning simplifies routing and power control. Taking advantage of all these opportunities facilitates effective and efficient radio resource management.

In the following, we present the pulse-based UWB radio resource management solutions in the avenues of multiple access, overhead reduction, and power/rate allocation.

9.3 Multiple Access

For a UWB wireless network, multiple access is achieved by medium access control (MAC) mechanism, whose function is to coordinate the access among competing nodes in an orderly and efficient manner.

Similar to the traditional IEEE 802.11 wireless local area networks (WLANs), multiple access with a single channel can be achieved in UWB networks. In the *single channel* case, each node and its neighbors share the same channel. At the receiver, the received signals from multiple nodes may collide. However, because of the inherent spread-spectrum nature in UWB transmission, simultaneous transmissions can be supported by proper pseudorandom code design and call admission control (CAC), referred to as *multichannel* case. In this section, we place emphasis on the multichannel case, which is more relevant to UWB.

For multichannel multiple access, in the limit of infinite bandwidth ($W \to \infty$), the optimal scheme is to simply allow transmissions over all the links simultaneously, because interference becomes negligible [8]. However, for a practical UWB network, the bandwidth is large but finite, so uncontrolled simultaneous transmissions are not optimal [9, 10]. Hence, it is critical to determine when, where, and how to allow simultaneous transmissions, and how to alleviate the induced interference in order to achieve desired performance.

9.3.1 Exclusive versus Concurrent Transmissions

In the IEEE 802.11 MAC protocols, all transmissions are in the same channel. Hence, simultaneous transmissions in a nearby neighborhood collide with each other. One effective solution to eliminate or reduce the collision is to use temporally exclusive mechanisms, for example, collision resolution protocols [11, 12], time division–based schemes [13], or a combination of both [14]. However, when implemented in an ad hoc manner, these mechanisms may suffer from a large overhead due to control message exchanges and packet (or control message) collisions.

In fact, by properly managing interference, simultaneous transmissions can be allowed in wireless communication networks, especially in ad hoc networks. For example, in CDMA-based networks, simultaneous transmissions can be easily supported by using

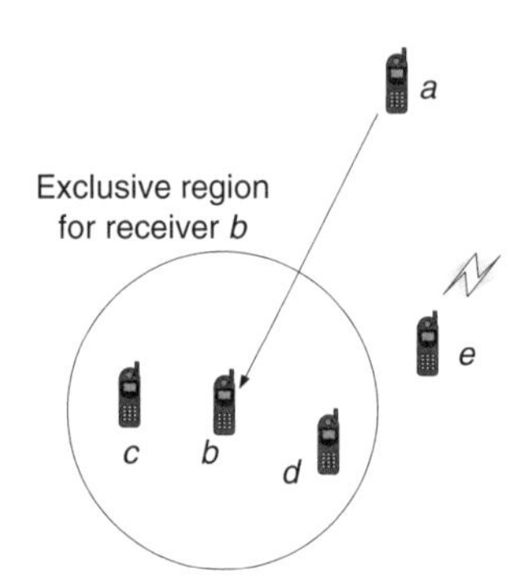

Figure 9.2 The exclusive region concept

power control in a cellular system, or by hybrid power control and some exclusion mechanism in an ad hoc network. This is particularly true for UWB networks with low transmission power, thus providing an efficient mechanism in UWB multiple access. Since transmission power in UWB networks is very low, two transmission pairs with a large separation in space will cause negligible interference to each other and therefore can work at the same time even when both of them use the same code channel. A concept, called *exclusive region* [7], is defined to clarify such large space separation. As shown in Figure 9.2, when transmitter a is sending to receiver b, the transmitters c and d (within the exclusive region) either keep silence or transmit with interference mitigation technique as discussed in Section 9.3.3 when the desired receiver b begins receiving information, while transmitter e (outside the exclusive region) is allowed to transmit at the same time. Note that the exclusive region is defined for receivers only. The exclusive region approach is optimal in terms of throughput to allow interfering sources to transmit simultaneously. Finding an optimal exclusive region is a challenge that should be addressed. A small exclusive region allows more simultaneous transmissions in the desired receiver's neighborhood but may result in a large transmission error probability due to large interference, while a large exclusive region improves the transmission accuracy but may lead to inefficiency due to a less extent of frequency reuse. Although some preliminary research work has been done in this area [7], it is still an open issue to obtain optimal exclusive region. In addition, joint power allocation and exclusive region determination can lead to high bandwidth efficiency and low power consumption [15].

9.3.2 Code Assignment

Both DS-UWB and TH-UWB have the potential to support concurrent transmissions. For DS-UWB, the transmission of each link is spread by a pseudorandom sequence, and the receiver despreads the received signal and recovers the original information. The spreading allows several independently coded signals to be transmitted simultaneously over the same frequency band. For TH-UWB, each link transmits one pulse per frame based on a distinct pulse-shift pattern called a time-hopping sequence (THS). Multiple access can be achieved if each link uses an independent pseudorandom THS [7]. However, for ad hoc networks, because there is no central controller, a code (or sequence) assignment protocol is necessary to determine the DSs or THSs used for traffic transmission and for monitoring any new traffic arrival over the channel. Currently, there are three basic types of code assignment protocols [16]:

- *Common code*: All the transmissions are assigned a common code. The packet header contains the address information. Each node monitors this information for any packets intended for it. As a common code is used for all transmissions, collision may occur in case of multiple simultaneous transmissions.
- *Receiver-based code*: Each node is assigned a unique receiving code. The code of the destination is used for any peer-to-peer transmission. Hence, for any intended traffic arrival, each node only needs to monitor its own receiving code. The main drawback of this scheme is the possible collision when multiple senders try to send packets to the same receiver simultaneously, as the same code is used.
- *Transmitter-based code*: Each node is assigned a unique sending code. Each transmitter uses its sending code for transmission to any receiver. The main advantage of this scheme is that multiple transmissions from multiple senders will not collide. However, a mechanism is needed to let the intended receiver be aware of an upcoming transmission.

In order to reduce the collision probability and make the handshaking procedure manageable, hybrid schemes should be more effective. For example, a combination of common code and transmitter-based code results in common-transmitter-based (C-T) protocols, while a combination of receiver-based code and transmitter-based code leads to receiver-transmitter-based (R-T) protocols [16]. However, in these schemes, as the sender transmits the data packets regardless of the reception status (i.e., collided or successfully received), wastage of bandwidth may be possible. Hence, the RTS (request-to-send) – CTS (clear-to-send) dialogue in multiple access with collision avoidance (MACA) [12] can be incorporated into C-T and R-T protocols, referred to as MACA/C-T and MACA/R-T protocols, respectively [17].

- *MACA/C-T*: Each node is assigned a sending code. The RTS-CTS dialogue uses the common code. If the dialogue is successfully exchanged, the subsequent data transmission is sent with the sender's sending code. As shown in Figure 9.3, nodes 1 and 3 intend to send packets to nodes 2 and 4, respectively. When node 1 sends a DATA packet to node 2, it does not collide with the overheard RTS or DATA exchange from node 3 to node 4, because different codes are used. Collisions only happen when multiple RTS-CTS dialogues exist in the same region, for example, both node 1 to 2 and node 3 to 4 pairs are in RTS-CTS exchange simultaneously in Figure 9.3. Multicast and broadcast can be inherently supported in MACA/C-T as all nodes tune to the common code for the RTS-CTS dialogue and the sender's sending code is used for DATA transmission.
- *MACA/R-T*: Each node is assigned a sending code and a receiving code. The RTS packet is sent using the destination's receiving code, while CTS and DATA packets are transmitted by the associated sending codes, respectively, or a code private to a source-destination pair [7], as shown in Figure 9.4 where nodes 1 and 3 intend to transmit to nodes 2 and 4, respectively. The channel code of RTS can use the lowest possible rate

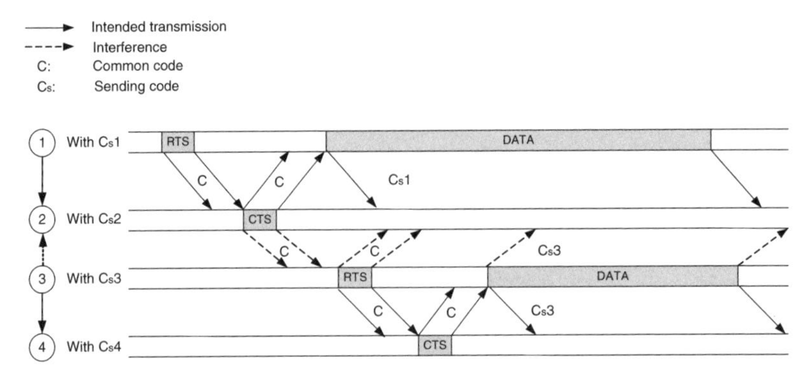

Figure 9.3 The MACA/C-T protocol [17]

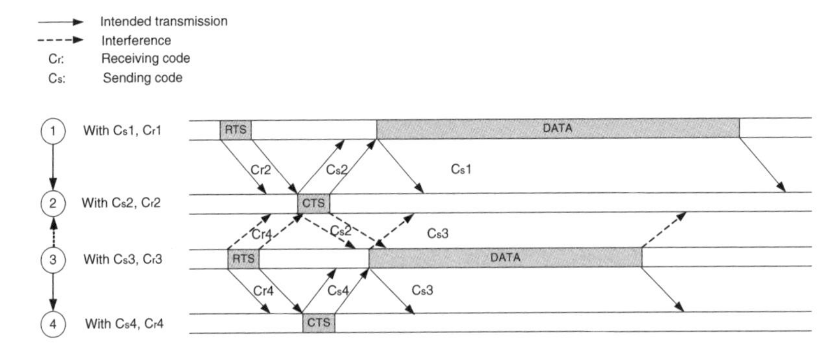

Figure 9.4 The MACA/R-T protocol [17]

so that all neighboring nodes that want to transmit to the same receiver can overhear it. In Figure 9.4, multiple RTS-CTS dialogues and DATA packets can be exchanged successfully without a collision due to the different codes used. Actually, collisions only happen when multiple senders attempt to send RTS to the same receiver, for example, when both nodes 1 and 3 use the same code $Cr2$ to send RTS to node 2. Therefore, MACA/R-T can achieve a higher channel throughput than that in MACA/C-T.

9.3.3 Interference Mitigation in TH-UWB

As mentioned in Section 9.3.1, it is desired to have simultaneous transmissions with the constraint of the exclusive region. Within an exclusion region, either no simultaneous transmissions are allowed, or interference from simultaneous transmissions can be combated. The exclusion region is difficult to determine. Further, coordination among nodes is needed to enforce the exclusion region, thus resulting in an information exchange overhead. If the effect of interference in an exclusion region can be mitigated, an exclusion region with a negligible size can be achieved [18]. Specifically, in addition to the traditional interference mitigation mechanisms used in conventional narrowband systems, the unique channelization in TH-UWB can be explored to implement a more flexible and efficient interference mitigation mechanism.

In TH-UWB, multiuser interference (MUI) is due to pulse collisions between the desired and interfering flows. Pulse collisions due to a near interfering node (very likely with strong interfering power at the receiver) greatly degrade the performance of the desired reception. In TH-UWB, a bit is modulated over a number (N_s) of pulses with a pseudo-random hopping sequence. At the receiver side, a matched filter is used, which has the input sampled at the desired hopping interval and generates the symbol decision variable. If one sample has a very high power level, it is likely that there exists a collision with a strong interferer. Hence, a chip discrimination principle can be effective where an acceptance level threshold is applied to each pulse sample prior to its entering the matched filter. A pulse with a larger power level (than the threshold) is skipped, and an erasure is declared. The remaining pulses should still be able to give an accurate detection decision. A substantial bit error rate (BER) improvement can be achieved for a large near/far power ratio [19]. The loss due to the erasure can be mitigated by rate control. If the ratio of pulse erasure is recorded by the receiver, it can be fed back to the sender so that the sender can determine the minimum pulse rate per bit (i.e., N_s) to meet the required BER [20]. In addition, the loss due to pulse erasures can be recovered by channel coding, thus leading to a bit rate reduction. Dynamic channel coding can be used to improve the system throughput performance [18], similar to methods discussed in Section 9.5.3.

9.4 Overhead Reduction

Overhead, such as frame headers, control messages, and so on, inevitably exists in any wireless system. Effects of overhead on the system throughput is more severe in UWB networks supporting very high data rate, since overhead is normally transmitted at a low rate to guarantee reliable detection at the receiver. For example, the physical layer and MAC layer headers are usually transmitted at a low rate for robustness, thus requiring a relatively large bandwidth in a high-rate system. The interframe spaces also consume channel bandwidth [14, 21]. In UWB networks, one source of overhead results from the long acquisition time required by the high precision synchronization, which usually varies from tens of microseconds to tens of milliseconds, compared to microseconds in narrowband systems. For example, consider a time-division multiple access (TDMA)-based UWB network with a 50 Mbps channel. Acquisition time is assumed to be 1 ms and packet size is 1500 bytes. Neglecting other timing components and overhead, the transmission efficiency, defined as the fraction of time used for actual data transmission, can be roughly calculated as [22]

$$\frac{1500 \text{ bytes}/50 \text{ Mbps}}{1 \text{ ms} + 1500 \text{ bytes}/50 \text{ Mbps}} = 19\ \%, \tag{4}$$

which is too low.

The relatively large overhead and long acquisition time in UWB transmissions may limit UWB radio resource management design. Therefore, it is critical to design an efficient radio resource management protocol that keeps the system overhead as low as possible, in order to fully explore the high-rate transmission.

9.4.1 ACK Mechanisms

Acknowledgement (ACK) and retransmission are usually adopted by the MAC layer to correct transmission errors. In the IEEE 802.15.3 [14], three types of ACK mechanisms are defined for MAC – no-ACK, immediate ACK (Imm-ACK), and delayed ACK (Dly-ACK), as shown in Figure 9.5. Not using any acknowledgement, no-ACK is suitable for transmissions not requiring reliable delivery. Two successive frames are separated by a minimum interframe space (MIFS). In the Imm-ACK mechanism, each data frame is always followed by an ACK frame from the receiver to indicate its correct reception. A short interframe space (SIFS) is used between the transmitted frame and ACK. In the Dly-ACK mechanism, instead of acknowledging each data frame, after a burst of frames are received, the whole burst is acknowledged by one ACK frame. The sender retransmits (in the next burst) the frames not ACKed in the previous ACK frames.

As shown in Figure 9.5, the average total transmission time t_m of one frame for Imm-ACK and Dly-ACK can be calculated as $t_p + t_{\mathrm{ACK}} + 2t_{\mathrm{SIFS}}$ and $(nt_p + t_{\mathrm{ACK}} + (n-1)t_{\mathrm{MIFS}} + 2t_{\mathrm{SIFS}})/n$, respectively, where t_p, t_{ACK}, t_{SIFS}, and t_{MIFS} denote the transmission times of the data frame, the ACK frame, the SIFS, and the MIFS, respectively; and n is the burst size in Dly-ACK. The values of t_m in different ACK mechanisms are given in Table 9.1, where 5-Dly-ACK and 10-Dly-ACK represent Dly-ACK mechanism with burst size 5 and 10, respectively [21].

It can be observed that, at a high data rate, Imm-ACK results in bandwidth inefficiency because the time to transmit the payload (i.e., t_p) is only a small portion of t_m. Therefore, Dly-ACK becomes a more suitable ACK mechanism for UWB networks, taking advantage of the reduced number of ACK frames and the associated interframe spaces (IFSs).

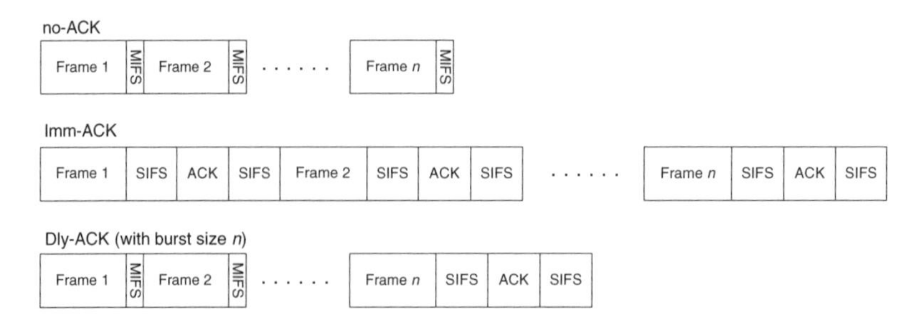

Figure 9.5 The no-ACK, Imm-ACK, and Dly-ACK mechanisms in IEEE 802.15.3

Table 9.1 Transmission time under various data rates [9]

Data rate (Mbps)	Imm-ACK		5-Dly-ACK		10-Dly-ACK	
	t_m(μs)	t_p/t_m (%)	t_m(μs)	t_p/t_m (%)	t_m(μs)	t_p/t_m (%)
110	112.7	73.4	90.3	91.6	87.5	94.5
200	80	62.5	57.9	86.4	55.0	90.9
480	56.7	47.1	34.5	77.2	31.7	81.6

Traditional Dly-ACK uses a fixed burst size, which may lead to severe *local information problem* [21]. On the basis of the IEEE 802.15.3 specification, each MAC Service Data Unit (MSDU) is divided into single or multiple smaller parts, termed MAC Protocol Data Units (MPDUs), and all MPDUs must be delivered to the upper layer in order. In transmission, it is possible that some MPDUs with higher IDs are correctly received while those with lower IDs are not. The ordered delivery requires that such MPDUs with higher IDs must wait until all MPDUs with lower IDs have already been correctly received. However, since the receiver does not have the information of the number of retransmissions that the erroneous MPDUs have already experienced, it may keep waiting for an MPDU that has been discarded owing to it exceeding the maximum retransmission time. This is the local information problem. In addition, because of the bursty nature of traffic, the source MAC queue may be empty from time to time. With a fixed burst size, when there are not enough MPDUs (because of an empty MAC queue) to trigger an ACK at the receiver, the sender cannot retransmit erroneous frames in time because it keeps waiting for additional MPDUs to fill the burst before it requests an ACK from the destination.

A possible solution to resolve the local information problem is to introduce a retransmission counter at the receiver. If an expected MPDU is not received successfully, the receiver activates a counter to count the number of the ACK frames indicating the erroneous reception of the MPDU, which is equivalently the retransmission time of the MPDU at the sender. As soon as the counter reaches a preset threshold, the MPDU is considered to be discarded and the receiver will deliver the previously received MPDUs with higher IDs to the upper layer immediately. For the fixed burst size problem, a possible solution is that the source MAC can request the Dly-ACK frame when the source MAC queue becomes empty [21].

Although Dly-ACK can achieve better bandwidth utilization (than Imm-ACK), its impact on delay performance is twofold. The total delay consists of the queuing delay at the sender, transmission delay, and reordering delay at the receiver. On one hand, the queuing delay at the sender can be reduced, as the data frames are transmitted faster. On the other hand, the reordering delay at the receiver can degrade the end-to-end delay performance. To trade off the two conflicting effects, there exists an optimal burst size that is heavily dependent on the input traffic volume and is insensitive to the channel error rate within a normal error rate range [23]. The end-to-end delay performance is particularly critical to real-time applications that are usually delay-sensitive.

9.4.2 Long Acquisition Time

From (4), it can be seen that the low channel utilization is because of the dominant acquisition time in the packet transmission duration. Hence, to reduce the overhead introduced by the acquisition time, an intuitive solution is to enlarge the packet size. However, as the wireless channel is usually error-prone, a large packet size leads to a high packet error probability, thus introducing a different kind of overhead due to retransmissions of the erroneous packets and the induced delay. How to balance the trade-off between these two effects is challenging. Further, a large packet size may result in a large packetization delay, which may not be suitable for real-time applications such as voice-over IP (VoIP) and video streaming.

A general approach of 'packet packing' can be an effective way to compensate for the long acquisition time [24]. The approach consists of 5 policies: (1) packet classification policy, (2) buffer management policy, (3) packet assembly policy, (4) acknowledgement policy, and (5) packet error control policy. The basic idea is to assemble multiple upper-layer packets into one burst frame at the MAC layer. In each transmission, a whole burst frame is sent, rather than delivering each packet individually as in traditional approaches. Transmitting multiple packets in one frame can significantly reduce the synchronization overhead. It can be seen that this approach tries to enlarge the transmitted payload size in each transmission. Hence, it has a similar principle as in the 'large packet size' approach discussed earlier, and also has similar drawbacks.

Another possible solution to the long acquisition time is to use a link maintenance scheme [25]. The physical link is not torn down when there is no data to transmit. Rather, low-rate control packets are transmitted to maintain the physical link for the lifetime of a user's call so that the reacquisition overhead for future transmissions can be reduced. This solution has its inherent drawbacks. First, the transmission time is enlarged, thereby increasing interference to other links. The extra power consumption for the transmission of low-rate control packets is also critical for UWB devices that usually operate with limited power supply and low power consumption requirement. In addition, for bidirectional UWB communications, a node generally cannot send and receive at the same time. To address this problem, full-duplex is achieved by blanking the receiver at a node during pulse transmissions of its transmitter. The complexity will increase significantly when a node keeps multiple full-duplex links simultaneously.

9.5 Power/Rate Allocation

For UWB networks, to efficiently utilize the bandwidth and achieve desired QoS, an effective resource allocation scheme is needed to determine at what power level and rate a node can access the wireless medium. Basic power allocation and rate guarantee mechanisms are discussed in Section 9.5.1 and Section 9.5.2, respectively. The unique pulse transmission in UWB networks provides flexibility in rate control, as shown in Section 9.5.3. For resource allocation, benefits can also be obtained from cross-layer design approaches, as presented in Section 9.5.4.

9.5.1 Power Allocation

For multichannel UWB networks, code assignment can deal with the primary collisions due to the same code being used in simultaneous transmissions. However, the well-known near-far problem may induce the secondary collisions because of the intolerable interference experienced from simultaneous transmissions (spread by different codes), that is, MUI in the vicinity [26]. The transmission power of each link should be managed to make the network stable and to achieve the desired system performance.

For TH-UWB transmission, the combined MUI can be approximated by additive white Gaussian noise (AWGN) in a multiuser environment, if the number of users is large and different terminals use independent pseudorandom codes [27, 28]. In the following, a power allocation strategy for TH-UWB is discussed [29], and a similar principle can be applied to DS-UWB networks as well.

Consider a UWB network with N active links. The achieved signal to interference-plus-noise ratio (SINR) at the receiver of the ith link is

$$\mathrm{SINR}_i = \frac{P_i h_{ii}}{R_i(\eta_i + T_f \sigma^2 \sum_{j=1, j\neq i}^{N} P_j h_{ji})}, \qquad i = 1, \ldots, N, \tag{5}$$

where P_i denotes the average transmission power of link i's transmitter, h_{ij} the path gain from link i's transmitter to link j's receiver, R_i the ith link bit rate, η_i the background noise energy plus interference from other non-UWB systems, T_f the pulse repetition time, and σ^2 a parameter depending on the shape of the pulse.

It is interesting that, if only the best-effort service is to be considered, each sender should either transmit with the allowed maximum power level, or not transmit at all. From the view point of a transmission link i, an increase in transmission power leads to a higher SINR at the receiver side, therefore resulting in a higher achievable rate. It also increases the interference level to other simultaneous links. However, the loss in other links can be compensated by the gain obtained by link i [10, 29].

On the other hand, for services with QoS requirements, generally the physical layer should provide an upper bound of the BER, which can be translated into a prespecified SINR threshold, say γ_i for link i. In addition, each transmitter should maintain an upper bound of average power level, say $P_{\max}$. The $P_{\max}$ value can be determined by the emission regulation and the energy consumption of the terminals. Thus, the power levels of the N active links should comply with the following constraints:

$$\begin{cases} \dfrac{P_i h_{ii}}{R_i(\eta_i + T_f \sigma^2 \sum_{j=1, j\neq i}^{N} P_j h_{ji})} \geq \gamma_i, & i = 1, \ldots, N \\ 0 \leq P_i \leq P_{\max}, & i = 1, \ldots, N. \end{cases}$$

To meet these constraints, a centralized power allocation controller (if available) can take the advantage of system-level information such as the transmission power level of every node and the path gain of every link. Combined with rate allocation, the power allocation issue can lead to a joint optimization problem, to optimize network metrics such as system throughput or energy consumption [29].

For centralized resource allocation in UWB, the system-level information exchange usually imposes heavy signaling overhead, and a central controller (maybe a selected node) is necessary to (a) broadcast synchronization information; (b) collect the traffic request of every node and status of every link; and (c) determine active links with the allowed power levels, their transmission time duration, and transmission rates. However, a centralized controller may not always be available, especially in ad hoc networks. A suboptimal but distributed power/rate allocation scheme is more realistic for UWB networks.

For a distributed power/rate allocation scheme, each node performs admission decision for each request and determines the transmission power and rate if the request is admitted, according to its local measurements of the system and information obtained from the control message exchanges. In order to avoid frequent power reconfiguration after each new request is admitted, each link keeps an interference margin, also referred to as maximum sustainable interference (MSI), denoting the additional tolerable interference while not violating the SINR requirement. For link i,

$$\frac{P_i h_{ii}}{R_i(\eta_i + T_f\sigma^2 \sum_{j=1, j\neq i}^{N} P_j h_{ji} + \text{MSI}_i)} = \gamma_i, \tag{6}$$

which leads to

$$\text{MSI}_i = \frac{P_i h_{ii}}{\gamma_i R_i} - \eta_i - T_f\sigma^2 \sum_{j=1, j\neq i}^{N} P_j h_{ji}. \tag{7}$$

The MSIs of all the links should be nonnegative.

Each active link periodically announces its MSI over a control channel. When a new call request arrives at one node, according to local measurements of interference and noise levels, and MSI information of other links, the node determines whether it is feasible to assign a power level and a rate such that: (a) its own MSI is nonnegative; and (b) the interference (because of the new transmission if admitted) to any other existing active link does not exceed its MSI [29, 30].

9.5.2 Rate Guarantee

For multiple access in a multichannel case, if a flow MSI is honored by all the neighboring flows, its transmission rate can be guaranteed. Consider a UWB network with N active flows with rate requirements from R_1 to R_N. Upon a new transmission flow request i with required rate R_i, the following procedure can be implemented [29, 30]:

- *Step 1*: Calculate

$$P_i = \min\left\{P_{\max}, \min_{1\leq j\leq N}\left\{\frac{\text{MSI}_j}{T_f\sigma^2 h_{ij}}\right\}\right\}. \tag{8}$$

  If $P_i = 0$, reject the flow request; otherwise, continue to the next step;
- *Step 2*: Check whether or not

$$\frac{P_i h_{ii}}{\gamma_i R_i} - \eta_i - T_f\sigma^2 \sum_{j=1}^{N} P_j h_{ji} \geq 0. \tag{9}$$

  If it is true, assign power P_i and rate R_i; otherwise, reject the flow request.

The first step is to guarantee that the MSIs of all the existing flows are honored, while the second step is to guarantee that a newly admitted flow can obtain a nonnegative MSI. It can be seen that this rate mechanism is similar to the circuit-switching channel reservation in cellular networks. It is not efficient to meet the different QoS requirements of various traffic types in a packet-switching environment.

9.5.3 Rate Control

Although power control is usually considered as an effective way in CDMA-based systems to combat MUI, guarantee the required SINR at the receiver, and lengthen the battery life, it is not always true, or at least not efficient under some conditions. For example, when a link experiences deep fading, power control significantly increases its transmission power to keep the same SINR at the receiver. This large transmission power introduces large interference to other links and may reduce the interference-limited system capacity. As an extreme case of CDMA, UWB encounters a similar problem. Rate control is effective to compensate for such shortcomings in the power control. Instead of changing the transmission power, rate control adapts the data transmission rate such that more or less redundancy is introduced for compensating channel fading and interference. In UWB networks, rate control can be achieved by adapting the channel coding rate as discussed in Section 9.5.3.1. On the other hand, collocated UWB WPANs interfere with each other. For TH-UWB, an effective way to reduce such interference is to control the 'pulse rate' (i.e., the number of pulses transmitted per second) in each WPAN, as discussed in Section 9.5.3.2.

9.5.3.1 Adaptive Channel Coding

Channel encoder is a basic component of a wireless system to overcome channel errors at the receiver by inserting some redundancy at the transmitter. The selection of the channel code is determined by the trade-off between the error correction capacity and the introduced transmission overhead. Since a wireless channel is time variant due to user mobility, an adaptive channel encoder should be more efficient in such an environment. Research in [10, 31, 32] also indicates that the optimal radio resource management should make use of the allowed maximum power at each active link, and that power control does not provide a significant gain when dynamic channel coding is used. Basically, adaptive channel encoding at the transmitter should consist of following four steps:

- Based on channel statistics, a channel code is selected at the sender side;
- After adding a Cyclic Redundancy Check (CRC) and encoding, the sender transmits the encoded packet;
- At the receiver side, CRC is checked and ACK (positive or negative) is fed back to the sender;
- Based on the ACK feedback, the channel coding rate is adjusted such that a lower (higher) rate channel code is used for the next transmission if the channel becomes worse (better).

From the practical implementation point of view, a new channel code with different coding rate can be transmitted at each adaptation step. In order to reduce the overhead and the transmission delay, a more efficient way, called *incremental redundancy* [7], can be applied. In incremental redundancy, a special channel coding scheme is adopted such that a high-rate code is the subset of the lower-rate codes. In transmission, if the current channel code cannot provide sufficient protection for decoding at the receiver end, only the redundant bits (which are the different bits between the current channel code and the lower-rate code) are transmitted. In other words, none of the coded bits already transmitted are transmitted again.

Many convolutional codes have been developed to provide variable encoding rates as well as incremental redundancy [33]. One example of such a code is Rate Compatible Punctured Convolutional (RCPC) code [7, 18, 34]. In RCPC codes, a high-rate code is created by puncturing coded bits from the lowest rate block of coded bits. It is easy to prove that RCPC codes have compatibility property. For instance, given a set of codes with rates $R_0 = 1 > R_1 > R_2 > \cdots > R_N$ (where '1' means uncoded case and R_N represents the lowest coding rate), the code with rate R_n is the subset of the code with rate R_{n+1}. In addition, since the encoder only needs to generate the code with the lowest coding rate, one encoder/decoder pair is enough for encoding and decoding the coded bits with all coding rates, thus further reducing the system complexity [7].

To implement adaptive channel coding, it is critical to determine the initial highest code rate for reliable transmission. For RCPC codes, as an example, the initial highest code rate can be determined by the following procedure [7]:

- At the beginning, the first packet is coded using the most powerful code, that is, the lowest rate code.
- At the receiver, the decoding is carried out by step-wise traversal of the trellis of the Viterbi decoder. If the outcome of a decoding step for a higher rate code differs from that of the actual code, the higher rate codes are eliminated.
- The remaining highest rate code is the code powerful enough for decoding.

9.5.3.2 TH Sequence Parameters Adaptation

On the basis of the IEEE 802.15.3, WPANs work in both coordinated and uncoordinated ways. Well designed radio resource management can coordinate multiple transmissions within each WPAN by using contention-free techniques, random access, or a combination of both. In addition, adapting the parameters of TH in each WPAN is effective to combat the mutual interference among uncoordinated WPANs. One way to achieve rate control is to adapt TH sequence parameters such that the spreading gain is changed with respect to the channel and interference variance [35]. Such parameters include number of pulses for each bit (N_s), maximum TH shift (N_h), and TH unit (T_c). It is a unique control method inherent in UWB networks.

Effects of TH parameters on system performance in terms of throughput and BER have been studied in [36] through simulation. Consider the case when TH is allowed over the whole frame time T_f, that is, $T_f = N_h T_c$. The basic observation is that N_h should be increased with the number of WPANs in order to reduce the amount of generated interference and therefore reduce collisions. This *hard link adaptation* changes the TH sequence used and requires explicit information exchanges. To avoid the overhead, a *soft link adaptation* scheme can be applied, which varies the values of N_h and N_s, while keeping $N_h \cdot N_s$ constant such that the bit rate remains unchanged. Let h_0 denote the possible minimum value of N_h. For the soft link adaptation, the TH sequence with parameter h_0 is first generated. Since $T_b = N_h N_s T_c$, for a fixed chip duration T_c, the maximum processing gain is $N_s^{\max} = \frac{T_b}{h_0 T_c}$. Define a probability parameter q. Different values of N_h can be achieved by the so-called 'chip puncturing': If a pulse should be transmitted in a certain chip, the node transmits a pulse with probability q and keeps silent with probability $1 - q$ in that chip interval. After chip puncturing, the average number of pulses transmitted per bit is $q \cdot N_s^{\max}$ such that each chip of the TH sequence can hop on a wider range, that is, a virtually larger N_h. Since the chip puncturing does not need to be coordinated with the receiver, it can be applied autonomously by the sender in the WPAN without the overhead of control packet exchanges. By chip puncturing, the transmitted pulse rate is reduced, thus generating less interference to neighboring WPANs, at the cost of a smaller processing gain. It is still an open issue as to how to determine a proper probability q to achieve the best trade-off between the gain from less inter-WPAN interference and the loss due to a smaller processing gain.

9.5.4 Cross-Layer Design

UWB wireless system performance should benefit from cross-layer design approaches, taking advantage of information exchanges across the protocol layers, which may not be available in the traditional layered architecture, that is, the open system interconnection (OSI) protocol stack. Specifically, designing efficient radio resource management can utilize information from both the physical layer and the network layer, such as channel status, location information, and routing information, as shown in Figure 9.6.

9.5.4.1 Joint Routing/Resource Allocation in Multihop UWB Transmissions

In multihop UWB networks, more challenges will be encountered than those in the single-hop case: (a) An optimal route should be chosen appropriately by considering the traffic load distribution, power consumption, and system overhead; (b) Resource allocation parameters in each link should be determined, such as rate and power, and how to control them with a fluctuating interference/contention level; (c) Most importantly, routing and resource allocation interact with each other. Therefore, designers need to determine how the routing and resource allocation should interact, and jointly design them accordingly. Generally, joint design of routing and resource allocation can achieve performance improvement at the cost of complexity [37]. Although radio resource allocation in UWB networks has been shown to be insensitive to route selection strategies for best-effort

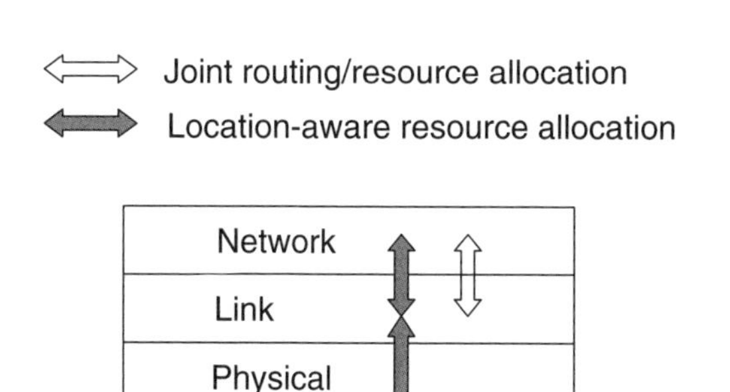

Figure 9.6 Cross-layer design for UWB MAC

services [10], further research efforts are needed to investigate how routing and resource allocation interact for multimedia traffic with various QoS requirements.

9.5.4.2 Location-Aware Resource Allocation

One advantage of UWB technology is its potential to provide accurate distance information. In addition, nodes in UWB indoor networks usually have low mobility. If distance information is exchanged among nodes (with limited overhead due to low mobility), a node may know the (accurate or coarse) location information of other nodes [38–40]. All these can be beneficial to the radio resource management design.

- Routing in UWB networks can benefit greatly from location information. Signaling overhead can be reduced significantly. In a search for a route from a source node (say a) to the destination node (say b), instead of the flooding used in traditional approaches, a smaller forwarding zone can be selected based on location information of nodes a and b [38, 41–44]. In addition, on the basis of location information of the nodes, the geographic area can be divided into *grids*, each with a grid leader. The grid mechanism can offer efficient route discovery and resilient route maintenance [43]. With an effective routing mechanism, complexity can be reduced in the joint routing/resource allocation design.
- Power/rate control can take advantage of the location information. On the basis of the location information of the communication pair and a signal propagation model, power level, or the UWB channelization parameters (such as TH sequence parameters, or variable spreading factors) can be selected appropriately, to achieve the required SINR [45].
- The exclusion region can be implemented more conveniently with the help of location information.
- Given the location information, each node may estimate the traffic density in its vicinity. A node in a sparse area can use different parameters (e.g., power levels and backoff parameters) in its transmission from those of a node in a dense neighborhood [46].

9.6 Conclusions

Radio resource management plays a very important role in UWB wireless networks to support effective and efficient communications. Unique characteristics of the UWB physical layer and network layer provide both challenges and opportunities for the radio resource management design. This chapter provides a comprehensive overview of the state-of-the-art research in pulse-based UWB radio resource management, in the avenues of three major research topics, multiple access, overhead reduction, and power/rate allocation. More flexibility can be obtained from the inherent support of simultaneous transmissions in UWB technologies, but at the same time this leads to complexity in terms of power, rate, and interference control. Relatively large overhead is a significant challenge in UWB transmissions, and should be suppressed to as low a level as possible. For pulse transmissions in UWB, its unique channelization features can benefit rate control and interference control. Moreover, the cross-layer approach should be exploited in UWB system design for better performance.

References

[1] R. C. Qiu, H. Liu, and X. Shen, "Ultra-Wideband for Multiple-Access Communications," *IEEE Commun. Mag.*, vol. 43, no. 2, pp. 80–87, Feb. 2005.

[2] W. Zhuang, X. Shen, and Q. Bi, "Ultra-Wideband Wireless Communications," *Wireless Commun. Mobile Comput.*, vol. 3, no. 6, pp. 663–685, Sept. 2003.

[3] *First report and order in the matter of revision of part 15 of the commission's rules regarding ultra-wideband transmission systems*, Federal Communications Commission (FCC 02-48), Std., Apr. 2002, ET Docket 98–153

[4] S. Roy, J. R. Foerster, V. S. Somayazulu, and D. G. Leeper, "Ultra-Wideband Radio Design: The Promise of High-Speed, Short Range Wireless Connectivity," *Proc. IEEE*, vol. 92, no.2, pp. 295–311, Feb. 2004.

[5] G. R. Aiello and G. D. Rogerson, "Ultra-Wideband Wireless Systems," *IEEE Microw. Magn.*, vol. 4, no. 2, pp. 36–47, Jun. 2003.

[6] S. Aedudodla, S. Vijayakumaran, and T. F. Wong, "Rapid Ultra-Wideband Signal Acquisition," *Proc. IEEE WCNC'04*, Atlanta, GA, vol. 2, pp. 1148–1153, 2004.

[7] J.-Y. Le Boudec, R. Merz, B. Radunovic, and J. Widmer, "A MAC Protocol for UWB Very Low Power Mobile Ad Hoc Networks Based on Dynamic Channel Coding with Interference Mitigation," EPFL-DI-ICA Technical Report IC/2004/02, Ecole Polytechnique Fédérale de Lausanne, Switzerland, Jan. 2004.

[8] R. Negi and A. Rajeswaran, "Capacity of Power Constrained Ad Hoc Networks," *Proc. IEEE INFOCOM'04*, Hong Kong, China, vol. 1, pp. 443–453, 2004.

[9] R. Negi and A. Rajeswaran, "Scheduling and Power Adaptation for Networks in the Ultra-Wideband Regime," *Proc. IEEE GLOBECOM'04*, Dallas, TX, vol. 1, pp. 139–145, 2004.

[10] B. Radunovic and J.-Y. Le Boudec, "Optimal Power Control, Scheduling, and Routing in UWB Networks," *IEEE J. Select. Areas Commun.*, vol. 22, no. 7, pp. 1252–1270, Sept. 2004.

[11] IEEE 802.11 WG, Part 11: Wireless LAN Medium Access Control (MAC) and Physical Layer (PHY) Specification, Standard, IEEE, Aug. 1999.

[12] P. Karn, "MACA – A New Channel Access Method for Packet Radio," *Proc. ARRL/CRRL Amateur Radio 9th Computer Networking Conference*, London, Ontario, Canada, pp. 134–140, 1990.

[13] S. Jiang, J. Rao, D. He, X. Ling, and C. C. Ko, "A Simple Distributed PRMA for MANETs," *IEEE Trans. Veh. Technol.*, vol. 51, no. 2, pp. 293–305, Mar. 2002.

[14] IEEE Std 802.15.3™-2003: Wireless Medium Access Control (MAC) and Physical Layer (PHY) Specifications for High Rate Wireless Personal Area Networks (WPANs), Sept. 2003.

[15] J. Cai, K. H. Liu, X. Shen, J. W. Mark, and T. D. Todd, "Power Allocation and Scheduling for MAC Layer Design in UWB Networks", *Proc. 2nd International Conference on Quality of Service in Heterogeneous Wired/Wireless Networks (QShine'05)*, Orlando, FL, 2005.

[16] E. S. Sousa and J. A. Silvester, "Spreading Code Protocols for Distributed Spread-Spectrum Packet Radio Networks," *IEEE Trans. Commun.*, vol. 36, no. 3, pp. 272–281, Mar. 1988.

[17] M. Joa-Ng and I-T. Lu, "Spread Spectrum Medium Access Protocol with Collision Avoidance in Mobile Ad Hoc Wireless Network," *IEEE INFOCOM'99*, New York, NY, vol. 2, pp. 776–783, 1999.

[18] R. Merz, J.-Y. Le Boudec, J. Widmer, and B. Radunovic, "A Rate-Adaptive MAC Protocol for Low-Power Ultra-Wide Band Ad Hoc Networks," *Proc. 3rd International Conference on Ad Hoc Networks and wireless*, Vancouver, British Columbia, Canada, 2004.

[19] W. M. Lovelace and J. K. Townsend, "Chip Discrimination for Large Near Far Power Ratios in UWB Networks," *Proc. IEEE MILCOM'03*, Boston, MA, vol. 2, pp. 868–873, 2003.

[20] W. M. Lovelace and J. K. Townsend, "Adaptive Rate Control with Chip Discrimination in UWB Networks," *Proc. IEEE Conference on Ultra Wideband Systems and Technologies*, Reston, VA, pp. 195–199, 2003.

[21] H. Chen, Z. Guo, R. Yao, and Y. Li, "Improved Performance with Adaptive Dly-ACK for IEEE 802.15.3 WPAN Over UWB PHY," *IEICE Trans. Fundamentals*, to appear.

[22] J. Ding, L. Zhao, S. R. Medidi, and K. M. Sivalingam, "MAC Protocols for Ultra-Wide-Band (UWB) Wireless Networks: Impact of Channel Acquisition Time," *Proc. SPIE ITCOM Conference*, Boston, MA, 2002.

[23] H. Chen, Z. Guo, R. Yao, X. Shen, and Y. Li, "Performance Analysis of Delayed Acknowledgement Scheme in UWB Based High Rate WPAN," *IEEE Trans. Veh. Technol.*, to appear.

[24] K. Lu, D. Wu, Y. Fang, and R. C. Qiu, "On Medium Access Control for High Data Rate Ultra-Wideband Ad Hoc Networks," *Proc. IEEE WCNC'05*, New Orleans, LA, vol. 2, pp. 795–800, 2005.

[25] S. S. Kolenchery, J. K. Townsend, and J. A. Freebersyser, "A Novel Impulse Radio Network for Tactical Military Wireless Communications," *Proc. IEEE MILCOM'98*, Boston, MA, vol. 1, pp. 59–65, 1998.

[26] A. Muqattash and M. Krunz, "CDMA-Based MAC Protocol for Wireless Ad Hoc Networks," *Proc. 4th ACM International Symposium on Mobile Ad Hoc Networking & Computing*, Annapolis, MD, pp. 153–164, 2003.

[27] M. Z. Win and R. A. Scholtz, "Impulse Radio: How it Works," *IEEE Commun. Lett.*, vol. 2, no. 2, pp. 36–38, Feb. 1998.

[28] M. Z. Win and R. A. Scholtz, "Ultra-Wide Bandwidth Time-Hopping Spread-Spectrum Impulse Radio for Wireless Multiple-Access Communications," *IEEE Trans. Commun.*, vol. 48, no. 4, pp. 679–691, Apr. 2000.

[29] F. Cuomo, C. Martello, A. Baiocchi, and F. Capriotti, "Radio Resource Sharing for Ad Hoc Networking with UWB," *IEEE J. Select. Areas Commun.*, vol. 20, no.9, pp. 1722–1732, Dec. 2002.

[30] Y. Chu and A. Ganz, "MAC Protocols for Multimedia Support in UWB-Based Wireless Networks," *International Workshop on Broadband Wireless Multimedia (BroadWim 2004)*, San José, CA, Oct. 2004.

[31] S. Raj, E. Telatar, and D. Tse. "Job Scheduling and Multiple Access," *DIMACS Ser. Discrete Math. Theor. Comput. Sci.*, vol. 66, pp.127–137, 2003.

[32] S. Toumpis and A. J. Goldsmith, "Capacity Regions for Wireless Ad Hoc Networks," *IEEE Trans. Wireless Commun.*, vol. 2, no. 4, pp. 736–748, Jul. 2003.

[33] J. G. Proakis, *Digital Communications*, 4th edition, McGraw-Hill, New York, 2001.

[34] J. Hagenauer, "Rate-Compatible Punctured Convolutional Codes (RCPC Codes) and Their Applications," *IEEE Trans. Commun.*, vol. 36, no. 4, pp. 389–400, Apr. 1988.

[35] S. S. Kolenchery, J. K. Townsend, J. A. Freebersyser, and G. Bilbro, "Performance of Local Power Control in Peer-to-Peer Impulse Radio Networks with Bursty Traffic," *Proc. IEEE GLOBECOM'97*, Phoenix, AZ, vol. 2, pp. 910–916, 1997.

[36] H. Yomo, P. Popovski, C. Wijting, I. Z. Kovács, N. Deblauwe, A. F. Baena, and R. Prasad, "Medium Access Techniques in Ultra-Wideband Ad Hoc Networks," *Proc. 6th National Conference of Society for Electronic, Telecommunication, Automatics, and Informatics (ETAI) of the Republic of Macedonia*, Ohrid, Macedonia, Sept. 2003.

[37] P. Baldi, L. De Nardis, and M.-G. Di Benedetto, "Modeling and Optimization of UWB Communication Networks Through a Flexible Cost Function," *IEEE J. Select. Areas Commun.*, vol. 20, no. 9, pp. 1733–1744, Dec. 2002.

[38] S. Basagni, I. Chlamtac, V. R. Syrotiuk, and B. A. Woodward, "A Distance Routing Effect Algorithm for Mobility (DREAM)," *Proc. ACM MOBICOM'98*, Dallas, TX, pp. 76–84, 1998.

[39] C. T. Cheng, H. L. Lemberg, S. J. Philip, E. van den Berg, and T. Zhang, "SLALoM: A Scalable Location Management Scheme for Large Mobile Ad-Hoc Networks," *Proc. IEEE WCNC'02*, Orlando, FL, vol. 2, pp. 574–578, 2002.

[40] J. Li, J. Jannotti, D. S. J. De Couto, D. R. Karger, and R. Morris, "A Scalable Location Service for Geographic Ad Hoc Routing," *Proc. ACM MOBICOM'00*, Boston, MA, pp. 120–130, 2000.

[41] L. De Nardis, G. Giancola, and M.-G. Di Benedetto, "A Position Based Routing Strategy for UWB Networks," *Proc. 2003 IEEE Conference on Ultra Wideband Systems and Technologies*, Reston, VA, pp. 200–204, 2003.

[42] F. Legrand, I. Bucaille, S. Héthuin, L. De Nardis, G. Giancola, M.-G. Di Benedetto, L. Blazevic, and P. Rouzet, "U.C.A.N.'s Ultra Wide Band System: MAC and Routing Protocols," *Proc. International Workshop on Ultra Wideband Systems(IWUWBS)*, Oulu, Finland, 2003.

[43] W.-H. Liao, Y.-C. Tseng, and J.-P. Sheu, "GRID: A Fully Location-Aware Routing Protocol for Mobile Ad Hoc Networks," *Telecommun. Syst.*, vol. 18. pp. 37–60, 2001.

[44] Y.-C. Tseng, S.-L. Wu, W.-H. Liao, and C.-M. Chao, "Location Awareness in Ad Hoc Wireless Mobile Networks," *Comput.*, vol. 34, no. 6, pp. 46–52, Jun. 2001.

[45] W. Horie and Y. Sanada, "Novel Packet Routing Scheme Based on Location Information for UWB Ad-Hoc Network," *Proc. IEEE Conference on Ultra Wideband Systems and Technologies*, pp. 185–189, 2003.

[46] T. Issariyakul, E. Hossain, and D. I. Kim, "Medium Access Control Protocols for Wireless Mobile Ad Hoc Networks: Issues and Approaches," *Wireless Commun. Mobile Comput.*, vol. 3, no. 8, pp. 935–958, Dec. 2003.

10

Pulsed UWB Interference to Narrowband Receivers

Jay E. Padgett

10.1 Introduction

Narrowband receivers typically have a structure similar to that shown in Figure 10.1. The incoming signal is amplified and filtered at radio frequency (RF), then down-converted with a (generally tunable) local oscillator (LO) to a first intermediate frequency (IF), at which point there may be more gain adjustments and filtering. There will often be another conversion to a second IF. The output of the final IF stage is a replica of the received RF signal, but shifted in frequency.

The receiver-specific processing occurs after that point, and might consist of something as simple as envelope detection (for analog AM) or limiting with frequency discrimination (for analog FM). For digital systems, the processing could include a combination of coherent or noncoherent symbol detection, deinterleaving, decoding, diversity combining, and despreading.

This chapter focuses primarily on analysis of the output of the final IF stage when the input to the receiver is a UWB signal, because that analysis is fairly general and applicable to a broad range of pulsed UWB signals and narrowband receiver types. Detailed analysis of the UWB interference impact for specific detectors must be done on a case-by-case basis. However, for pulse rates that are high compared to the receiver bandwidth, the UWB interference at the final IF output appears as a combination of a continuous wave (CW) tone and Gaussian noise, and the effects of those types of interference on receiver performance are generally well understood. If the pulse rate is less than the receiver bandwidth, the IF output is a sequence of distinct IF impulse responses and the interference appears impulsive.

Linearity is assumed here. While it is conceivable that a UWB signal from a nearby and/or high power source could generate enough power into the receiver front end to saturate the low-noise amplifier (LNA) or first mixer, such interference would usually be well above the levels required to cause noticeable performance degradation, such as

Ultra-wideband Wireless Communications and Networks
Edited by Xuemin (Sherman) Shen, Mohsen Guizani, Robert Caiming Qiu and Tho Le-Ngoc

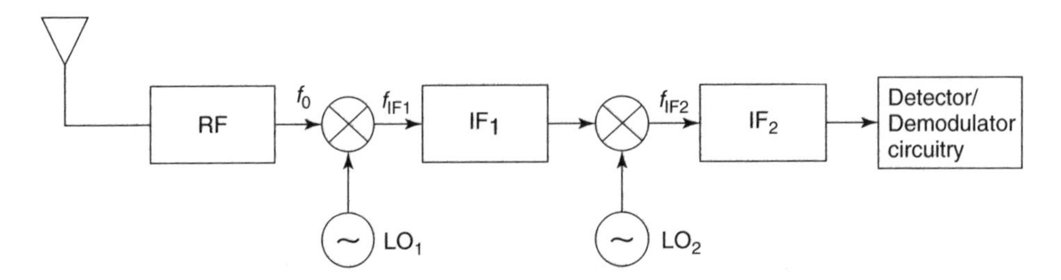

Figure 10.1 Generic narrowband receiver structure

increased bit error rate. Therefore, the focus here is on linear modeling that can be used to support signal-to-interference analysis. If necessary, the techniques developed here can easily be extended to explore conditions required for receiver front-end compression by UWB signals.

It is useful to look at the UWB signal and its relationship to the narrowband receiver from two different perspectives. The first is a time-domain view that leads to an expression for the waveform at the output of the final IF, in response to a UWB input at the front end. The second is a frequency-domain view that uses the power spectral density (PSD), which is the average power per Hertz as a function of frequency. These two approaches tend to be complementary.

The PSD approach provides a single expression that gives the distribution of power over frequency, and allows its sensitivity to changes in system parameters to be explored. PSD analysis can also provide good insights into the effects of periodicities and structure in signals. However, PSD analysis is limited to *average* power within a given receiver pass band, and may not give an adequate accounting of temporal and statistical effects that can be important in determining the impact on the detector/demodulator performance.

The strength of the temporal approach is that it supports detailed waveform-level analysis and simulation of the final IF output, which can be used to explore in detail the effect of the UWB interference on a specific receiver. Time-based waveform simulations can also be used to corroborate PSD analysis, by operating the simulation in a frequency-stepped mode, whereby the simulation is performed repeatedly at a series of center frequencies, and the average power is recorded at each frequency. This effectively simulates the operation of a spectrum analyzer. In the limit as the filter bandwidth decreases, the output approaches the PSD.

Both approaches are developed here, and examples are provided to illustrate their application and compare results. The intent is to provide the reader with the understanding and tools necessary to analyze the effect of pulsed UWB interference on any particular type of receiver.

10.2 Pulsed UWB Signal Model

The UWB signal of interest here is a sequence of very short pulses. If the basic pulse waveform is $p(t)$, the UWB signal can be described as

$$w(t) = \sum_k a_k p(t - T_k), \tag{1}$$

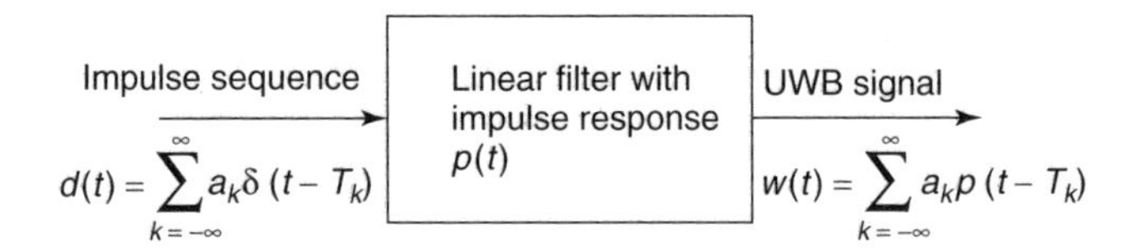

Figure 10.2 Relationship between impulse sequence and actual UWB signal

where T_k and a_k are the transmit time and amplitude modulation of the kth pulse, respectively. For analysis, it is often useful to define

$$d(t) \equiv \sum_k a_k \delta(t - T_k) \qquad D(f) = \sum_k a_k e^{-j2\pi f T_k}, \tag{2}$$

where $\delta(t)$ is the Dirac delta function. The UWB signal is then $w(t) = p(t) * d(t)$, where $*$ denotes convolution. Figure 10.2 shows the relationship between $d(t)$, $p(t)$, and $w(t)$.

The Fourier transform of the UWB signal is

$$W(f) = P(f)D(f) = P(f) \sum_k a_k e^{-j2\pi f T_k}. \tag{3}$$

Since the $\{a_k\}$ and $\{T_k\}$ are in general random, the UWB signal must be modeled as a random process, and a more useful frequency-domain description of the UWB signal is the PSD, which represents the average power per Hertz as a function of frequency. The PSD of a process $w(t)$ is denoted here by $\overline{S}_{ww}(f)$, and has units of watts per Hertz. In the case of the UWB signal,

$$\overline{S}_{ww}(f) = |P(f)|^2 \overline{S}_{dd}(f). \tag{4}$$

The term $|P(f)|^2$ is the magnitude squared of the Fourier transform of the pulse waveform $p(t)$:

$$P(f) = \int_{-\infty}^{\infty} p(t) e^{-j2\pi f t}\, dt = |P(f)| e^{j\psi(f)}, \tag{5}$$

and $|P(f)|^2$ represents the energy spectral density (ESD) of a single pulse, in units of joules per Hertz. The PSD of the UWB signal therefore can be represented as the product of two components: the ESD of the pulse itself, which provides the overall large-scale 'shape' of the spectrum, and the PSD of the process $d(t)$, which determines the fine structure that depends on how the pulse is modulated in amplitude and positioned in time. A detailed analysis of the PSD is provided later in this chapter.

The Gaussian pulse or one of its derivatives is often used to represent $p(t)$. While the pulse shape *per se* is normally not of great importance in interference analysis, it is worthwhile to have some examples available. Table 10.1 expresses the Gaussian pulse $g_0(t)$ and its first two derivatives $g_1(t)$ and $g_2(t)$ as well as their energy spectra, in terms of the associated energy per pulse and equivalent rectangular (one-sided) bandwidth.

Figure 10.3 shows the energy- and bandwidth-normalized versions of these pulses; for example, $x_0(t) = 2e^{-8\pi t^2}$ and $g_0(t) = \sqrt{E_0 B_0} x_0(B_0 t)$. Figures 10.4 and 10.5 show the normalized energy spectra on linear and dB scales, respectively. Note that f_1 and f_2 are

Table 10.1 Waveforms and energy spectra for the Gaussian pulse and its first two derivatives

$g_0(t) = 2\sqrt{E_0 B_0} e^{-8\pi(B_0 t)^2}$	$\lvert G_0(f)\rvert^2 = \frac{E_0}{2B_0} e^{-\pi\left(\frac{f}{2B_0}\right)^2}$
$g_1(t) = -\frac{4}{e}\sqrt{E_1 B_1}\frac{t}{t_1} e^{\frac{1}{2}\left[1-(t/t_1)^2\right]}$ $t_1 = \frac{e}{8\sqrt{\pi} B_1} = \frac{1}{2\pi f_1}$	$\lvert G_1(f)\rvert^2 = \frac{E_1}{2B_1}\left(\frac{f}{f_1}\right)^2 e^{1-(f/f_1)^2}$ $f_1 = \frac{4}{e\sqrt{\pi}} B_1 = \frac{1}{2\pi t_1}$
$g_2(t) = \frac{16}{3e}\sqrt{E_2 B_2} e^{-(\pi f_2 t)^2}[2(\pi f_2 t)^2 - 1]$ $f_2 = \frac{32}{3e^2}\sqrt{\frac{2}{\pi}} B_2$	$\lvert G_2(f)\rvert^2 = \frac{E_2}{2B_2}\left(\frac{f}{f_2}\right)^4 e^{2[1-(f/f_2)^2]}$

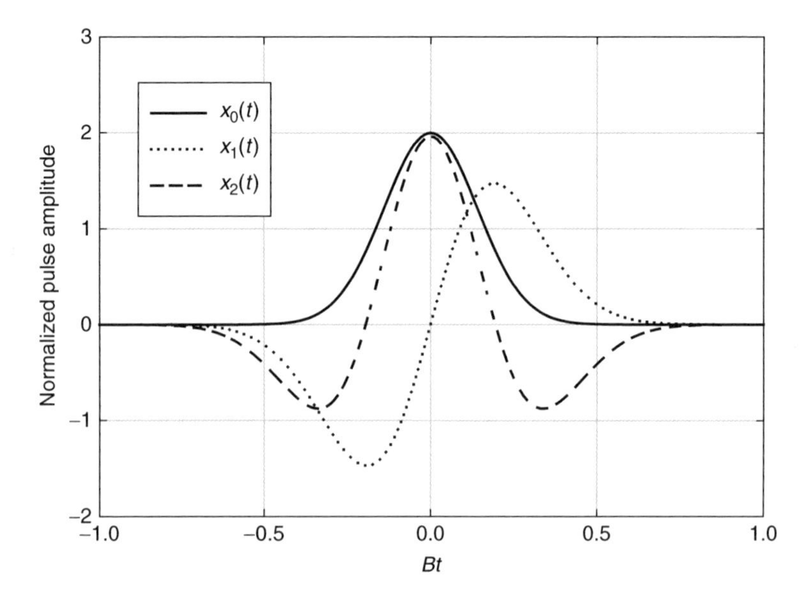

Figure 10.3 The normalized Gaussian pulse $x_0(t)$ and its first two derivatives $x_1(t)$ and $x_2(t)$

the frequencies at which the energy spectra of $G_1(f)$ and $G_2(f)$ reach their respective maximum values, and $\pm t_1$ are the times at which $g_1(t)$ reaches its maximum power.

Often, the UWB timing structure consists of intervals of duration T, with one pulse transmitted in each interval. The position of the pulse within the frame may vary according to pulse-position modulation (PPM) and/or pseudorandom dithering of the pulse position. Hence, $T_k = kT + \gamma_k$, where γ_k includes the combined effects of PPM and dithering. The, average pulse repetition frequency (PRF) is $R = 1/T$.

The energy in a single pulse is $E_p = \int_{-\infty}^{\infty} |P(f)|^2\, df$ joules, and the total average power of the UWB signal is $P_{\text{uwb}} = \left\langle a_k^2 \right\rangle R E_p$ watts, where $\langle\cdot\rangle$ denotes expectation. An

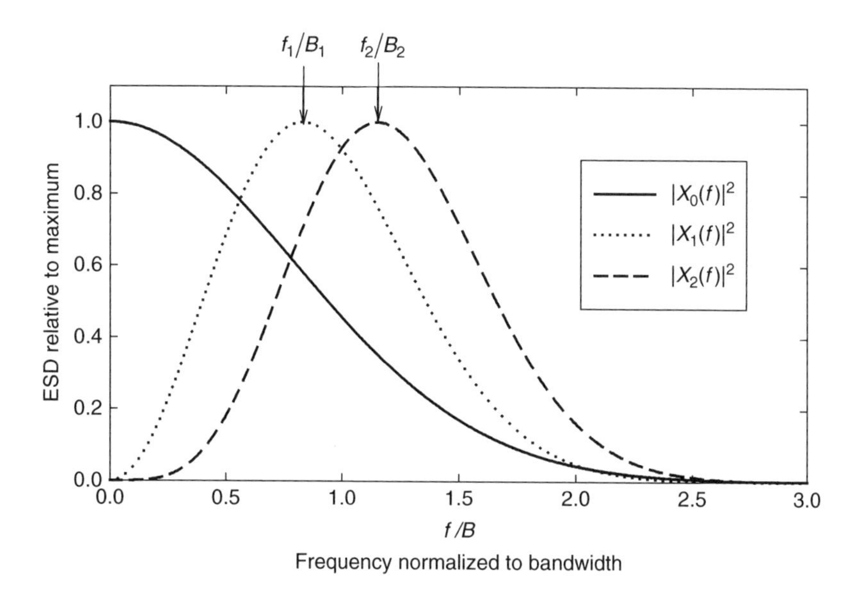

Figure 10.4 Normalized energy spectral densities of the Gaussian pulses

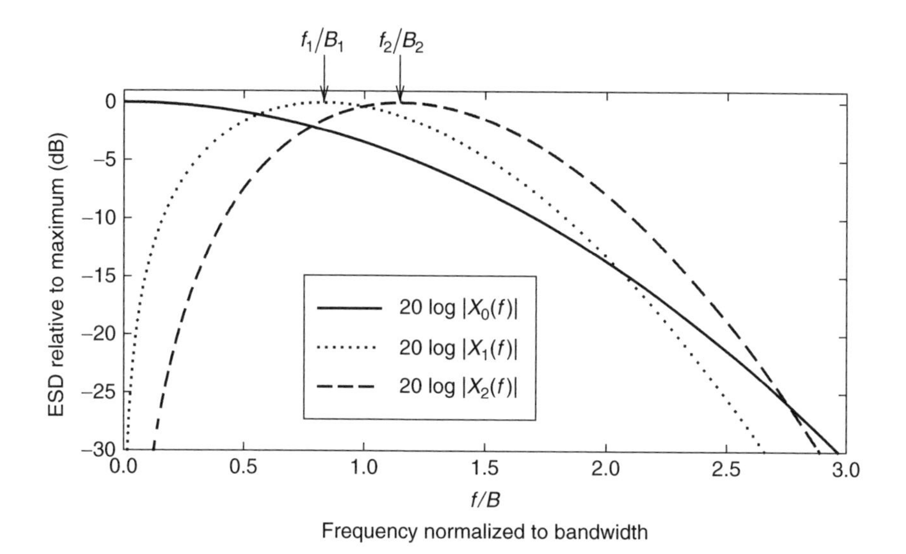

Figure 10.5 Normalized energy spectra of the Gaussian pulses, decibel scale

equivalent rectangular UWB pulse bandwidth can be defined as

$$B_p = \frac{E_p}{2|P(f)|^2_{\max}} \text{ Hz}, \tag{6}$$

where the factor of 2 in the denominator reflects the two-sided definition (positive and negative frequencies) of $P(f)$. It should be noted that since the pulse waveform $p(t)$ is real

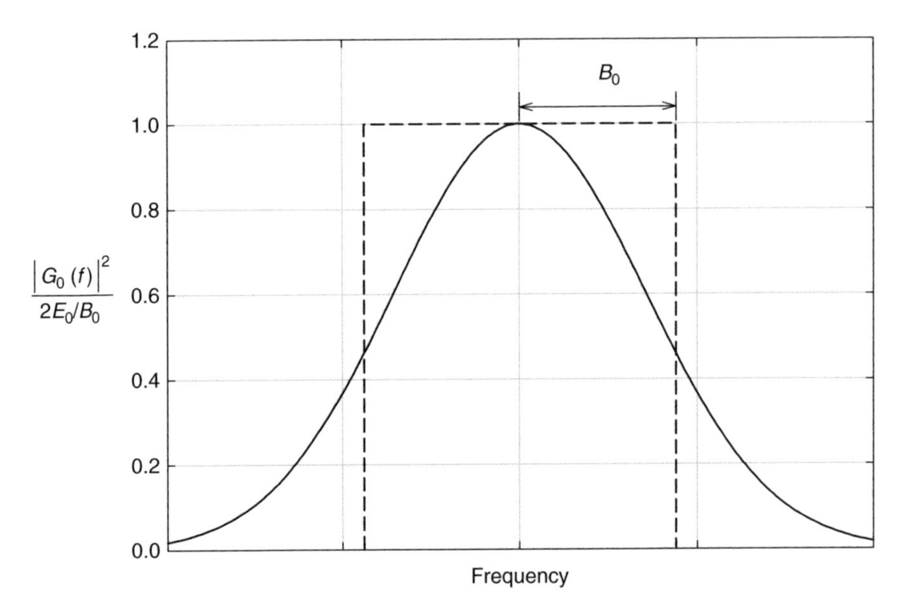

Figure 10.6 Energy spectral density and equivalent rectangular bandwidth for a Gaussian pulse

(has no imaginary component), $P(f)$ is conjugate-symmetric; that is $P(-f) = P^*(f)$; therefore, $|P(f)| = |P(-f)|$. Figure 10.6 shows the ESD and rectangular bandwidth for the Gaussian pulse $g_0(t)$.

10.3 Narrowband Receiver Model

Figure 10.7 shows a linear system model that represents the receive chain, where $H_0(f)$, $H_1(f)$, and $H_2(f)$ are linear band-pass filters. If the receiver can tune to multiple channels, the channel filtering is typically provided by the IF filters $H_1(f)$ and $H_2(f)$; the RF filter $H_0(f)$ will often have a bandwidth that spans all the channels. The objective is to find an expression for the output of the final IF filter, denoted $g(t)$, given that the input is a pulse sequence.

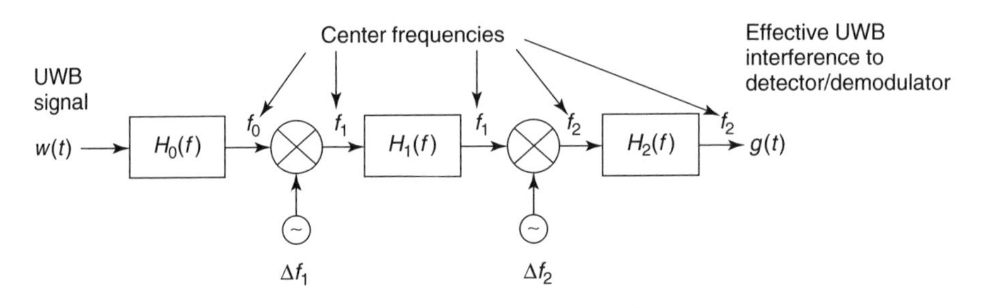

Figure 10.7 Linear system model of receiver

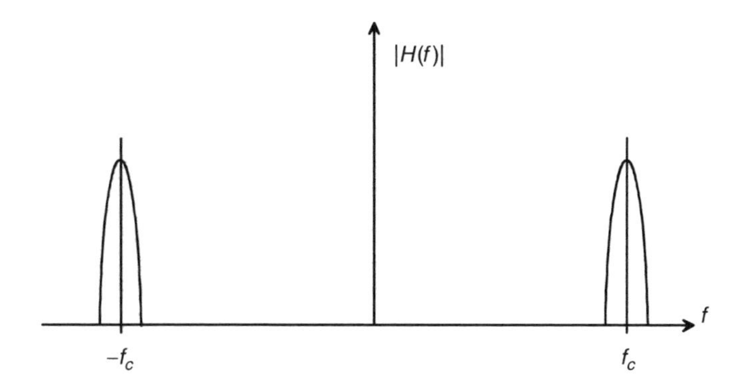

Figure 10.8 Illustration of a band-pass filter

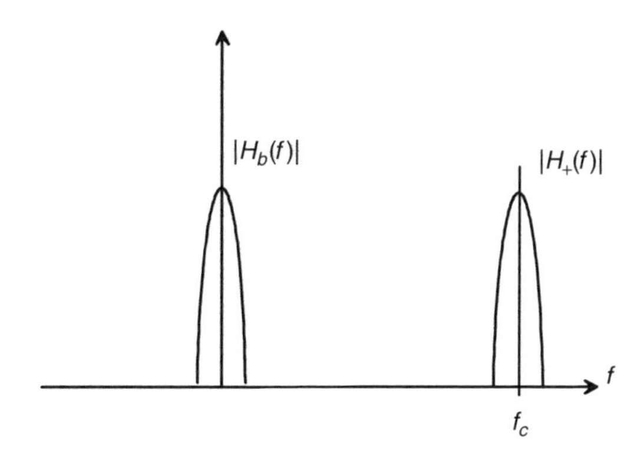

Figure 10.9 Analytic and baseband (translated) spectra of $H(f)$

The use of baseband-equivalent representations will facilitate this objective, so a brief summary of baseband–band-pass relationships is provided here (see, for example, [1], Section 4.2 and [2], Section 4.1 for more detail). Suppose that $H(f)$ is the frequency response of a band-pass filter with center frequency f_c, as shown in Figure 10.8.

Defining $H_+(f) = H(f)U(f)$, where $U(f)$ is the unit step function: $U(f) = 1$ for $f \geq 0$, $U(f) = 0$ for $f < 0$, the baseband equivalent is $H_b(f) = H_+(f + f_c)$, as shown in Figure 10.9. Hence, $H_+(f) = H_b(f - f_c)$. Since the filter impulse response $h(t)$ is assumed real, its transform is conjugate-symmetric, that is, $H(-f) = H^*(f)$. Therefore,

$$\begin{aligned} H(f) &= H_+(f) + H_+^*(-f) \\ &= H_b(f - f_c) + H_b^*(-f - f_c), \end{aligned} \tag{7}$$

which gives the band-pass transfer function $H(f)$ in terms of its baseband equivalent $H_b(f)$ and center frequency f_c.

Taking the inverse Fourier transform gives the impulse response

$$h(t) = h_b(t)e^{j2\pi f_c t} + h_b^*(t)e^{-j2\pi f_c t} = 2\text{Re}\,[h_b(t)e^{j2\pi f_c t}], \tag{8}$$

and $h_b(t) = e^{-j2\pi f_c t} h_+(t)$.

In general, $h_b(t)$ is complex, and $2h_b(t)$ is sometimes called the *complex envelope* of $h(t)$. Also, $2h_+(t) = h(t) + j\hat{h}(t)$, where $2h_+(t)$ is the analytic part of $h(t)$ and $\hat{h}(t)$ is the Hilbert transform of $h(t)$.

Band-pass *signals* can be similarly represented. Let $a(t)$ and $b(t)$ be the input and output of the linear system with impulse response $h(t)$, and define $A_+(f) = 2A(f)U(f)$ and $A_b(f) = A_+(f + f_c)$. Therefore, $A(f) = \frac{1}{2}[A_b(f - f_c) + A_b^*(-f - f_c)]$ and $a(t) = \text{Re}\,[a_b(t)e^{j2\pi f_c t}]$. The transform of the output signal is $B(f) = A(f)H(f)$. Hence, following [1], $B_+(f) = 2B(f)U(f) = A_+(f)H_+(f)$ and

$$\begin{aligned} B_b(f) &= B_+(f + f_c) = A_+(f + f_c)H_+(f + f_c) \\ &= A_b(f)H_b(f). \end{aligned} \tag{9}$$

The total signal energy of $a(t)$ is half that of $a_b(t)$:

$$\begin{aligned} E_a &= \int_{-\infty}^{\infty} |A(f)|^2\,df = \frac{1}{4}\int_{-\infty}^{\infty} [|A_b(f - f_c)|^2 + |A_b^*(-f - f_c)|^2]\,df \\ &= \frac{1}{2}\int_{-\infty}^{\infty} |A_b(f)|^2\,df. \end{aligned} \tag{10}$$

This assumes that $A_b(f - f_c)A_b^*(-f - f_c) = 0$.

10.4 Equivalent Receiver Model and Response to a Pulse

To analyze the receiver response to the UWB signal, it is useful to simplify the receiver model. Consider the output of the band-pass RF filter $H_0(f)$, which has a center frequency f_0 and baseband-equivalent $H_{b0}(f)$. For an input $P(f)$, the output is $g_0(t)$, and its transform is $G_0(f) = P(f)H_0(f)$. The corresponding baseband-equivalent signal is

$$\begin{aligned} G_{b0}(f) &= G_{0+}(f + f_0) = 2P(f + f_0)H_{0+}(f + f_0) \\ &= 2P(f + f_0)H_{b0}(f). \end{aligned} \tag{11}$$

and its inverse transform is $g_{b0}(t)$. The response of $H_0(f)$ to a pulse arriving at time T_k is therefore

$$g_{0,k}(t) = g_0(t - T_k) = \text{Re}\,[g_{b0}(t - T_k)\,e^{j2\pi f_0(t-T_k)}]. \tag{12}$$

The first mixer multiplies this signal by $2\cos(2\pi\Delta f_1 t + \alpha) = e^{j(2\pi\Delta f_1 t+\alpha)} + e^{-j(2\pi\Delta f_1 t+\alpha)}$ where $\Delta f_1 = f_0 - f_1$ and α is an arbitrary phase offset. Ignoring double-frequency terms, which will be rejected by $H_1(f)$, the signal at the first mixer output and its transform are

$$g_{01,k}(t) = \text{Re}\,[g_{b0}(t - T_k)\,e^{-j2\pi f_0 T_k}\,e^{j2\pi f_1 t}\,e^{-j\alpha}], \tag{13}$$

$$\begin{aligned} G_{01,k}(f) &= \frac{1}{2}e^{-j\alpha}e^{-j2\pi f_0 T_k}G_{b0}(f - f_1)\,e^{-j2\pi(f-f_1)T_k} \\ &\quad + \frac{1}{2}e^{j\alpha}e^{j2\pi f_0 T_k}G_{b0}^*(-f - f_1)\,e^{j2\pi(-f-f_1)T_k}. \end{aligned} \tag{14}$$

Multiplying by $H_1(f) = H_{b1}(f - f_1) + H_{b1}^*(-f - f_1)$ gives the filtered signal

$$
\begin{aligned}
G_{11,k}(f) = {} & \tfrac{1}{2} e^{-j\alpha} e^{-j2\pi f_0 T_k} G_{b1}(f - f_1)\, e^{-j2\pi(f-f_1)T_k} \\
& + \tfrac{1}{2} e^{j\alpha} e^{j2\pi f_0 T_k} G_{b1}^*(-f - f_1)\, e^{j2\pi(-f-f_1)T_k},
\end{aligned} \quad (15)
$$

where $G_{b1}(f) = G_{b0}(f)H_{b1}(f)$.

Down-converting to f_2 and filtering by $H_2(f) = H_{2b}(f - f_2) + H_{2b}^*(-f - f_2)$ gives the final output as

$$
\begin{aligned}
G_k(f) = {} & \tfrac{1}{2} e^{-j(\alpha+\beta)} e^{-j2\pi f_0 T_k} G_b(f - f_2)\, e^{-j2\pi(f-f_2)T_k} \\
& + \tfrac{1}{2} e^{j(\alpha+\beta)} e^{j2\pi f_0 T_k} G_b^*(-f - f_2)\, e^{j2\pi(-f-f_2)T_k},
\end{aligned} \quad (16)
$$

where

$$
\begin{aligned}
G_b(f) &= G_{b1}(f)H_{b2}(f) \\
&= 2P(f + f_0)H_{b0}(f)H_{b1}(f)H_{b2}(f),
\end{aligned} \quad (17)
$$

and β is the arbitrary phase shift introduced by the second mixer. The time-domain output is

$$g_k(t) = \text{Re}\,[e^{-j(\alpha+\beta)} g_b(t - T_k)\, e^{-j2\pi f_0 T_k} e^{j2\pi f_2 t}]. \quad (18)$$

It is clear from this result that the receive chain, up to the final IF output, can be represented using a single band-pass filter and a single down-conversion as shown in Figure 10.10, where

$$
\begin{aligned}
H(f) &= H_b(f - f_0) + H_b^*(-f - f_0) \\
H_b(f) &= H_{b0}(f)H_{b1}(f)H_{b2}(f).
\end{aligned} \quad (19)
$$

The down-conversion is assumed to include sufficient filtering to reject the double-frequency terms. The output of the system in Figure 10.10 in response to an input of $p(t - T_k)$ is the same as (18). For purposes of interest here, any phase shift introduced by the mixers can be ignored and the output can be represented as

$$g_k(t) = \text{Re}\,[g_b(t - T_k)e^{-j2\pi f_0 T_k} e^{j2\pi f_{\text{if}} t}]. \quad (20)$$

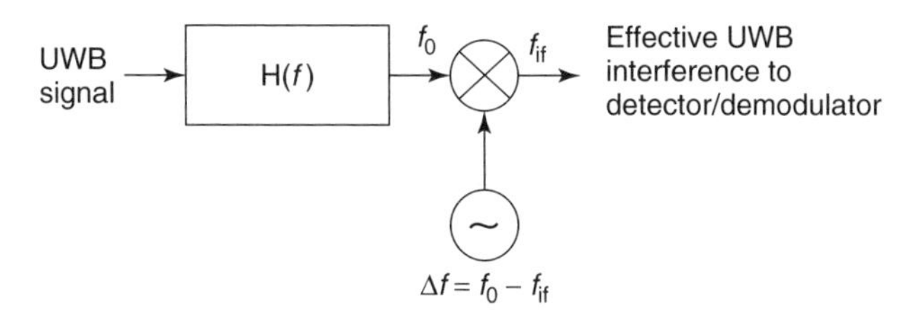

Figure 10.10 Equivalent narrowband receiver

The exact value of the final IF f_{if} usually is unimportant for interference analysis, which often will focus on the complex envelope at the IF output and the sum of the complex envelopes from a sequence of pulses:

$$g(t) = \sum_k g_k(t) = \text{Re}\left[e^{j2\pi f_{\text{if}} t} \sum_k a_k g_b(t - T_k) e^{-j2\pi f_0 T_k} \right], \tag{21}$$

that is, the complex envelope is the sum of the complex envelopes from the individual pulses. The next section develops a more specific relationship for $g(t)$, which can be used for simulation and analysis.

An important point here is that the phase relationships among the receive chain responses to individual pulses depend on the RF f_0 as well as the $\{T_k\}$, and are independent of the IFs. It is these phase relationships that determine the characteristics of the IF output when $R \gg B_{\text{if}}$ and the IF output is the sum of many overlapping impulse responses, where R is the pulse rate and B_{if} is the noise bandwidth of $H(f)$ in Figure 10.10.

10.5 Response to a Pulse Sequence

Figure 10.11 illustrates the general relationship between the effective receiver pass-band filtering, as represented by $H(f)$, and the UWB pulse spectrum. The receive chain sees only the portion of $P(f)$ near $\pm f_0$. It is assumed that the pulse bandwidth is much greater than the bandwidth of $H(f)$, and that: (1) the magnitude of $P(f)$ is roughly flat across the victim receiver pass band; that is, $|P(f)| \cong |P(f_0)|$; and (2) the phase is approximately linear. That is, letting $\psi_0 = \psi(f_0)$ and $\psi_0' = \psi'(f_0) = d\psi/df|_{f=f_0}$, then within the receiver pass band,

$$\begin{aligned} \psi(f) &\cong \psi_0 + (f - f_0)\psi_0' \qquad && f > 0 \\ \psi(f) &\cong -\psi_0 + (f + f_0)\psi_0' \qquad && f < 0. \end{aligned} \tag{22}$$

Within the pass band of $H(f)$, the UWB pulse spectrum can be approximated as

$$P(f) \cong \begin{cases} |P(f_0)| e^{j[\psi_0 + \psi_0'(f - f_0)]} & f > 0 \\ |P(f_0)| e^{j[-\psi_0 + \psi_0'(f + f_0)]} & f < 0 \end{cases} \tag{23}$$

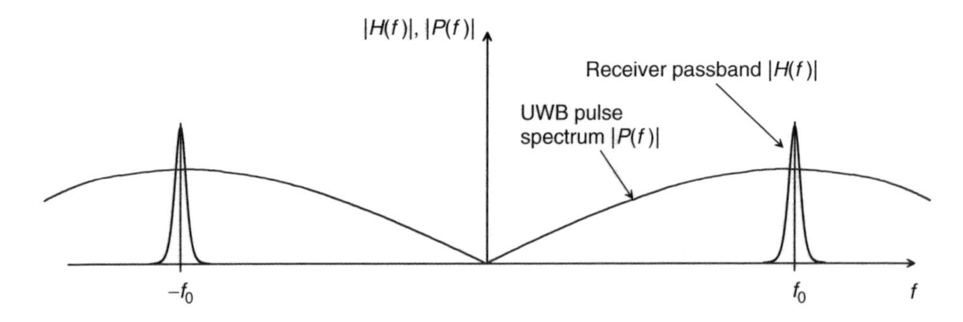

Figure 10.11 UWB pulse spectrum and narrowband receiver pass band (illustrative)

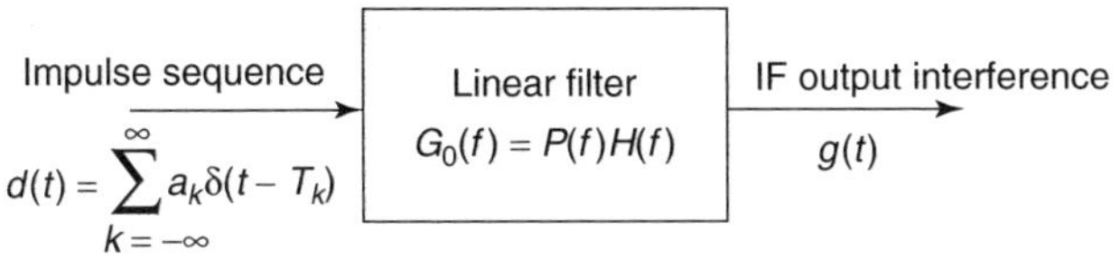

Figure 10.12 Relationship between input impulse sequence and IF output interference

Ignoring the frequency conversion for the moment, the output of the effective band-pass filter is

$$G_0(f) = P(f)H(f) \\ = |P(f_0)| \left\{ e^{j[\psi_0 + \psi_0'(f - f_0)]} H_b(f - f_0) + e^{-j[\psi_0 + \psi_0'(-f - f_0)]} H_b^*(-f - f_0) \right\}. \quad (24)$$

Figure 10.12 shows the relationship between the impulse sequence $d(t)$, $G_0(f)$, and the IF output interference $g(t)$.

From (17),

$$G_b(f) \equiv 2|P(f_0)|\, e^{j(\psi_0 + \psi_0' f)} H_b(f), \quad (25)$$

and (24) becomes $G_0(f) = \frac{1}{2} G_b(f - f_0) + \frac{1}{2} G_b^*(-f - f_0)$, and $g_0(t) = \mathrm{Re}\,[g_b(t) e^{j2\pi f_0 t}]$. From (25),

$$g_b(t) = 2|P(f_0)| e^{j\psi_0} h_b \left(t + \frac{\psi_0'}{2\pi} \right). \quad (26)$$

Thus,

$$g_0(t) = 2|P(f_0)| \mathrm{Re} \left[h_b \left(t + \frac{\psi_0'}{2\pi} \right) e^{j(2\pi f_0 t + \psi_0)} \right]. \quad (27)$$

The term $-\psi_0'/2\pi$ represents the *group delay* (see, e.g., [1], pp. 123–4). Letting $t_g = -\psi_0'/2\pi$, (27) can be written as

$$g_0(t) = 2|P(f_0)| \mathrm{Re}\,[h_b(t - t_g) e^{j(2\pi f_0 t + \psi_0)}]. \quad (28)$$

The final IF output, in response to a pulse sequence, is therefore

$$g(t) = 2|P(f_0)| \mathrm{Re} \left[e^{j2\pi f_{\mathrm{if}} t} e^{j\psi_0} \sum_k a_k e^{-j2\pi f_0 T_k} h_b(t - t_g - T_k) \right]. \quad (29)$$

For most purposes related to interference analysis, t_g and ψ_0 can be ignored, and the IF output can be represented as

$$g(t) = 2|P(f_0)| \mathrm{Re} \left[e^{j2\pi f_{\mathrm{if}} t} \sum_k a_k e^{-j2\pi f_0 T_k} h_b(t - T_k) \right]. \quad (30)$$

At this point, it is clear that the important factors in determining the characteristics of the interference presented to the detector/demodulator stage are as follows:

1. The ESD $|P(f)|^2$ of the pulse at the center frequency f_0 of the narrowband receiver, is represented by $|P(f_0)|$. Note that the *shape* of the pulse is unimportant.
2. The relationship between the pulse arrival times $\{T_k\}$ and the receiver center frequency f_0. As a simple illustration, assume that the pulses are transmitted at a regular interval T; that is, $T_k = kT$. The spectrum of the UWB signal then consists of discrete tones at frequencies that are harmonics of $1/T$. If f_0 is at one of these harmonics, then the receiver responses to successive pulses add constructively (perfectly in phase). From a frequency-domain perspective, a tone appears within the receiver pass band.
3. The effective impulse response of the cascaded filtering sections, as represented by $h_b(t)$, and its time duration relative to the interval between pulses. In the frequency domain, this corresponds to the relationship between the pulse rate and the bandwidth of $H(f)$, denoted B_{if}.

The term $|P(f_0)|$ is simply a scaling factor that can be related to the received pulse energy and bandwidth by

$$|P(f_0)| = A(f_0)\sqrt{\frac{E_p}{2B_p}}, \tag{31}$$

where

$$A(f) = \frac{|P(f)|}{|P(f)|_{\max}}, \tag{32}$$

is the magnitude of the pulse spectrum relative to its maximum. Thus, $A_{\max}(f) = 1$.

The relationships among the $\{T_k\}$, the $\{a_k\}$, f_0, and B_{if}, and their effect on the IF output interference are more involved and often require PSD analysis and simulation to fully understand. However, some initial qualitative discussion is useful as a preface.

In the time domain, each UWB pulse effectively acts as an impulse input to the filter $H(f)$, so the filter output in response to each UWB pulse is a scaled, time-shifted version of the filter impulse response $h(t)$ as shown in Figure 10.13. What is actually shown is $h_b(t)$, which is the *envelope* of the impulse response. The impulse response is the carrier f_0 amplitude-modulated by the envelope $h_b(t)$.

The actual UWB signal is a sequence of pulses, and the IF response will be a sequence of impulse responses as shown in Figure 10.14. If response envelopes overlap, the composite output signal will depend on the phase relationships among the successive responses. If the pulse rate is low compared to the IF bandwidth, there will be no significant overlap and the receiver will respond to each pulse individually. As will be seen, in this case the peak envelope power output of the IF stage varies as the square of the IF bandwidth.

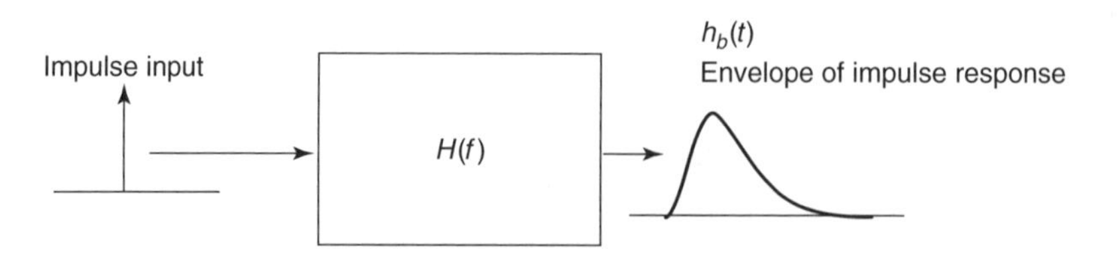

Figure 10.13 Response of the receive chain to a single pulse

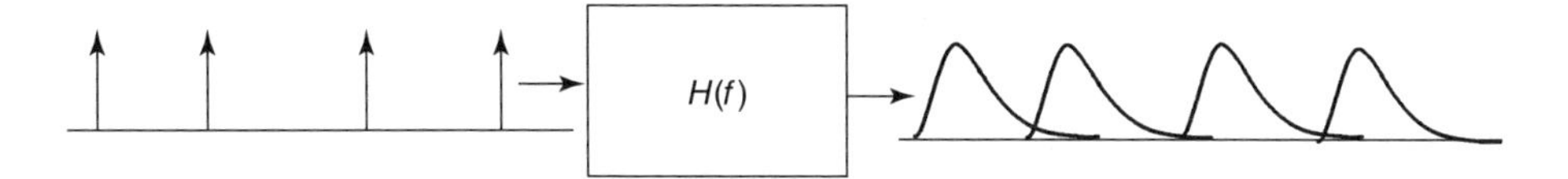

Figure 10.14 Response of the chain to a pulse sequence

10.6 Simulating the Response to a Pulse Sequence

10.6.1 I/Q Component Formulation

It is useful to further develop the expression in (30) for the time-domain IF output interference, both for simulation purposes and to support some physical insights into the characteristics of the interference. To those ends, (30) can be expressed in terms of in-phase and quadrature (I/Q) components. Letting

$$x_k(t) = \mathrm{Re}\{a_k e^{-j2\pi f_0 T_k} h_b(t - T_k)\} \qquad x(t) = \sum_k x_k(t)$$
$$y_k(t) = \mathrm{Im}\{a_k e^{-j2\pi f_0 T_k} h_b(t - T_k)\} \qquad y(t) = \sum_k y_k(t), \tag{33}$$

(30) becomes

$$g(t) = 2|P(f_0)|[x(t)\cos(2\pi f_{\mathrm{if}} t) - y(t)\sin(2\pi f_{\mathrm{if}} t)]. \tag{34}$$

The rms envelope of the interference at the IF output is

$$r(t) = \sqrt{2}|P(f_0)|\sqrt{x^2(t) + y^2(t)}, \tag{35}$$

and the envelope power is $r^2(t)$. In terms of $r(t)$, the final IF output is

$$g(t) = \sqrt{2} r(t) \cos\left[2\pi f_{\mathrm{if}} t + \tan^{-1}\frac{y(t)}{x(t)}\right]. \tag{36}$$

In general, $h_b(t)$ is complex and can be written as $h_b(t) = |h_b(t)|e^{j\theta(t)}$. Hence,

$$x_k(t) = a_k|h_b(t - T_k)|\cos[\theta(t - T_k) - 2\pi f_0 T_k]$$
$$y_k(t) = a_k|h_b(t - T_k)|\sin[\theta(t - T_k) - 2\pi f_0 T_k]. \tag{37}$$

If $H_b(f)$ is conjugate-symmetric, that is, $H_b(f) = H_b^*(-f)$, then $h_b(t)$ is real and $\theta(t) = 0$, in which case

$$x_k(t) = a_k h_b(t - T_k)\cos(2\pi f_0 T_k)$$
$$y_k(t) = -a_k h_b(t - T_k)\sin(2\pi f_0 T_k). \tag{38}$$

In either case, given the algorithms to generate the $\{a_k\}$, $\{T_k\}$, and an expression for the baseband-equivalent IF response $h_b(t)$, the envelope of the receiver IF response can be readily simulated.

10.6.2 Simulation Parameters

The simulation must compute sampled versions of $x(t)$ and $y(t)$; that is, sequences $x_n = x(n\Delta t)$ and $y_n = y(n\Delta t)$. The simulation sampling rate $R_S(= 1/\Delta t)$ must be high enough to capture the IF filter response with reasonable resolution. This translates to the constraint $R_S \geq J_{\min} B_{\text{if}}$ where $J_{\min}$ is the minimum number of samples per IF filter time constant.

Letting K_S represent the number of simulation samples per UWB frame, and R the average UWB pulse rate, then $R_S = K_S R$, so $K_S \geq J_{\min} B_{\text{if}}/R$. Adding the constraint $K_S \geq 1$ (i.e., there must be at least one sample per UWB frame) gives the sampling rate as

$$R_S = \left\lceil J_{\min} \frac{B_{\text{if}}}{R} \right\rceil R, \tag{39}$$

where $\lceil \cdot \rceil$ denotes the smallest integer greater than or equal to the argument.

Therefore, if $B_{\text{if}} \ll R$, then $R_S = R$ (one sample per UWB frame); conversely, if $B_{\text{if}} \gg R$, then $R_S = J_{\min} B_{\text{if}}$. Thus, the constraint on the sampling rate depends on the relationship between the IF bandwidth and the pulse rate.

The IF impulse response has significant magnitude for some time interval $T_{\text{if}} = K_{T_{\text{if}}}/B_{\text{if}}$ (for an n-pole filter, $K_{T_{\text{if}}}$ is of the order of 2), so each sample of the IF output must account for the additive effect of RT_{if} UWB pulses. Each simulation sample is therefore the sum of L delayed and weighted IF impulse response samples, where $L = \lceil RK_{T_{\text{if}}}/B_{\text{if}} \rceil$, or in the case of the n-pole filter discussed below, $L = \lceil 2R/B_{\text{if}} \rceil$. The simulated I/Q components are then

$$x_n = \sum_{l=1}^{L} a_l |h_b(n\Delta t - t_l)| \cos[\theta(n\Delta t - t_l) - 2\pi f_0 t_l]$$

$$y_n = \sum_{l=1}^{L} a_l |h_b(n\Delta t - t_l)| \sin[\theta(n\Delta t - t_l) - 2\pi f_0 t_l], \tag{40}$$

where n is the sample index, $\Delta t = 1/R_S$ is the simulation sample interval, t_L is the arrival time of the most recent UWB pulse, t_{L-1} is the arrival time of the pulse prior to that, and so on.

10.6.3 Normalization

To make the simulation as general as possible, it is useful to normalize simulation time to the effective IF bandwidth B_{if}; that is, simulation time is $\tau = B_{\text{if}} t$. The normalized (unity-bandwidth) baseband impulse response is

$$h_1(t) = \frac{h_b(t/B_{\text{if}})}{B_{\text{if}}}. \tag{41}$$

Thus, $h_b(t) = B_{\text{if}} h_1(B_{\text{if}} t)$, where $h_1(t)$ and its argument are dimensionless and $h_b(t)$ has dimensions of 1/sec as should be the case for an impulse response. It is easily seen that $H_b(f) = H_1(f/B_{\text{if}})$, and $\int_{-\infty}^{\infty} |h_b(t)^2\, dt| = B_{\text{if}} \int_{-\infty}^{\infty} |h_1(t)^2\, dt|$. Equivalently, $\int_{-\infty}^{\infty} |H_b(f)|^2\, df = B_{\text{if}} \int_{-\infty}^{\infty} |H_1(f)|^2\, df$. Clearly, $H_b(0)$ is independent of B_{if}, so the energy captured by $H(f)$, assuming a spectrally flat input, is proportional to B_{if}. If $|H_b(0)| = 1$, then $\int_{-\infty}^{\infty} |H_1(f)|^2\, df = 1$ and $\int_{-\infty}^{\infty} |H_b(f)|^2\, df = \int_{-\infty}^{\infty} |h_b(t)^2\, dt| = B_{\text{if}}$.

In the simulation, the timescale and center frequency f_0 must also be scaled by B_{if}. Defining the normalized simulation timescale as $\tau = B_{\text{if}}t$, with $\tau_k = B_{\text{if}}T_k$, the normalized I/Q components are

$$c_k(\tau) = a_k|h_1(\tau - \tau_k)|\cos\left[\theta_1(\tau - \tau_k) - 2\pi\frac{f_0}{B_{\text{if}}}\tau_k\right]$$
$$s_k(\tau) = a_k|h_1(\tau - \tau_k)|\sin\left[\theta_1(\tau - \tau_k) - 2\pi\frac{f_0}{B_{\text{if}}}\tau_k\right], \tag{42}$$

where $\theta_1(t) = \theta(t/B_{\text{if}})$. Thus, $x_k(t) = B_{\text{if}}c_k(B_{\text{if}}t)$ and $y_k(t) = B_{\text{if}}s_k(B_{\text{if}}t)$. The normalized I/Q terms computed in the simulation are

$$c_n = \sum_{l=1}^{L} a_l|h_1(n\Delta\tau - \tau_l)|\cos[\theta_1(n\Delta\tau - \tau_l) - 2\pi f_0\tau_l/B_{\text{if}}]$$
$$s_n = \sum_{l=1}^{L} a_l|h_1(n\Delta\tau - \tau_l)|\sin[\theta_1(n\Delta\tau - \tau_l) - 2\pi f_0\tau_l/B_{\text{if}}]. \tag{43}$$

The envelope is then

$$r\left(\frac{n\Delta\tau}{B_{\text{if}}}\right) = \sqrt{2}|P(f_0)|B_{\text{if}}\sqrt{c_n^2 + s_n^2}. \tag{44}$$

10.6.4 Example Filter Response: The ***n*****-Pole Filter*

An expression is needed for the baseband-equivalent IF response $h_b(t)$. Consider the low-pass n-pole filter with the transfer function and impulse response:

$$H_b(f) = \frac{1}{(1 + j2\pi f/\alpha)^n} \qquad h_b(t) = \frac{1}{(n-1)!}\alpha^n t^{n-1}e^{-\alpha t}, \tag{45}$$

where the parameter α sets the bandwidth. Specifically, if B_{if} is the noise-equivalent bandwidth[1]

$$B_{\text{if}} = \frac{\int_{-\infty}^{\infty}|H_b(f)|^2\,df}{|H_b(0)|^2}. \tag{46}$$

With $|H_b(0)| = 1$,

$$B_{\text{if}} = \int_{-\infty}^{\infty}|H_b(f)|^2\,df = \frac{\alpha}{2\pi}\int_{-\infty}^{\infty}\frac{d\xi}{(1+\xi^2)^n}$$
$$= \frac{\alpha}{2\pi}\frac{\Gamma\left(\frac{1}{2}\right)\Gamma\left(n - \frac{1}{2}\right)}{(n-1)!} = \frac{\alpha}{2}\prod_{k=1}^{n-1}\frac{2k-1}{2k}, \tag{47}$$

then $\alpha = B_{\text{if}}x_n$ where $x_n = \frac{\alpha}{B_{\text{if}}} = 2\prod_{k=1}^{n-1}\frac{2k}{2k-1}$.

[1] Note that bandwidth is defined differently for the baseband-equivalent transfer function than for the UWB pulse $P(f)$. The reason is that the bandwidth definition for $H_b(f)$ will yield a definition for the band-pass function $H(f)$ that is consistent with that used for $P(f)$.

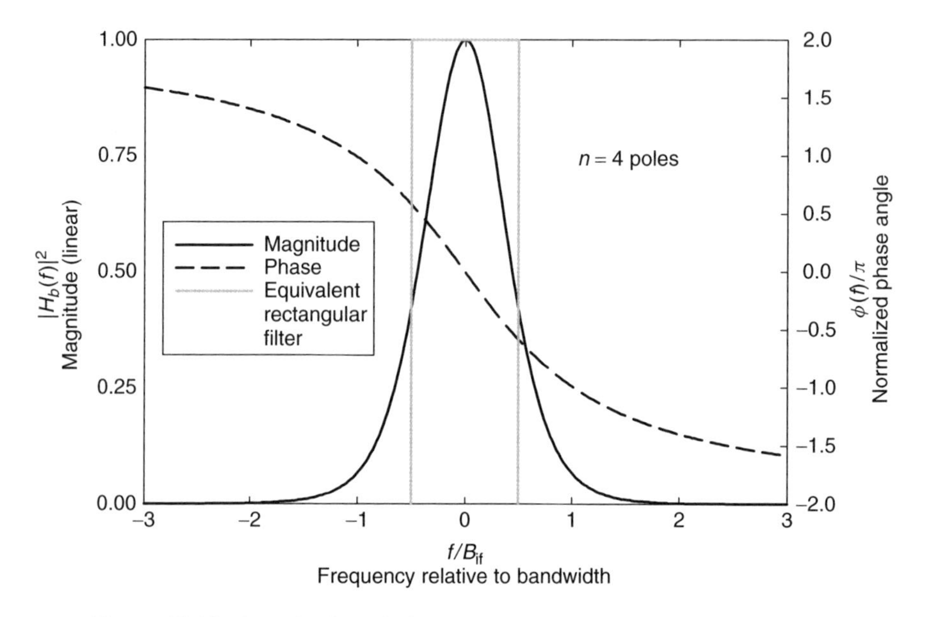

Figure 10.15 Magnitude and phase response of the 4-pole baseband filter

The magnitude and phase are, respectively, $|H_b(f)| = [1 + (2\pi f/\alpha)^2]^{-n/2}$ and $\phi(f) = -n\tan^{-1}(2\pi f/\alpha)$. Figure 10.15 shows the magnitude $|H_b(f)|^2$ and $\phi(f)$ for a 4-pole filter versus normalized frequency f/B_{if}. This filter is a useful and relevant example because it is real, causal, symmetric, and physically realizable (for example, using R-C sections). Note that within the filter pass band, the phase is approximately linear.

Expressed in terms of B_{if}, the impulse response of the filter is

$$h_b(t) = \frac{B_{\text{if}}x^n}{(n-1)!}(B_{\text{if}}t)^{n-1}e^{-xB_{\text{if}}t}, \tag{48}$$

and the normalized impulse response is therefore

$$h_1(t) = \frac{h_b(t/B_{\text{if}})}{B_{\text{if}}} = \frac{x^n}{(n-1)!}t^{n-1}e^{-xt}. \tag{49}$$

Hence, $h_b(t) = B_{\text{if}} \cdot h_1(B_{\text{if}}t)$. Note that $\int_0^\infty h_1(t)\,dt = \int_0^\infty h_b(t)\,dt = H_b(0)$. Figure 10.16 shows $h_b(t)/B_{\text{if}}$ versus $B_{\text{if}}t$. Note that since $h_b(t)$ is real, the n-pole filter introduces no phase modulation in the impulse response ($\theta(t) = 0$).

The band-pass n-pole filter impulse and frequency responses are

$$h(t) = 2h_b(t)\cos 2\pi f_0 t$$

$$H(f) = \frac{1}{[1 + j2\pi(f - f_0)/\alpha]^n} + \frac{1}{[1 + j2\pi(f + f_0)/\alpha]^n}. \tag{50}$$

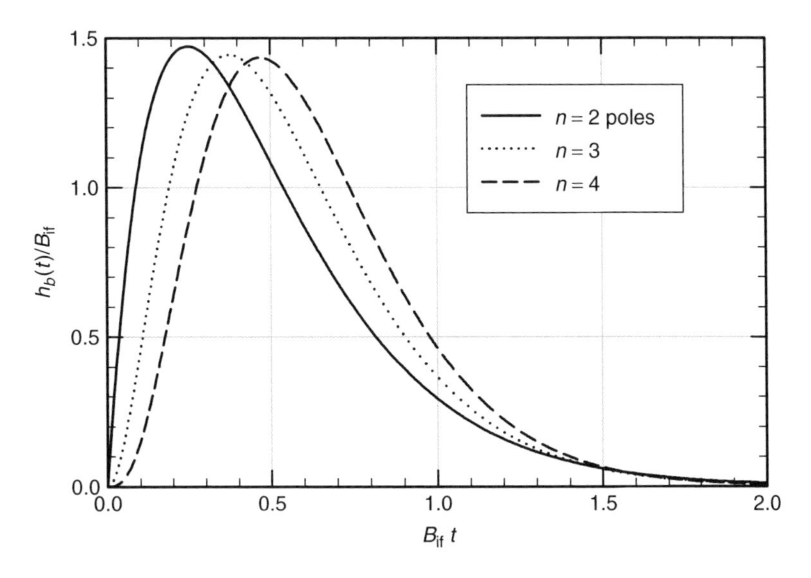

Figure 10.16 Normalized baseband-equivalent impulse response of n-pole filter

10.7 General Properties of the IF Output

While a detailed characterization of the final IF output requires simulation and/or PSD analysis, there are some basic observations that are useful for an initial assessment of interference effects.

10.7.1 Case 1: Pulse Rate Less than IF Bandwidth

If $R < B_{\text{if}}$, then the response time of $h_b(t)$ is less than the time between pulses, and the IF output is a sequence of isolated impulse responses. The envelope of the kth response can be written as

$$\begin{aligned} r_k(t) &= \sqrt{2}|P(f_0)||a_k||h_b(t - T_k)| \\ &= \sqrt{2}B_{\text{if}}|P(f_0)||a_k||h_1(t - T_k)|. \end{aligned} \tag{51}$$

With $|P(f_0)| = A(f_0)\sqrt{E_p/2B_p}$, the peak envelope power is

$$P_{\text{peak}} = B_{\text{if}}^2 A^2(f_0)\frac{E_p}{B_p}|a_k|^2|h_1(t)|_{\max}^2. \tag{52}$$

With $P_{\text{uwb}} = \left\langle a_k^2 \right\rangle RE_p$, and assuming $a_k = 1$,

$$P_{\text{peak}} = \frac{B_{\text{if}}^2 P_{\text{uwb}}}{RB_p}A^2(f_0)|h_1(t)|_{\max}^2. \tag{53}$$

Thus, for a given composite filter impulse response form as dictated by $h_1(t)$, the peak power varies as the square of the IF bandwidth. For the n-pole filter, $h_{1,\max} \cong \sqrt{2}$.

The energy in the IF output pulse is

$$E_{\text{if}} = 2|P(f_0)|^2|a_k|^2 \int_{-\infty}^{\infty} |h_b(t - T_k)|^2\, dt$$
$$= |a_k|^2 A^2(f_0) \frac{B_{\text{if}}}{B_p} E_p, \tag{54}$$

which is not a surprising result. This assumes that $|H_b(0)| = 1$, so that $\int_{-\infty}^{\infty} |h_b(t)|^2\, dt = B_{\text{if}}$. If $|a_k| = 1$ and the receiver center frequency f_0 is at the spectral peak of the pulse, then $E_{\text{if}} = E_p \cdot B_{\text{if}}/B_p$ as would be expected. Note that $P_{\text{peak}}/E_{\text{if}} = B_{\text{if}}|h_1(t)|^2_{\max}$.

If the UWB pulse has the same shape factor $|h_1(t)|_{\max}$, then $P_{p,\text{peak}}/E_p = B_p|h_1(t)|^2_{\max}$, and

$$\frac{P_{\text{if,peak}}}{P_{p,\text{peak}}} = \frac{P_{\text{if,peak}}/E_{\text{if}}}{P_{p,\text{peak}}/E_p} \cdot \frac{E_{\text{if}}}{E_p} = \left(\frac{B_{\text{if}}}{B_p}\right)^2 |a_k|^2|A^2(f_0)|. \tag{55}$$

The average power output of the IF filter is

$$\overline{P}_{\text{if}} = E_{\text{if}} R = \left\langle |a_k|^2 \right\rangle A^2(f_0) \frac{B_{\text{if}}}{B_p} E_p R$$
$$= A^2(f_0) \frac{B_{\text{if}}}{B_p} P_{\text{uwb}}, \tag{56}$$

which again, is not a surprising result.

10.7.2 Case 2: Pulse Rate Greater than IF Bandwidth

If $R > B_{\text{if}}$, the IF responses overlap and the resulting IF output waveform depends on the phase relationships among them as determined by the offsets $\{2\pi f_0 T_k\}$. While a detailed understanding of this is best provided by the PSD analysis, it is useful to illustrate the basic principles for two simplified extreme cases. Representing the complex envelope of the IF output as

$$z(t) = x(t) + jy(t) = \sum_k a_k e^{-j2\pi f_0 T_k} h_b(t - T_k), \tag{57}$$

the envelope power is $r^2(t) = 2|P(f_0)|^2|z(t)|^2$. The average power is therefore

$$\overline{P}_{\text{if}} = \left\langle r^2(t) \right\rangle = 2|P(f_0)|^2 \left\langle |z(t)|^2 \right\rangle, \tag{58}$$

where

$$\left\langle |z(t)|^2 \right\rangle = \sum_k \sum_n \left\langle a_k a_n^* e^{-j2\pi f_0 (T_k - T_n)} \right\rangle \left\langle h_b(t - T_k) h_b^*(t - T_n) \right\rangle. \tag{59}$$

At one extreme, $\left\langle a_k a_n^* e^{-j2\pi f_0(T_k - T_n)}\right\rangle = 0$, $k \neq n$, in which case

$$\begin{aligned}\left\langle |z(t)|^2\right\rangle &= \sum_k \left\langle |a_k|^2\right\rangle\left\langle |h_b(t-T_k)|^2\right\rangle \\ &= \left\langle |a_k|^2\right\rangle \sum_k \left\langle |h_b(t-T_k)|^2\right\rangle,\end{aligned} \tag{60}$$

where the $\left\langle |a_k|^2\right\rangle$ is taken outside the sum because of the expectation, assuming the $\{a_k\}$ and the $\{T_k\}$ are uncorrelated. Assuming that the process $\sum_k |h_b(t-T_k)|^2$ is ergodic, a time average can be used to compute the expectation:

$$\begin{aligned}\left\langle \sum_k |h_b(t-T_k)|^2 \right\rangle &= \lim_{\tau\to\infty} \frac{1}{2\tau}\int_{-\tau}^{\tau} \sum_k |h_b(t-T_k)|^2\, dt \\ &= \lim_{\tau\to\infty} \sum_k \frac{1}{2\tau}\int_{-\tau}^{\tau} |h_b(t-T_k)|^2\, dt.\end{aligned} \tag{61}$$

In the interval 2τ, there are an average of $2\tau R$ pulses. Therefore,

$$\left\langle \sum_k |h_b(t-T_k)|^2 \right\rangle = R\int_{-\infty}^{\infty} |h_b(t)|^2\, dt = RB_{\text{if}}, \tag{62}$$

and hence, $\langle |z(t)|^2\rangle = \langle |a_k|^2\rangle RB_{\text{if}}$ and

$$\begin{aligned}\langle r^2(t)\rangle &= 2|P(f_0)|^2\left\langle |a_k|^2\right\rangle RB_{\text{if}} = A^2(f_0)\frac{E_p}{B_p}\left\langle |a_k|^2\right\rangle RB_{\text{if}} \\ &= A^2(f_0)P_{\text{uwb}}B_{\text{if}}/B_p.\end{aligned} \tag{63}$$

In other words, the UWB interference appears noiselike with an effective two-sided PSD of

$$N_{0\text{eff}} = \frac{A^2(f_0)}{2B_p}P_{\text{uwb}}. \tag{64}$$

The I/Q components $x(t)$ and $y(t)$ are each the sum of independent random variables. If enough terms contribute significantly to the sum, the central limit theorem suggests that $x(t)$ and $y(t)$ approach Gaussian processes. If they are independent and zero mean, the IF output resembles Gaussian noise in its statistical properties (Gaussian instantaneous voltage, Rayleigh-distributed envelope, exponentially distributed power) and $\left\langle x^2(t)\right\rangle = \left\langle y^2(t)\right\rangle = \left\langle |a_k|^2\right\rangle RB_{\text{if}}/2$. If they are nonzero mean, then there is a discrete spectral component as well as a noiselike component, and the envelope will approach a Rician distribution. Note that if the $\{a_k\}$ are independent of the $\{T_k\}$ and $\left\langle a_k\right\rangle = 0$, there can be no spectral line.

At the other extreme, if $T_k = kT$ and f_0 is a harmonic of the pulse rate R $(= 1/T)$, then $e^{-j2\pi f_0 T_k} = e^{j\phi}$ (a constant), and

$$z(t) = e^{j\phi} \sum_k a_k h_b(t - kT). \tag{65}$$

Letting $a_k = 1 \forall k$ and assuming $B_{\text{if}} \ll R$,

$$z(t) = e^{j\phi} \sum_k h_b(t - T_k) \cong e^{j\phi} R \int_{-\infty}^{\infty} h_b(t)\, dt = e^{j\phi} R H_b(0) = e^{j\phi} R. \tag{66}$$

Therefore,

$$g(t) = 2|P(f_0)| R \cos(2\pi f_{\text{if}} t + \phi), \tag{67}$$

which is a CW tone at frequency f_{if} with power

$$p_g = \frac{E_p}{B_p} \cdot R^2 A^2(f_0). \tag{68}$$

This is not surprising, because in this case

$$w(t) = \sum_k p(t - kT) = \frac{1}{T} \sum_n P\left(\frac{n}{T}\right) e^{j2\pi nt/T}, \tag{69}$$

and

$$W(f) = \sum_n e^{-j2\pi fkT} P(f) = P(f) \cdot \frac{1}{T} \sum_n \delta\left(f - \frac{n}{T}\right), \tag{70}$$

where the second equality in each case follows from the Poisson sum formula.

10.8 Power Spectral Density

The PSD of the UWB signal is a key component to understanding the effect of UWB signals on narrowband receivers. The PSD is the average power per Hertz of the UWB signal as a function of frequency, and is typically computed as the Fourier transform of the autocorrelation function. The PSD, integrated over the pass band (channel bandwidth) of the victim narrowband receiver, gives the average interference power from the UWB transmitter that is seen by the narrowband receiver.

While this average interference power is a useful measure of the interference impact, the average alone is not adequate to completely characterize the effect of the UWB signal on the narrowband receiver. The temporal and statistical properties of the interference, after passing through the IF filtering stages of the receiver, are also important. As is shown here, PSD analysis can be used to provide insight into these properties.

A number of techniques are available in the literature for computing PSD in general (e.g., [2, 3]), and specifically for UWB signals (e.g., [4], which uses the general approach developed in [3]). Chapter 3 of [5] discusses the development of a technique for computing PSD based on the derivative of the average energy spectrum, and also derives comparable results based on the techniques of [3] and [6]. The PSD of a UWB signal for which the PRF itself is directly modulated is derived in [7].

A simple approach to UWB PSD computation is demonstrated here using various forms of the basic UWB signal. General expressions for the autocorrelation and PSD are developed and applied to some specific signals, and example spectra are shown and compared to simulation results.

The autocorrelation can be defined as

$$R_{ww}(t,\tau) = \left\langle w(t)w^*(t-\tau)\right\rangle. \tag{71}$$

Communications and radar signals generally have some periodic structure and are therefore modeled as cyclostationary random processes, meaning that

$$\begin{aligned} \overline{w}(t) &= \overline{w}(t-nT_p) \\ R_{ww}(t,\tau) &= R_{ww}(t-nT_p,\tau), \end{aligned} \tag{72}$$

where T_p is the period. Of most interest is usually the time-averaged autocorrelation

$$\overline{R}_{ww}(\tau) = \frac{1}{T_p}\int_0^{T_p} R_{ww}(t,\tau)\,dt. \tag{73}$$

The PSD is the Fourier transform of $\overline{R}_{ww}(\tau)$:

$$\overline{S}_{ww}(f) = \int_{-\infty}^{\infty} \overline{R}_{ww}(\tau)e^{-j2\pi f\tau}\,d\tau, \tag{74}$$

which gives the average power per Hertz generated by the UWB signal, as a function of frequency, and[2]

$$\overline{S}_{ww}(f) = |P(f)|^2\overline{S}_{dd}(f). \tag{75}$$

Letting $T_k = kT + \gamma_k$ the autocorrelation of $d(t)$ can be written as

$$\begin{aligned} R_{dd}(t,\tau) &= \left\langle d(t)d^*(t-\tau)\right\rangle = \left\langle \sum_n\sum_k a_k\delta(t-kT-\gamma_k)a_n^*\delta(t-\tau-nT-\gamma_n)\right\rangle \\ &= \left\langle \sum_m\sum_k a_{m+k}^*a_k\delta[t-\tau-(m+k)T-\gamma_{m+k}]\delta[t-kT-\gamma_k]\right\rangle. \end{aligned} \tag{76}$$

Since $d(t)$ is cyclostationary with period T_p, variations over t must be averaged out, giving

$$\begin{aligned} \overline{R}_{dd}(\tau) &= \frac{1}{T_p}\int_0^{T_p} R_{dd}(t,\tau)\,dt \\ &= \frac{1}{T_p}\sum_m\left\langle \sum_k a_{m+k}^*a_k\int_0^{T_p}\delta[t-\tau-(m+k)T-\gamma_{m+k}]\delta[t-kT-\gamma_k]\,dt\right\rangle. \end{aligned} \tag{77}$$

[2] Since $\delta(t)$ is the identity element for convolution, it has units of 1/s or Hz. Therefore, $R_{dd}(t,\tau)$ and $\overline{R}_{dd}(\tau)$ have units of Hz and $\overline{S}_{dd}(f)$ has units of Hz. Since $|P(f)|^2$ has units of joules per Hz, $\overline{S}_{ww}(f)$ has units of watts per Hz.

After the integration, the sum over k will include only terms for which $0 < kT + \gamma_k < T_p$. If the sequences $\{a_n\}$ and $\{\gamma_n\}$ are jointly stationary, then $T_p = T$. If γ_k is constrained such that there is always exactly one pulse in the integration interval, then the integral will be nonzero for $k = 0$ only. However, any integration interval of duration T is valid, and some arbitrary intervals may include two pulses while others might include none, but because of the averaging operation, (77) becomes

$$\begin{aligned}\overline{R}_{dd}(\tau) &= \frac{1}{T}\sum_m \left\langle a^*_{m+k} a_k \delta(\tau + mT + \gamma_{m+k} - \gamma_k)\right\rangle \\ &= \frac{1}{T}\sum_m \left\langle a^*_{m+k} a_k \delta(\tau + mT + \Delta_m)\right\rangle, \end{aligned} \tag{78}$$

where $\Delta_m = \gamma_{m+k} - \gamma_k$ has statistics that are independent of k, because of the assumed stationarity.

The PSD is

$$\begin{aligned}\overline{S}_{dd}(f) &= \frac{1}{T}\sum_m \left\langle a^*_{m+k} a_k \int_{-\infty}^{\infty} \delta(\tau + mT + \Delta_m) e^{-j2\pi f\tau}\, d\tau\right\rangle \\ &= \frac{1}{T}\sum_m \left\langle a^*_{m+k} a_k e^{j2\pi f(mT+\Delta_m)}\right\rangle \\ &= \frac{1}{T}\sum_m e^{j2\pi f mT}\left\langle a^*_{m+k} a_k e^{j2\pi f\Delta_m}\right\rangle. \end{aligned} \tag{79}$$

It is useful to explore special cases of (78). Often, the amplitude and shift sequences $\{a_n\}$ and $\{\gamma_n\}$ will be independent, in which case the autocorrelation can be written as

$$\overline{R}_{dd}(\tau) = \frac{1}{T}\sum_m R_a[m]\langle\delta(\tau + mT + \Delta_m)\rangle, \tag{80}$$

where $R_a[m] = \left\langle a^*_{m+k} a_k\right\rangle$. If $p_{\Delta_m}(\alpha)$ is the probability density function (PDF) of Δ_m, then the autocorrelation can be expressed as

$$\begin{aligned}\overline{R}_{dd}(\tau) &= \frac{1}{T}\sum_m R_a[m]\int_\alpha \delta(\tau + mT + \alpha) p_{\Delta_m}(\alpha)\, d\alpha \\ &= \frac{1}{T}\sum_m R_a[m] p_{\Delta_m}(-\tau - mT) \\ &= \frac{1}{T}\sum_m R_a[m] p_{\Delta_m}(\tau + mT). \end{aligned} \tag{81}$$

For the final equality, the PDF has been assumed symmetric $p_{\Delta_m}(\alpha) = p_{\Delta_m}(-\alpha)$. The corresponding PSD is

$$\overline{S}_{dd}(f) = \frac{1}{T}\sum_m R_a[m]\int_{-\infty}^{\infty} p_{\Delta_m}(\tau + mT) e^{-j2\pi f\tau}\, d\tau$$

$$= \frac{1}{T} \sum_m R_a[m] \int_{-\infty}^{\infty} p_{\Delta_m}(\xi) e^{-j2\pi f(\xi - mT)} \, d\xi$$

$$= \frac{1}{T} \sum_m R_a[m] e^{j2\pi f mT} \Phi_{\Delta_m}(-2\pi f)$$

$$= \frac{1}{T} \sum_m R_a[m] e^{j2\pi f mT} \Phi_{\Delta_m}(2\pi f), \tag{82}$$

where $\Phi_{\Delta_m}(\omega) = \left\langle e^{j\omega\Delta_m} \right\rangle$ is the characteristic function of Δ_m. If $p_{\Delta_m}(\alpha)$ is real and symmetric, then so is $\Phi_{\Delta_m}(\omega)$, accounting for the last equality in (82).

Note that if the $\{\gamma_k\}$ are i.i.d., then

$$p_{\Delta_m}(\tau) = \begin{cases} \delta(\tau), & m = 0 \\ p_\gamma(\tau) * p_\gamma(-\tau), & m \neq 0 \end{cases} \qquad \Phi_{\Delta_m}(\omega) = \begin{cases} 1, & m = 0 \\ |\Phi_\gamma(\omega)|^2, & m \neq 0 \end{cases}. \tag{83}$$

With $\Delta T_m = T_{m+k} - T_k = mT + \gamma_{m+k} - \gamma_k$, the autocorrelation and PSD can also be written as

$$\overline{R}_{dd}(\tau) = \frac{1}{T} \sum_m R_a[m] p_{\Delta T_m}(\tau)$$

$$\overline{S}_{dd}(f) = \frac{1}{T} \sum_m R_a[m] \Phi_{\Delta T_m}(2\pi f). \tag{84}$$

Note that if $p_{\Delta_m}(\tau)$ is discrete (Δ_m can take on a finite number of values), then the autocorrelation $\overline{R}_{dd}(\tau)$ is discrete, and therefore $\overline{S}_{dd}(f)$ is periodic in the frequency domain. The actual UWB PSD is $|P(f)|^2 \overline{S}_{dd}(f)$, which generally is not periodic, because of the shaping by the pulse ESD $|P(f)|^2$.

The next section gives an example for which the PDF of Δ_m is discrete, and the section after that gives two examples for which the PDF is continuous.

10.9 Discrete PDF PSD Example: Equally Spaced, Equally Likely Time Offsets

An important example is the case in which the offsets $\{\gamma_k\}$ are i.i.d. and can take on N equally likely values separated by a bin interval T_c. In that case, the PDF and characteristic function are

$$p_\gamma(\tau) = \frac{1}{N} \sum_{k=0}^{N-1} \delta(\tau - kT_c) \tag{85}$$

$$\Phi_\gamma(2\pi f) = \frac{1}{N} \sum_{k=0}^{N-1} e^{-j2\pi f k T_c} = e^{-j\pi f(N-1)T_c} \frac{\sin \pi f N T_c}{N \sin \pi f T_c}. \tag{86}$$

Hence,

$$\Phi_{\Delta_m}(2\pi f) = \begin{cases} 1, & m = 0 \\ \left(\dfrac{\sin \pi f N T_c}{N \sin \pi f T_c} \right)^2, & m \neq 0 \end{cases} \tag{87}$$

and the PSD is

$$\overline{S}_{dd}(f) = \frac{1}{T}\left[R_a[0] + \left(\frac{\sin \pi f N T_c}{N \sin \pi f T_c}\right)^2 \sum_{m \neq 0} R_a[m] e^{j2\pi f m T}\right]. \tag{88}$$

The autocorrelation itself is also of interest, and that requires finding $p_{\Delta_m}(\tau)$ explicitly. For $m = 0$, $\Delta_m = 0$, and $p_{\Delta m}(\tau) = \left\langle \delta(\tau + \Delta_m) \right\rangle = \delta(\tau)$. For $m \neq 0$, $p_{\Delta_m}(\tau)$ can be found by simply finding the total number of combinations of γ_{m+k} and γ_k, which result in a value of Δ_m that is a given multiple of T_c, and dividing by the total number of combinations, which is N^2:

$$p_{\Delta m}(\tau) = \left\langle \delta(\tau - \Delta_m) \right\rangle = \frac{1}{N^2} \sum_{n=0}^{N-1} \sum_{l=0}^{N-1} \delta[\tau - (n-l)T_c] \quad m \neq 0. \tag{89}$$

This can be simplified by recognizing that there are $N - |k|$ combinations that give $\Delta_m = kT_c$, leading to the identity

$$\sum_{n=0}^{N-1} \sum_{l=0}^{N-1} f(n-l) = \sum_{k=-(N-1)}^{N-1} (N - |k|) f(k), \tag{90}$$

which gives

$$p_{\Delta_m}(\tau) = \frac{1}{N^2} \sum_{k=-(N-1)}^{N-1} \delta(\tau - kT_c)(N - |k|) \quad m \neq 0, \tag{91}$$

as shown in Figure 10.17 for $N = 10$.

Therefore, the time-averaged autocorrelation is

$$\overline{R}_{dd}(\tau) = \frac{1}{T}\left\{R_a[0]\delta(\tau) + \sum_{m \neq 0} R_a[m] p_{\Delta_m}(\tau + mT)\right\}. \tag{92}$$

Note that the Fourier transform of (92) yields (88).

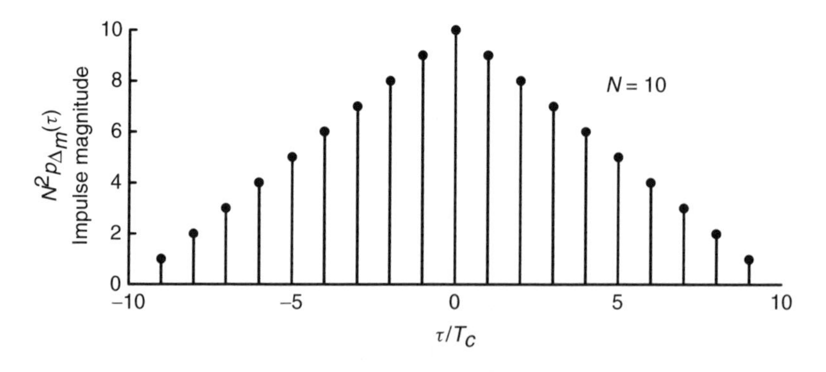

Figure 10.17 PDF of the time offset difference Δ_m, for $m \neq 0$ and $N = 10$

If $T_c = T/N$, then $\sum_{m=-\infty}^{\infty} p_{\Delta_m}(\tau + mT) = \frac{1}{N}\sum_{m=-\infty}^{\infty} \delta(\tau - mT_c)$, where $p_{\Delta_m}(\tau)$ is given by (91) for all m. Therefore, if $R_a[m] = R_{a,m\neq 0}, \forall m \neq 0$, then

$$\overline{R}_{dd}(\tau) = \frac{1}{T}\left\{R_a[0]\delta(\tau) + R_{a,m\neq 0}\left[\frac{1}{N}\sum_{m=-\infty}^{\infty} \delta(\tau - mT_c) - p_{\Delta_m}(\tau)\right]\right\} (T = NT_c), \tag{93}$$

which is shown in Figure 10.18, assuming $a_n = 1, \forall n$ and $T_c = T/N$. If $T_c \neq T/N$, the result will exhibit periodicity with period T. Figure 10.19 shows the autocorrelation for $T_c = T/5N$.

Assuming that $\{a_n\}$ is white; that is, $R_a[m] = |\mu_a|^2 + \sigma_a^2\delta[m]$, where $\delta[m]$ is the Kronecker delta function, defined as $\delta[0] = 1, \delta[m] = 0, m \neq 0$. The PSD is

$$\overline{S}_{dd}(f) = \frac{1}{T}\left[\left\langle |a|^2 \right\rangle - \left(\frac{\sin \pi f N T_c}{N \sin \pi f T_c}\right)^2 |\mu_a|^2\right] + \frac{1}{T}|\mu_a|^2 \left(\frac{\sin \pi f N T_c}{N \sin \pi f T_c}\right)^2 \sum_{m=-\infty}^{\infty} e^{j2\pi f m T}, \tag{94}$$

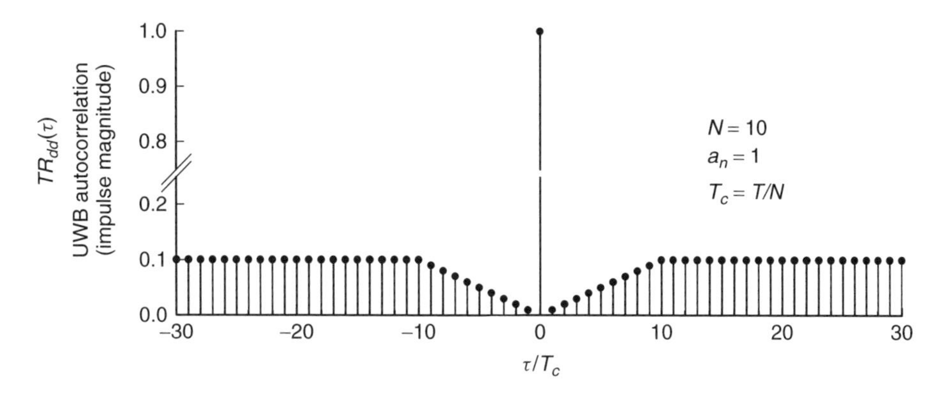

Figure 10.18 Autocorrelation function for 10 equally likely pulse positions with $T = NT_c$

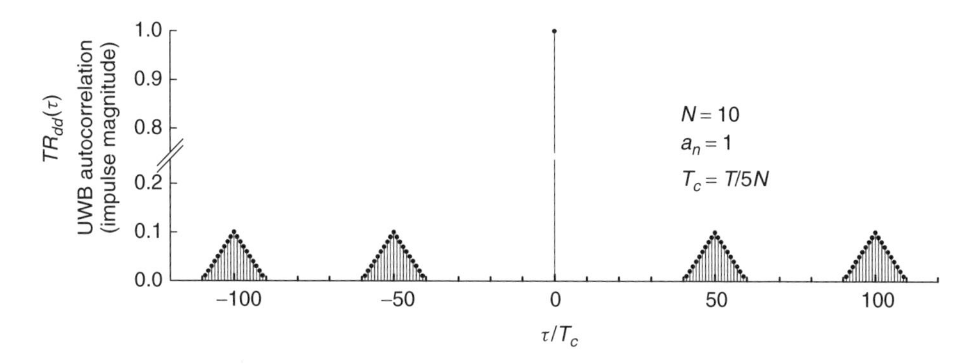

Figure 10.19 Autocorrelation function for 10 equally likely pulse positions with $T = 5NT_c$

and applying the Poisson sum formula gives

$$\overline{S}_{dd}(f) = \frac{1}{T}\left[\left\langle|a|^2\right\rangle - \left(\frac{\sin \pi f N T_c}{N \sin \pi f T_c}\right)^2 |\mu_a|^2\right] + \frac{1}{T^2}|\mu_a|^2 \left(\frac{\sin \pi f N T_c}{N \sin \pi f T_c}\right)^2 \sum_{m=-\infty}^{\infty} \delta\left(f - \frac{m}{T}\right). \tag{95}$$

In the special case for which $T = NT_c$, this becomes

$$\overline{S}_{dd}(f) = \frac{1}{T}\left[\left\langle|a|^2\right\rangle - \left(\frac{\sin \pi f T}{N \sin \pi f T_c}\right)^2 |\mu_a|^2\right] + \frac{1}{T^2}|\mu_a|^2 \sum_{m=-\infty}^{\infty} \delta\left(f - \frac{m}{T_c}\right). \tag{96}$$

The second equality follows because the term $\dfrac{\sin \pi m}{N \sin \pi m/N} = 0$ (for m an integer) unless m is an integral multiple of N, in which case the term is 1.

Finally, the PSD for the actual UWB signal, including the effect of the pulse waveform, is

$$\overline{S}_{ww}(f) = \underbrace{\frac{|P(f)|^2}{T}\left[\left\langle|a|^2\right\rangle - |\mu_a|^2 \left(\frac{\sin \pi f N T_c}{N \sin \pi f T_c}\right)^2\right]}_{\text{Continuous component}} + \underbrace{\frac{1}{T^2}|\mu_a|^2 \left(\frac{\sin \pi f N T_c}{N \sin \pi f T_c}\right)^2 \sum_{m=-\infty}^{\infty} \delta\left(f - \frac{m}{T}\right)\left|P\left(\frac{m}{T}\right)\right|^2}_{\text{Discrete component}}. \tag{97}$$

As can be seen, the PSD in general includes a discrete component consisting of spectral lines, and a continuous component which is noiselike. The discrete component arises from periodicities in the autocorrelation, which in turn are because of timing structure (constraints) in the signal (e.g., a pulse occurs every frame). The period is generally T, giving tones at harmonics of the average pulse rate $1/T$. However, as shown, if $N = T/T_c$, the tones occur only at harmonics of N/T. The continuous component corresponds to the aperiodic part of the autocorrelation, which is due to randomness in the signal (e.g., uncertainty in the transmission time of each pulse).

If $\mu_a = 0$ (the amplitude modulation is zero mean), then the discrete component vanishes and the PSD is simply

$$\overline{S}_{ww}(f) = \left\langle|a|^2\right\rangle \frac{|P(f)|^2}{T}. \tag{98}$$

The figures below show some examples of the PSD, assuming that the pulse ESD $|P(f)|^2$ is as shown in Figure 10.20, which is the ESD of the 'Gaussian monocycle' $g_1(t)$ (the first derivative of the basic Gaussian pulse). In this case, the spectral peak (f_1) is near 2 GHz. Since the PSD generally includes both continuous and discrete spectral components (which have units of watts and watts per Hertz, respectively), it is necessary to show the PSD as the power output of a resolution filter, similar to what would be seen

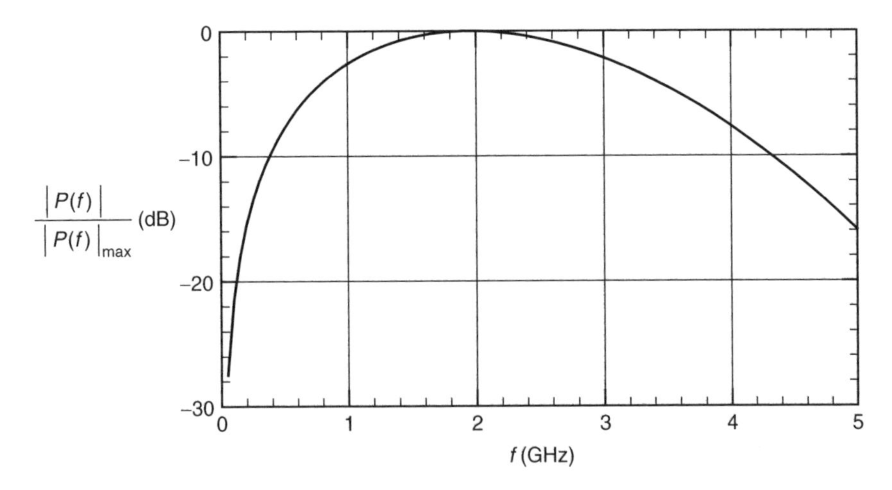

Figure 10.20 Pulse ESD used for the PSD examples

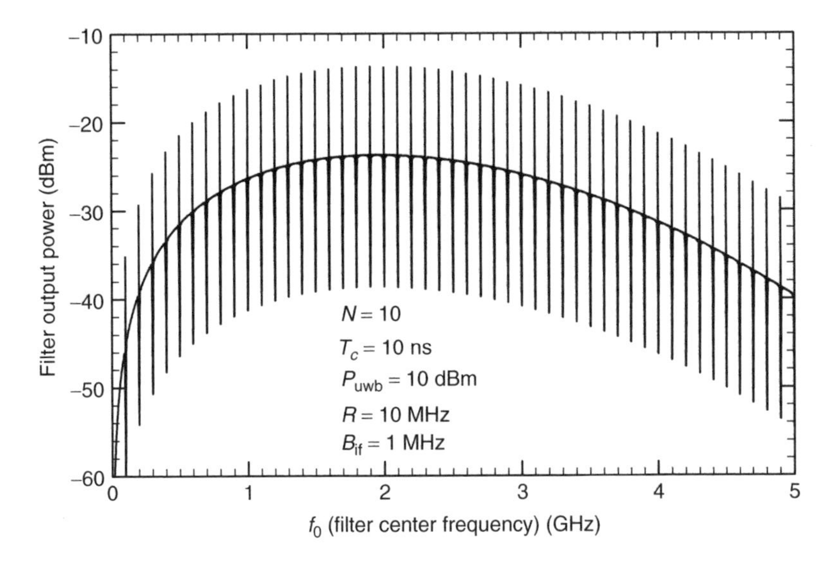

Figure 10.21 PSD (seen through a 1-MHz resolution filter) for a 10-MHz pulse rate, with 10 equally likely pulse positions separated by 10 ns

by a spectrum analyzer, averaging the resolution filter power output. The figures below assume a pulse rate of 10 MHz ($T = 100$ ns), and use a resolution filter bandwidth of 1 MHz.

Figure 10.21 shows the filter output assuming $N = 10$ equally likely pulse positions equally spaced $T_c = 10$ ns apart within the 100 ns frame, representing 'full-frame' time hopping. Figure 10.22 is a close-up of Figure 10.21. Note that the spectral lines are 100 MHz ($1/T_c$) apart, as would be expected from the above analysis. Figure 10.23 shows full-frame dithering with $N = 50$ and $T_c = 2$ ns. In this case, the lines are 500 MHz apart.

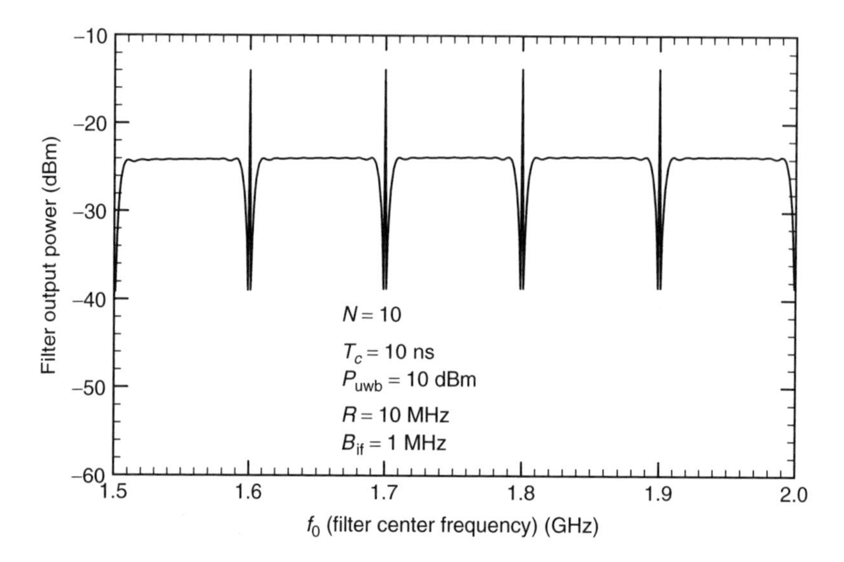

Figure 10.22 Close-up of Figure 10.21

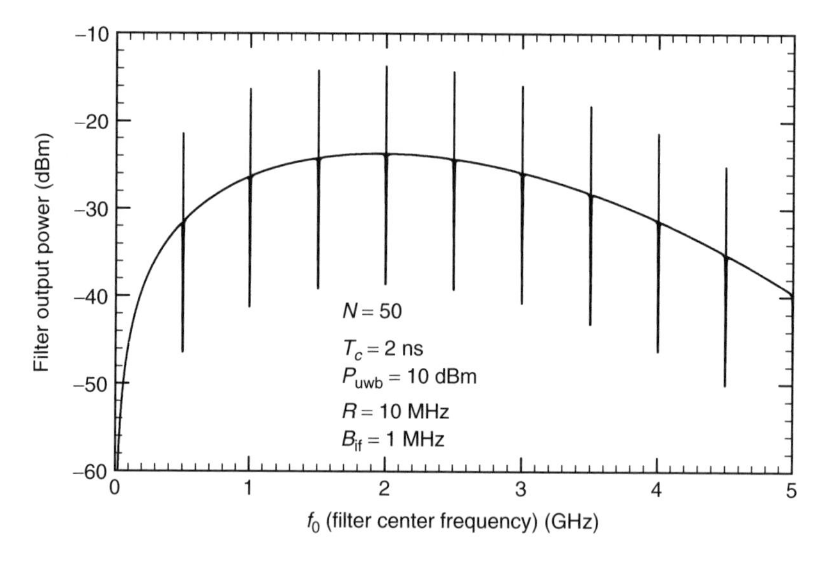

Figure 10.23 PSD with 50 equally likely pulse positions 2 ns apart

Figure 10.24 shows an example of partial-frame time hopping, where $T_c = 2$ ns, but there are only $N = 10$ possible pulse positions, so the pulse is restricted to 20 % of the nominal frame interval. Figure 10.25 shows a close-up of Figure 10.24. Note that there are some lines that are 10 MHz apart (i.e., harmonics of the frame rate), but the strongest lines are 500 MHz apart. Moreover, there is a clear periodicity in frequency, with a period of 500 MHz, since the minimum spacing between impulses in the autocorrelation function is 2 ns.

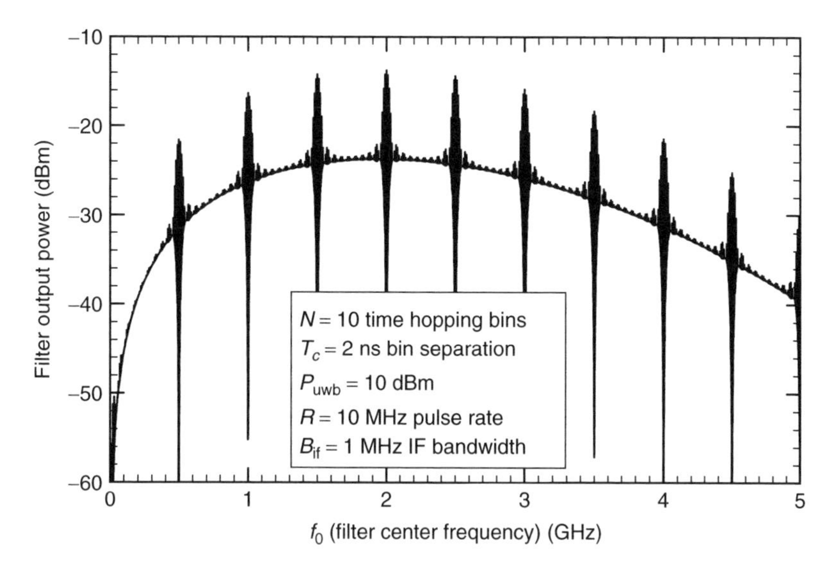

Figure 10.24 PSD with 10 pulse positions 2 ns apart (partial-frame time hopping)

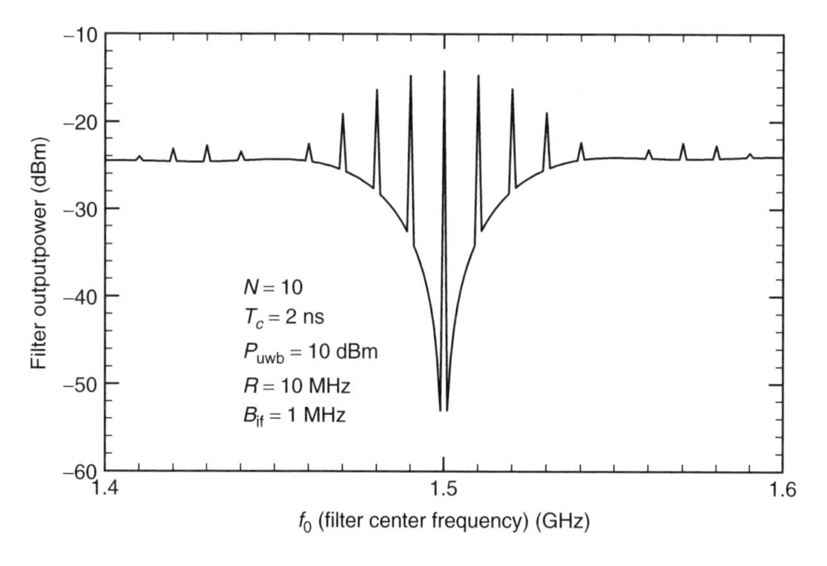

Figure 10.25 Close-up of Figure 10.24

10.10 Continuous PDF PSD Examples

10.10.1 The Poisson Process

The Poisson process can be written as

$$x(t) = \sum_n \delta(t - T_n), \tag{99}$$

with an average pulse rate of λ, so that $\overline{x}(t) = \lambda$ (see e.g. [8], p. 314). The Poisson process is stationary rather than cyclostationary. Since the autocorrelation is time-invariant, it should be unaffected by taking a time average. The averaging interval used here is $T = 1/\lambda$, which is the average time between pulses.

The PDF of the time between pulse k and pulse $k+m$ is

$$p_{\Delta T_m}(\tau) = \frac{\lambda^m}{(m-1)!}\tau^{m-1}e^{-\lambda\tau} \quad m > 0, \tau \geq 0. \tag{100}$$

Obviously, $p_{\Delta T_0}(\tau) = \delta(\tau)$. Applying (84) with $R_a[m] = 1$, the autocorrelation for $\tau \geq 0$ is

$$R_{xx}(\tau) = \frac{1}{T}\sum_m p_{\Delta T_m}(\tau) = \lambda\left[\delta(\tau) + \sum_{m=1}^{\infty}\frac{\lambda^m}{(m-1)!}\tau^{m-1}e^{-\lambda\tau}\right] \quad \tau \geq 0. \tag{101}$$

By symmetry, this also applies to $\tau \leq 0$, for which $m < 0$:

$$p_{\Delta T_m}(\tau) = \frac{\lambda^{-m}}{(-m-1)!}\tau^{-m-1}e^{\lambda\tau} \quad m < 0, \tau \leq 0. \tag{102}$$

However $\sum_{m=1}^{\infty}\frac{\lambda^m}{(m-1)!}\tau^{m-1} = \lambda\sum_{m=0}^{\infty}\frac{(\lambda\tau)^m}{m!} = \lambda e^{\lambda\tau}$, so

$$\begin{aligned} R_{xx}(\tau) &= \lambda\delta(\tau) + \lambda^2 \\ S_{xx}(f) &= \lambda + \lambda^2\delta(f), \end{aligned} \tag{103}$$

which agrees with results derived elsewhere by different means [8].

10.10.2 Continuous PDF Uniform Random Pulse Position

If γ_k is uniformly distributed over an interval T_γ, then the PDF can be written as

$$p_\gamma(\tau) = \frac{1}{T_\gamma}, |\tau| \leq \frac{T_\gamma}{2} \tag{104}$$

and the PDF of Δ_m is

$$p_{\Delta_m}(\tau) = \begin{cases} \delta(\tau), & m = 0 \\ \frac{1}{T_\gamma}q_{T_\gamma}(\tau), & m \neq 0 \end{cases}, \tag{105}$$

where

$$q_T(\tau) = \begin{cases} \left(1 - \frac{|\tau|}{T}\right), & |\tau| \leq T \\ 0, & |\tau| > T \end{cases} \tag{106}$$

The characteristic function of γ is

$$\Phi_\gamma(2\pi f) = \frac{1}{T_\gamma}\int_{-T_\gamma/2}^{T_\gamma/2} e^{j2\pi f\tau}\,d\tau = \frac{\sin\pi f T_\gamma}{\pi f T_\gamma} = \mathrm{sinc}(fT_\gamma), \tag{107}$$

and therefore,

$$\Phi_{\Delta_m}(2\pi f) = \begin{cases} 1, & m = 0 \\ \text{sinc}^2(fT_\gamma) & m \neq 0 \end{cases} \tag{108}$$

Recalling from (81) and (82) that

$$\overline{R}_{dd}(\tau) = \frac{1}{T}\sum_m R_a[m] p_{\Delta_m}(\tau + mT)$$
$$\overline{S}_{dd}(f) = \frac{1}{T}\sum_m R_a[m] e^{j2\pi fmT}\Phi_{\Delta_m}(2\pi f), \tag{109}$$

and assuming as before that $R_a[m] = |\mu_a|^2 + \sigma_a^2\delta[m]$, the autocorrelation and PSD are

$$\begin{aligned}\overline{R}_{dd}(\tau) &= \frac{1}{T}\left[\left\langle |a|^2\right\rangle\delta(\tau) + |\mu_a|^2\frac{1}{T_\gamma}\sum_{m\neq 0} q_{T_\gamma}(\tau + mT)\right] \\ &= \frac{1}{T}\left[\left\langle |a|^2\right\rangle\delta(\tau) - \frac{|\mu_a|^2}{T_\gamma}q_{T_\gamma}(\tau) + \frac{|\mu_a|^2}{T_\gamma}\sum_m q_{T_\gamma}(\tau + mT)\right],\end{aligned} \tag{110}$$

$$\begin{aligned}\overline{S}_{dd}(f) &= \frac{1}{T}\left[\left\langle |a|^2\right\rangle + |\mu_a|^2\text{sinc}^2(fT_\gamma)\sum_{m\neq 0} e^{j2\pi fmT}\right] \\ &= \frac{1}{T}\left[\left\langle |a|^2\right\rangle - |\mu_a|^2\text{sinc}^2(fT_\gamma) + |\mu_a|^2\text{sinc}^2(fT_\gamma)\sum_m e^{j2\pi fmT}\right] \\ &= \frac{1}{T}\left[\left\langle |a|^2\right\rangle - |\mu_a|^2\text{sinc}^2(fT_\gamma) + \frac{|\mu_a|^2}{T}\sum_m \delta\left(f - \frac{m}{T}\right)\text{sinc}^2\left(\frac{mT_\gamma}{T}\right)\right].\end{aligned} \tag{111}$$

Note that if $T_\gamma = T$ (full-frame dithering, for which the only constraint is that there be one pulse at some point in every T-second interval), then

$$\begin{aligned}\overline{R}_{dd}(\tau) &= \frac{1}{T}\left\{\left\langle |a|^2\right\rangle\delta(\tau) + \frac{|\mu_a|^2}{T}[1 - q_T(\tau)]\right\} \\ &= \frac{\left\langle |a|^2\right\rangle\delta(\tau)}{T} + \frac{|\mu_a|^2}{T^2}[1 - q_T(\tau)],\end{aligned} \tag{112}$$

$$\begin{aligned}\overline{S}_{dd}(f) &= \frac{\left\langle |a|^2\right\rangle}{T} + \frac{|\mu_a|^2}{T^2}[\delta(f) - T\text{sinc}^2(fT_\gamma)] \\ &= \frac{1}{T}[\left\langle |a|^2\right\rangle - |\mu_a|^2\text{sinc}^2(fT_\gamma)] + \frac{\delta(f)}{T^2}.\end{aligned} \tag{113}$$

It is interesting to compare these results to those of the Poisson process. If $a_n = 1$ and $\lambda = 1/T$, then for the uniform pulse distribution,

$$\begin{aligned} \overline{R}_{dd}(\tau) &= \lambda\delta(\tau) + \lambda^2[1 - q_T(\tau)] \\ \overline{S}_{dd}(f) &= \lambda[1 - \text{sinc}^2(f/\lambda)] + \lambda^2\delta(f), \end{aligned} \tag{114}$$

whereas for the Poisson process,

$$\begin{aligned} R_{xx}(\tau) &= \lambda\delta(\tau) + \lambda^2 \\ S_{xx}(f) &= \lambda + \lambda^2\delta(f). \end{aligned} \tag{115}$$

The difference is due to the fact that for the uniform pulse distribution, it has been assumed that there is exactly only pulse per T-second frame, while for the Poisson process, there is no constraint on when pulses can occur. In other words, the covariance of the Poisson process is $C_{xx}(\tau) = \lambda\delta(\tau)$, whereas the covariance of the uniform pulse distribution process is $C_{dd}(\tau) = \lambda\delta(\tau) - \lambda^2 q_T(\tau)$. The mean-square in both cases is λ^2.
Also note that in both cases, the PSD is not periodic in the frequency domain.

10.11 Comparison of PSD and Simulation Results

It is useful to test the simulation against a known analytic result to verify operation and correct power scaling. This was done using the previous example.

This case was duplicated using the simulation; the results are shown in Figure 10.26, with the results from the PSD analysis superimposed for comparison. The center frequency

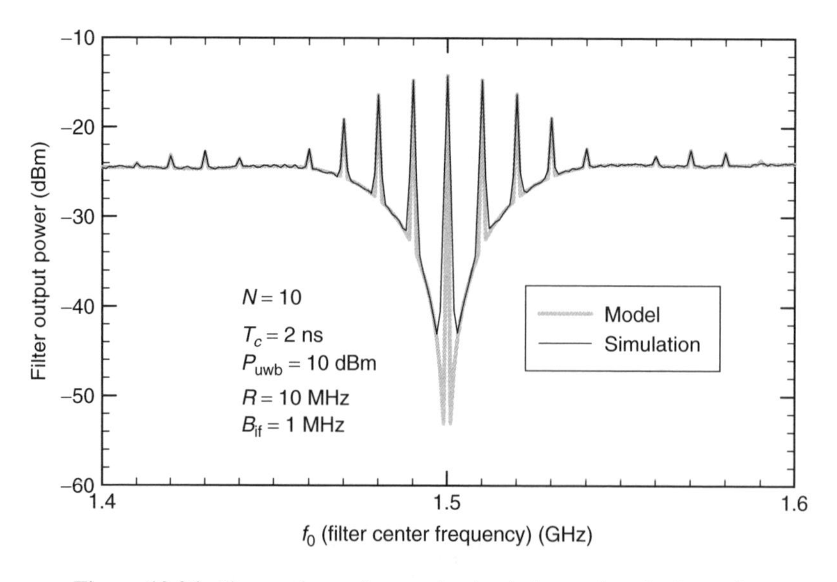

Figure 10.26 Comparison of example simulation and analysis results

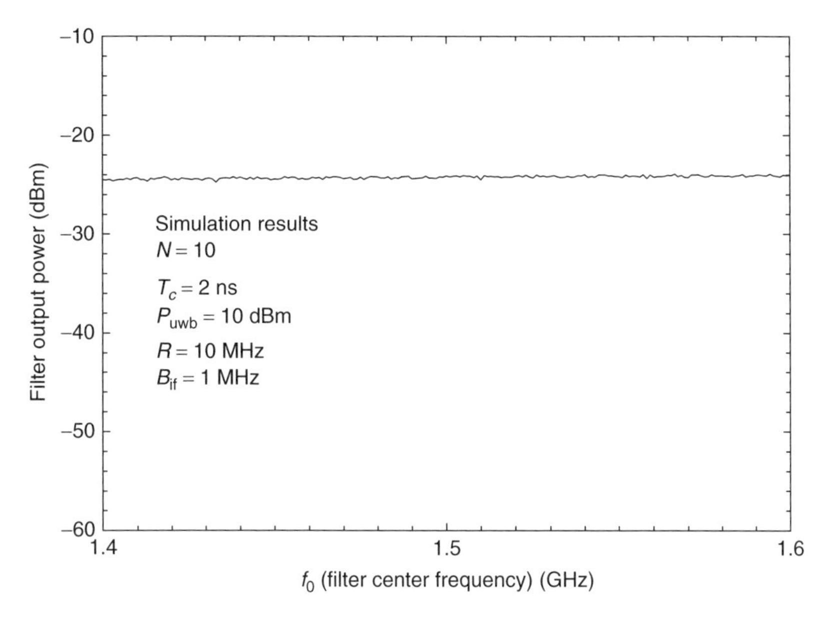

Figure 10.27 Effect of pulse polarity modulation

was stepped up from 1.4 to 1.6 GHz in 1-MHz steps. At each center frequency, a UWB signal consisting of 10,000 pulses was generated, with the pulse in each frame randomly positioned (i.e., dithering) in one of 10 positions, which are separated by 2 ns. There was no PPM. The output of the simulation was the average power at each center frequency. As can be seen, the simulation results agree well with the analysis.

The PSD analysis showed that if the pulse polarity is randomly modulated so that $\overline{a}_k = 0$, the spectral lines will vanish. This is easily verified with the simulation, as shown in Figure 10.27, which used exactly the same settings as those given in Figure 10.25, except that the pulse polarity was randomly varied from frame to frame. As can be seen, the tones are indeed gone.

Figure 10.28 shows the 1-MHz bandwidth filter power output versus time for a center frequency of 1.4 GHz. As can be seen, the signal appears very noiselike. Note that from Figure 10.25, there is no spectral line at 1.4 GHz.

Figure 10.29 shows the results for 1.46 GHz, which has a low-power spectral line, and the signal appears slightly less variable than for 1.4 GHz. However, as can be seen from Figure 10.30, Figure 10.31, and Figure 10.32, which show the output for 1.47, 1.48, and 1.50 GHz, respectively, the signal variability decreases markedly as the strength of the tone in the spectral representation increases. In fact, for 1.5 GHz, there is no significant variation, because the continuous component of the PSD within the 1-MHz filter bandwidth is very small at that center frequency. Finally, Figure 10.33 shows the filter output for $f_0 = 1.5$ GHz, but with the pulse polarity randomly varied. The variation in the envelope power is more pronounced than in the other cases; the reason is explained in the following section.

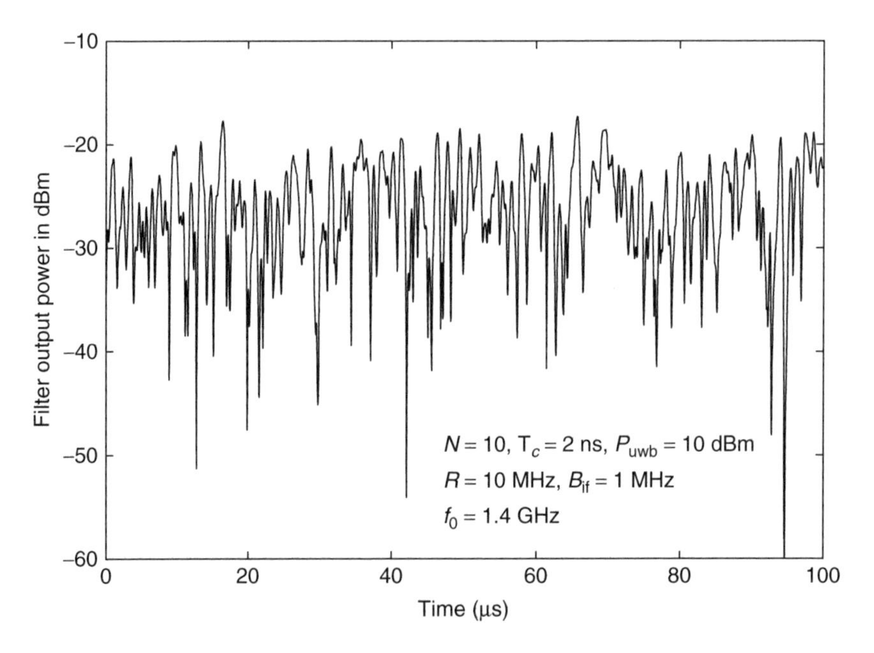

Figure 10.28 Simulation output for 1.4 GHz

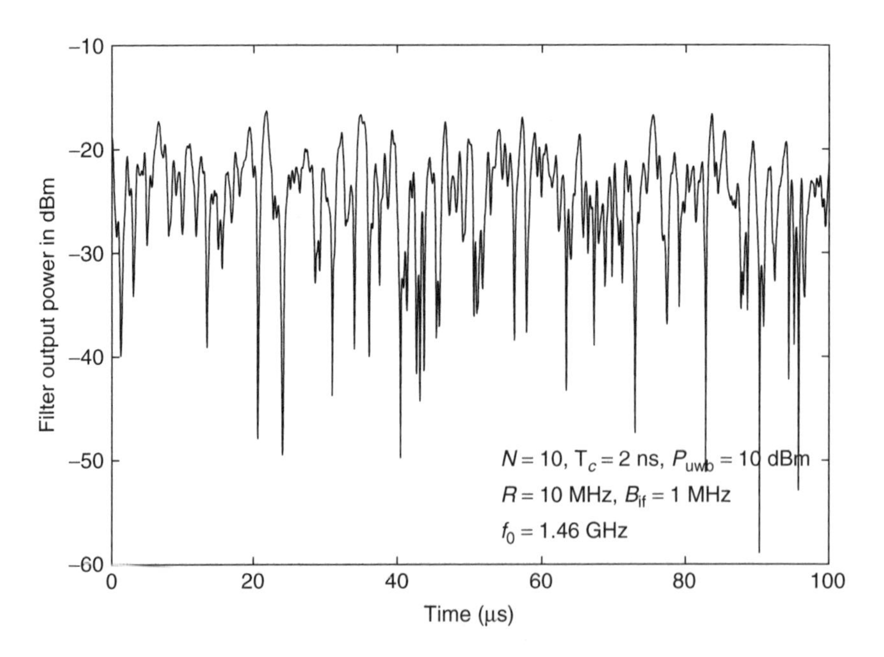

Figure 10.29 Simulation output for 1.46 GHz, which has a small spectral line

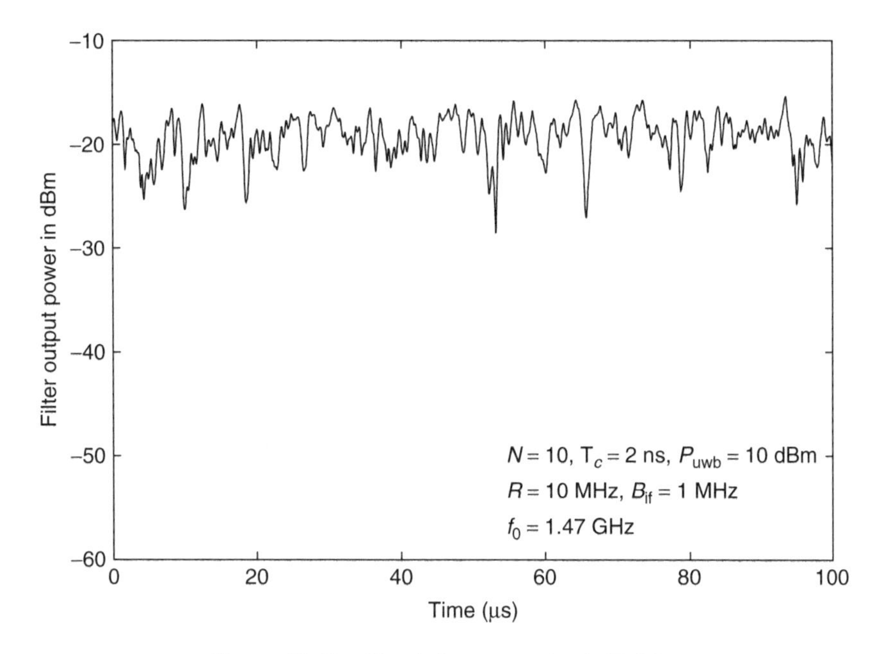

Figure 10.30 Simulation output for 1.47 GHz

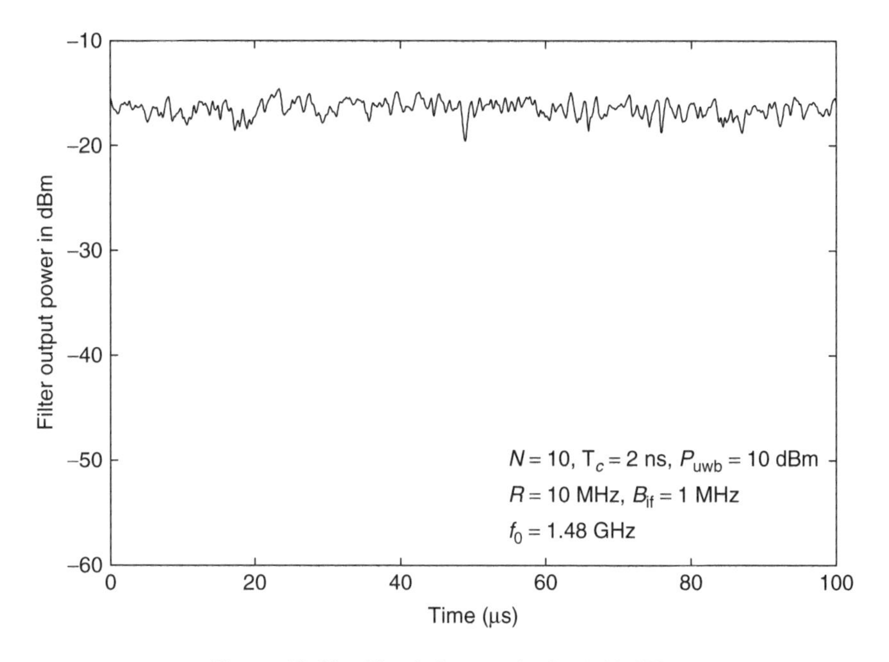

Figure 10.31 Simulation results for 1.48 GHz

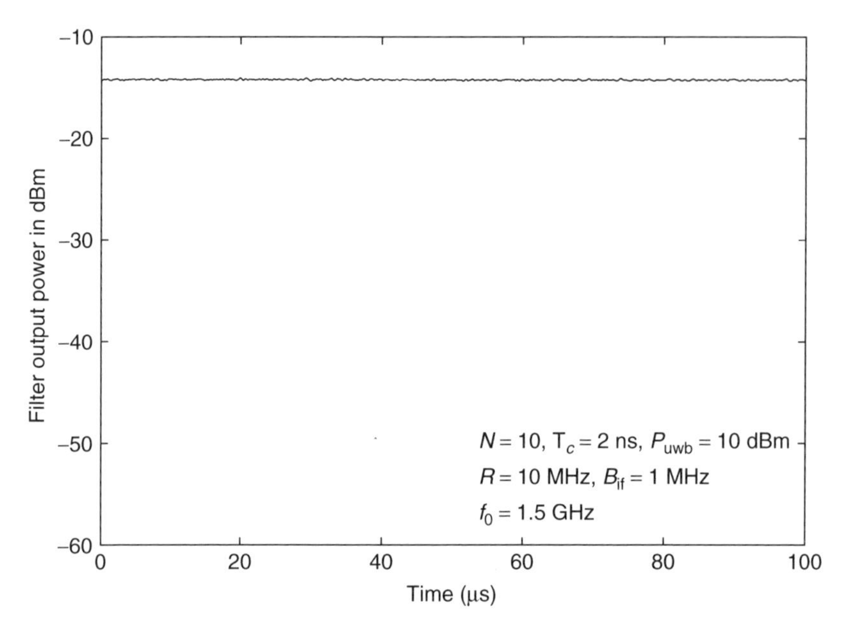

Figure 10.32 Simulation results for 1.5 GHz

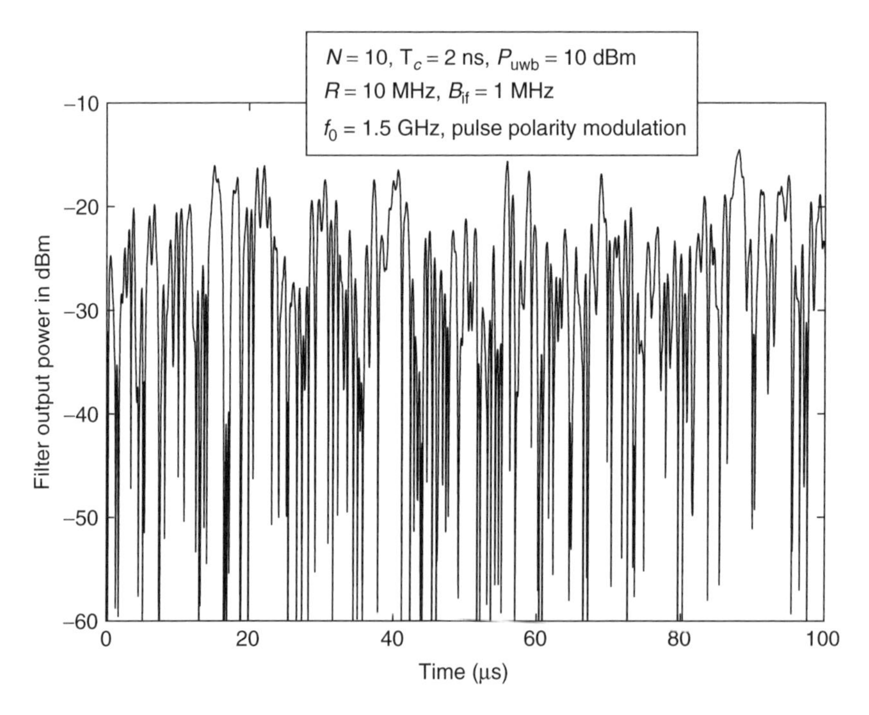

Figure 10.33 Simulation results for 1.5 GHz with pulse polarity modulation

10.12 Statistical Properties of the Output Envelope

The simulation can also be used to explore the statistical properties of the IF output due to the UWB interference. As noted, the continuous portion of the PSD corresponds to a noiselike signal, and if $R \gg B_{\text{if}}$, the I/Q components $x(t)$ and $y(t)$ should be approximately zero-mean Gaussian, each with average power $RB_{\text{if}}/2$, assuming $\langle |a_k|^2 \rangle = 1$. It is most straightforward to extract the distribution of the normalized I/Q components $c(\tau)$ and $s(\tau)$. Because $x(t) = B_{\text{if}} c(B_{\text{if}} t)$ and $y(t) = B_{\text{if}} s(B_{\text{if}} t)$, $c(\tau)$ and $s(\tau)$ should have average power

$$\sigma_c^2 = \sigma_s^2 = \frac{R}{2B_{\text{if}}}. \tag{116}$$

Figure 10.34 shows the cumulative distribution functions (CDFs) for $c(\tau)$ and $s(\tau)$ on a Gaussian scale (a Gaussian distribution appears as a straight line). The 90th and 10th percentile points correspond to 1.29σ above and below the median. For the example shown, $R/B_{\text{if}} = 10$, so $\sigma = \sqrt{5}$ and $1.29\sigma \cong 2.9$, which is consistent with the simulation results shown in Figure 10.34.

It is well known that if $c(\tau)$ and $s(\tau)$ are Gaussian with variance σ_{cs}, zero mean, and independent, the envelope $a(\tau) = \sqrt{c^2(\tau) + s^2(\tau)}$ is Rayleigh distributed, with PDF

$$f_a(x) = \frac{x}{\sigma^2} e^{-x^2/2\sigma^2}, \quad x \geq 0 \qquad \text{Rayleigh}, \tag{117}$$

where $\sigma = \sigma_{cs}$. The mean is $\overline{a} = \sigma\sqrt{\pi/2}$, and clearly $\langle a^2 \rangle = 2\sigma^2$, so the variance is $\sigma_a^2 = \sigma^2(2 - \pi/2)$.

If the filter pass band includes a discrete frequency component, then the envelope of the filter output should tend to be Rician distributed. In this case, if the (normalized) power

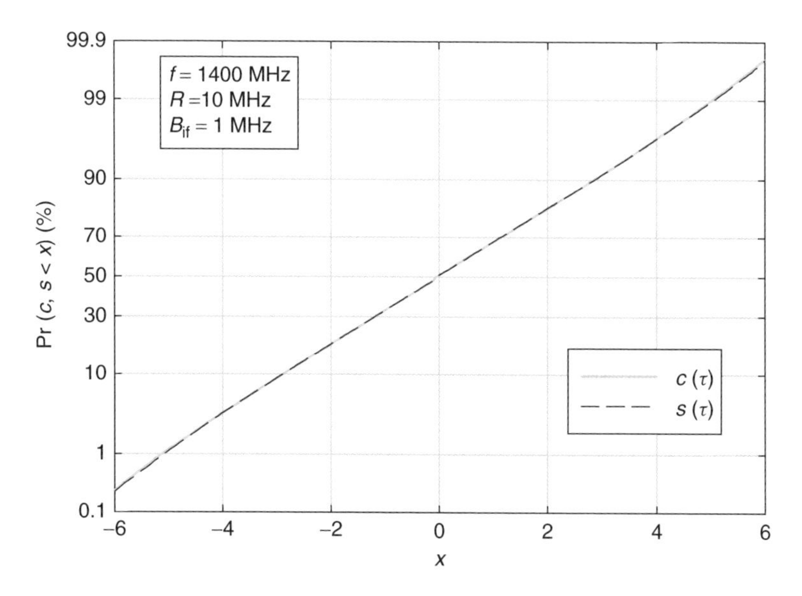

Figure 10.34 CDF of the normalized I/Q components (Gaussian scale)

in the discrete component is p_d, the envelope distribution is

$$f_a(x) = \frac{xe^{-(x^2+p_d)/2\sigma^2}}{\sigma^2} I_0\left(\frac{x\sqrt{p_d}}{\sigma^2}\right) \quad x \geq 0 \quad \text{Rician,} \tag{118}$$

where $I_0(\cdot)$ is the modified Bessel function of the first kind and zero order. The Rician distribution is sometimes characterized in terms of the parameter $k = p_d/2\sigma^2$, which is the ratio of the power in the CW component to that in the noise components.

For comparison of theoretical and simulation results, it is useful to express the Rayleigh and Rician PDFs in terms of the normalized envelope power. The envelope power is $\rho = a^2$, and its PDF is $f_\rho(y) = f_a(\sqrt{y})/2\sqrt{y}$. Applying this to the Rayleigh and Rician envelope voltage PDFs gives

$$f_\rho(y) = \frac{1}{2\sigma^2} e^{-y/2\sigma^2}, \qquad y \geq 0 \tag{119}$$

$$f_\rho(y) = \frac{e^{-(y+p_d)/2\sigma^2}}{2\sigma^2} I_0\left(\frac{\sqrt{yp_d}}{\sigma^2}\right), \qquad y \geq 0. \tag{120}$$

To normalize to an average power of unity, the transformation $f_v(\xi) = \overline{\rho} f_\rho(\overline{\rho}\xi)$ is used, where $v = \rho/\overline{\rho}$ is the normalized envelope power. The corresponding PDFs are

$$f_v(\xi) = e^{-\xi}, \qquad \xi \geq 0 \tag{121}$$

$$f_v(\xi) = (k+1)e^{-(k+1)\xi-k} I_0[2\sqrt{\xi(k^2+k)}], \qquad \xi \geq 0. \tag{122}$$

The advantage of expressing the variations this way is that the instantaneous envelope power is $v\overline{\rho}$, the product of a random variable and a constant. Note that $\sqrt{v} = a/a_{\text{rms}}$, where $a_{\text{rms}} = \sqrt{\langle a^2 \rangle}$, the rms value of the envelope.

Figure 10.35 shows the envelope CDF $\int_0^V f_v(\xi)\, d\xi$ versus $\sqrt{V}$ for different frequencies. Note that the envelope distribution for $f_0 = 1400$ MHz is very near the 'Rayleigh' curve, which is to be expected since there is no discrete component. The Rician curve for $k = 3$ is also shown, which should correspond to 1470 MHz because from Figure 10.26, the discrete component at that frequency is about 5 dB above the continuous component power (for $B_{\text{if}} = 1$ MHz). However, the envelope distribution does not closely match to the Rician distribution for $k = 3$, and the envelopes at frequencies with discrete components do not appear to be Rician-distributed.

An interesting case arises with polarity modulation, whereby a_k is randomly ± 1. As noted earlier, if $\mu_a = 0$, there are no spectral lines, and as seen in Figure 10.27, the average power does not vary with frequency except as dictated by the spectral shape of the pulse itself. However, the envelope distribution of the filter output does vary with frequency. Figure 10.33 shows the filter output envelope power versus time from the simulation, for $f_0 = 1.5$ GHz, but with pulse polarity modulation. As can be seen, the variations appear to be more pronounced than in Figure 10.28. The reason is that for $f_0 = 1.5$ GHz, the I/Q components have a fixed relationship, and in fact can be combined to form a single Gaussian process by simply adjusting the phase reference. Denoting this process by $z(t)$ and its variance σ_z^2, the envelope is $a_z(t) = |z(t)|$, so its distribution is

$$\Pr[a_z(t) < a] = \frac{2}{\sigma_z\sqrt{2\pi}} \int_0^a e^{-x^2/2\sigma_z^2}\, dx = \text{erf}\left(\frac{a}{\sigma_z\sqrt{2}}\right) \quad a \geq 0 \tag{123}$$

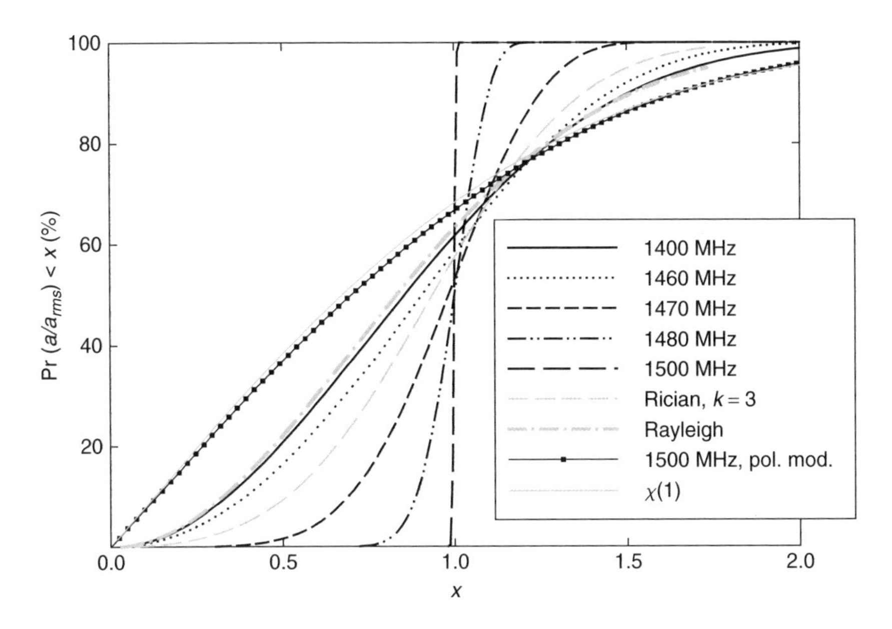

Figure 10.35 Normalized IF output envelope distributions for various cases

Table 10.2 Comparison of envelope statistics for the Rayleigh and $\chi(1)$ cases

	Rayleigh	$\chi(1)$
Mean envelope $\overline{a}$	$\sigma_{cs}\sqrt{\pi/2} \cong 1.25\sigma_{cs}$	$2\sigma_{cs}/\sqrt{\pi} \cong 1.13\sigma_{cs}$
Mean-squared envelope $\langle a^2 \rangle$	$2\sigma_{cs}^2$	$2\sigma_{cs}^2$
Envelope variance σ_a^2	$\sigma_{cs}^2(2 - \pi/2) \cong 0.43\sigma_{cs}^2$	$\sigma_{cs}^2(2 - 4/\pi) \cong 0.73\sigma_{cs}^2$

where $\text{erf}(x) = \frac{2}{\sqrt{\pi}} \int_0^x e^{-t^2} dt$ is the error function. This is a chi distribution with one degree of freedom ([8], p. 133) denoted here $\chi(1)$.

The average power is $\left\langle a_z^2(t) \right\rangle = \sigma_z^2 = 2\sigma_c^2$, and $\overline{a}_z(t) = \sigma_z\sqrt{2/\pi}$, so the variance is $\sigma_a^2 = \sigma_z^2(1 - 2/\pi)$. Table 10.2 compares the $\chi(1)$ and Rayleigh cases with respect to the envelope statistics. Consistent with Figure 10.33, the variance of the interference envelope is significantly greater in the $\chi(1)$ case than in the Rayleigh case, although the average power is the same $(2\sigma_{cs}^2)$ in both cases. The envelope distribution from the simulation, corresponding to the case of Figure 10.33, is shown in Figure 10.35, and compared with the $\chi(1)$ distribution of (123). As can be seen, agreement is excellent.

10.13 Summary

Models have been developed here for calculating and simulating the interference at the output of the final IF of a narrowband receiver in response to a UWB input signal. This signal will be the effective UWB interference as seen by the detector/demodulator. When the pulse rate is low compared to the receiver bandwidth, the IF output interference is essentially a sequence of isolated IF impulse responses. If the pulse rate exceeds the receiver bandwidth, then the impulse responses overlap and the composite signal depends on the phase relationships between the successive impulse responses, which in turn depends on the receiver RF center frequency f_0 and the interpulse intervals. The result is in general a combination of CW tones and a noiselike component.

Since the response of a receiver to pulsed, tone, and noiselike interference is generally known, an initial assessment of the impact of a particular UWB signal can often be made using PSD analysis and some of the basic relationships presented here. A more detailed assessment may require simulation, and the material developed here can be used to generate the simulated input to the detector/demodulator.

References

[1] A. Papoulis, *Signal Analysis*, McGraw-Hill, New York, 1977.

[2] J. G. Proakis, *Digital Communications*, 4th edition, McGraw-Hill, Boston, MA, 2001.

[3] M. K. Simon, S. M. Hinedi, and W. C. Lindsey, *Digital Communication Techniques – Signal Design and Detection*, Prentice-Hall, Upper Saddle River, NJ, 1995.

[4] M. Z. Win, "Spectral Density of Random Time-Hopping Spread Spectrum UWB Signals with Uniform Timing Jitter," *Proc. MILCOM*, Atlantic City, NJ, vol. 2, 1999.

[5] J. E. Padgett, J. C. Koshy, and A. A. Triolo, "Physical-Layer Modeling of UWB Interference Effects," Jan. 10, 2003, sponsored by DARPA under contract MDA972-02-C-0056.

[6] J. Romme and L. Piazzo, "On the Power Spectral Density of Time-Hopping Impulse Radio," *IEEE Conference on UWB Systems and Technologies (UWBST) 2002*, Baltimore, MD, May 20–23, 2002.

[7] J. E. Padgett, "The Power Spectral Density of a UWB Signal With Pulse Repetition Frequency (PRF) Modulation," *Proc. IEEE UWBST 2003*, Reston, VA, Nov. 2003.

[8] A. Papoulis and S. U. Pillai, *Probability, Random Variables, and Stochastic Processes*, 4th edition, McGraw-Hill, New York, 2002.

11

Digital-Carrier Spreading Codes for Baseband UWB Multiaccess

Liuqing Yang and Georgios B. Giannakis

11.1 Introduction

Conveying information over ultrashort waveforms, UWB has emerged as an exciting technology for commercial wireless communications with unique features such as low-complexity baseband transceivers, ample multipath diversity, and potential for major increase in user capacity.

To achieve these features, UWB radios have to overlay existing narrowband systems, and be able to accommodate multiple users in the presence of narrowband interference (NBI). Existing baseband (a.k.a. carrier-less) spreading schemes for multiple access rely on time-hopping (TH), or direct-sequence (DS) codes [1–3]. These codes can lead to constant-modulus transmissions, but they are not as flexible in handling multiuser interference (MUI) and NBI with low-complexity receivers – two critical factors limiting performance of UWB radios in the presence of multipath and coexisting narrowband services.

As an alternative, we introduce here two UWB multiaccess systems that utilize novel digital single-carrier (SC) or multicarrier (MC) spreading codes. These SC/MC codes are digital, lead to baseband operation, and offer flexibility in NBI mitigation by simply avoiding carriers residing on the contaminated band. On the basis of discrete cos/sin functions, these spreading codes also enjoy low-complexity implementation with standard discrete cosine transform (DCT) circuits.

We will see that, differing from orthogonal frequency-division multiple access (OFDMA) which has well-documented merits in conventional systems [4], these baseband SC- and MC-UWB spreading codes are real. The resulting baseband transceivers are carrier-free and thus immune to carrier frequency offset arising from the mismatch of transmit–receive oscillators. We show that unlike OFDMA, which has to resort to channel coding and/or frequency hopping to mitigate frequency-selective fading, UWB signaling

Ultra-wideband Wireless Communications and Networks
Edited by Xuemin (Sherman) Shen, Mohsen Guizani, Robert Caiming Qiu and Tho Le-Ngoc

with a single-carrier spreading code occupies multiple frequency bands, and the resulting multiband transmission enjoys multipath diversity gains.

Using a general discrete-time equivalent rake reception model, performance of various spreading codes will be compared in the presence of dense multipath channel and NBI. Their capability to mitigate MUI will also be compared by analysis as well as simulations.

Let us start with the construction of the digital-carrier codes.

11.2 Digital-Carrier Multiband User Codes

Consider a multiaccess UWB system with N_u users, where $s_u(n_s)$ denotes the n_sth information bearing symbol of user u. Every symbol is represented by N_f ultrashort pulses $p(t)$ of duration T_p, transmitted over N_f frames (one pulse per frame of duration T_f). The symbol transmitted during the kth frame can thus be written as $s_u(\lfloor k/N_f \rfloor)$, where $\lfloor \cdot \rfloor$ denotes the floor operation. With symbol duration $T_s := N_f T_f$, the symbol rate is $R := 1/T_s$. And with T_p in the order of nanoseconds, the transmission is UWB with bandwidth $B \approx 1/T_p$. Using binary pulse amplitude modulation (PAM), the uth user's transmitted signal is

$$v_u(t) = \sqrt{\frac{\mathcal{E}_u}{N_f}} \sum_{k=0}^{\infty} s_u(\lfloor k/N_f \rfloor) c_u(k) p(t - kT_f), \tag{1}$$

where $\mathcal{E}_u$ is the energy per symbol, and $c_u(k)$ denotes the spreading code of the uth user, $\forall u \in [0, N_u - 1]$. Differing from the well-known DS codes [1,2,5], $c_u(k)$ here will be SC or MC. Similar to DS though, these new codes will be periodic with period N_f, and with energy normalized so that $\sum_{k=0}^{N_f-1} c_u^2(k) = N_f$, $\forall u \in [0, N_u - 1]$.

11.2.1 Baseband Single-Carrier UWB

To introduce our SC-UWB user codes, let us first define $N_u = N_f$ digital carriers $\forall k \in [0, N_f - 1]$:

$$[\boldsymbol{g}_u]_k = \begin{cases} \sqrt{2}\cos(2\pi f_u k), & \text{if } u \in \left[0, \dfrac{N_f}{2} - 1\right], \\ \sqrt{2}\sin(2\pi f_u k), & \text{if } u \in \left[\dfrac{N_f}{2}, N_f - 1\right], \end{cases} \tag{2}$$

where $f_u := (u + 0.5)/N_f$, $\forall u \in [0, N_f - 1]$. This SC spreading code during the kth frame is then given by $c_u(k) = [\boldsymbol{g}_u]_k$, which means that the uth user relies on the *digital* frequency f_u to spread symbols. Stacking the N_f carriers into a matrix $\boldsymbol{G}_{sc} := [\boldsymbol{g}_0 \ \cdots \ \boldsymbol{g}_{N_f-1}]$, we construct the SC-UWB user codes as

$$\boldsymbol{c}_u = \boldsymbol{G}_{sc} \boldsymbol{e}_u, \quad \forall u \in [0, N_f - 1], \tag{3}$$

where $\boldsymbol{c}_u := [c_u(0), \ldots, c_u(N_f - 1)]^T$, T stands for transposition, and $\boldsymbol{e}_n$ denotes the $(n+1)$th column of the identity matrix $\boldsymbol{I}_{N_f}$. It can be easily verified that these digital SC spreading codes are orthogonal; that is, $\boldsymbol{c}_{u_1}^T \boldsymbol{c}_{u_2} = N_f \delta(u_1 - u_2)$; hence, the maximum number of users is $N_u = N_f$, as in DS-UWB using orthogonal DS sequences.

Differing from *narrowband* OFDMA, the codes in (3) are baseband real. More important, in *ultra-wideband* operation, these SC spreading codes result in multiband transmissions. To reveal this *multiband feature* of SC-UWB, we will next derive the power spectral density (PSD) of $v_u(t)$ in (1), when the SC codes of (3) are utilized.

For i.i.d. equi-probable binary PAM symbols, the PSD of $v_u(t)$ in (1) can be expressed as [6, Chapter 4]

$$\Phi_v^{(u)}(f) = \frac{\mathcal{E}_u}{N_f}\frac{1}{T_s}|P_{s,u}(f)|^2, \tag{4}$$

where $P_{s,u}(f) := \mathcal{F}\{p_{s,u}(t)\}$ is the Fourier Transform (FT) of the symbol level pulse shaper $p_{s,u}(t) := \sum_{k=0}^{N_f-1} c_u(k)p(t-kT_f)$. With the SC spreading codes in (3), it can be readily verified that

$$P_{s,u}(f) = P(f)\sum_{k=-\infty}^{+\infty} \mathcal{S}\left(f-\frac{k}{T_f}-\frac{f_u}{T_f}\right) \pm \mathcal{S}\left(f-\frac{k}{T_f}+\frac{f_u}{T_f}\right), \tag{5}$$

where $P(f) := \mathcal{F}\{p(t)\}$, and $\mathcal{S}(f) := (T_s/\sqrt{2})\exp(-j\pi T_s f)\text{sinc}(T_s f)$, with $\text{sinc}(f) := \sin(\pi f)/(\pi f)$. The '+' sign between the two $\mathcal{S}(\cdot)$ terms in (5) corresponds to users $u \in [0, N_f/2-1]$, while the '−' sign corresponds to users $u \in [N_f/2, N_f-1]$.

The nonzero frequency support of $P(f)$ is inversely proportional to the pulse duration T_p; whereas the sinc function has main lobe width $(2/T_s)$ Hz, and is repeated every $(1/T_f)$ Hz. Letting $N_p := T_f/T_p$ to be an integer without loss of generality, we deduce that there are $2N_p$ sinc main lobes over the bandwidth of $P(f)$. In UWB transmissions that typically have low duty-cycle, $T_f \gg T_p$ implies that the number of sinc main lobes $2N_p \gg 2$. In other words, utilizing a single digital 'carrier' f_u, each user's transmission occupies multiple frequency bands, as depicted in Figure 11.1. Also notice that due to the $0.5/N_f$ shift in the definition of f_u in (2), each user (subcarrier) occupies the same bandwidth.

The multiband feature of SC-UWB implies that each user's transmission is spread over the ultra-wide bandwidth, and enjoys the associated multipath diversity gains. In fact, we will later see that the baseband real SC-UWB codes in (3) enable full multipath

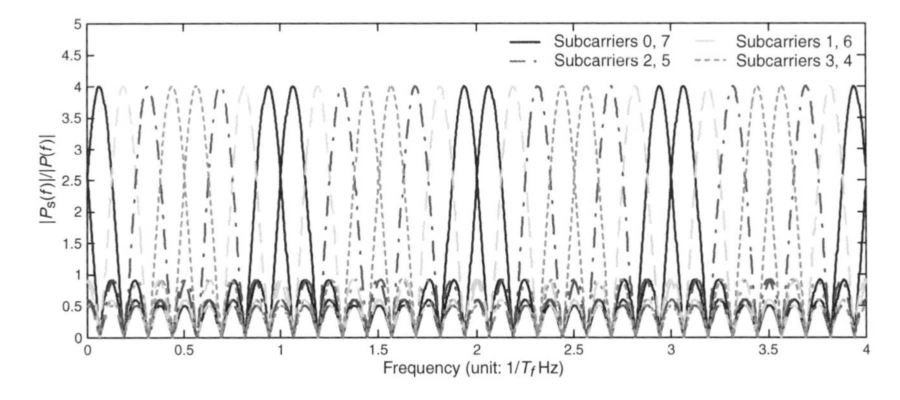

Figure 11.1 Subcarriers in baseband SC-UWB ($N_f = 8$, $N_p = 4$)

diversity, in contrast with narrowband OFDMA systems that have to resort to channel coding and/or frequency hopping to mitigate frequency-selective fading at the expense of (possibly considerable expense) bandwidth overexpansion.

Also implied by Figure 11.1 is the flexibility offered by SC-UWB in handling (e.g., GPS or WLAN induced) NBI. Since the transmit spectrum is distinctly determined by the digital carrier f_u, one gains resilience to NBI by simply avoiding usage of carriers residing on (or close to) these services. Such a flexibility in NBI avoidance is also shared by the MC-UWB spreading codes that we introduce next.

11.2.2 Baseband Multicarrier UWB

Instead of a SC, in MC-UWB, each user can utilize all digital carriers. To construct $N_u(= N_f)$ such user-specific codes $c_u(k)$, let us first define the following $N_f \times 1$ digital carriers $\forall k \in [0, N_f - 1]$:

$$[\bar{\boldsymbol{g}}_n]_k = \begin{cases} \cos(2\pi f_n k), & n = 0, \text{ or, } n = \dfrac{N_f}{2} \\ \sqrt{2}\cos(2\pi f_n k), & n \in \left[1, \dfrac{N_f}{2} - 1\right] \\ \sqrt{2}\sin(2\pi f_n k), & n \in \left[\dfrac{N_f}{2} + 1, N_f - 1\right] \end{cases}, \tag{6}$$

where $f_n := n/N_f$. Stacking the N_f carriers into a matrix $\boldsymbol{G}_{mc} := [\bar{\boldsymbol{g}}_0 \ \ldots \ \bar{\boldsymbol{g}}_{N_f-1}]$, we construct the MC-UWB spreading codes as

$$\boldsymbol{c}_u = \boldsymbol{G}_{mc}\boldsymbol{c}_u^{(o)}, \quad \forall u \in [0, N_f - 1], \tag{7}$$

where $\{\boldsymbol{c}_u^{(o)}\}_{u=0}^{N_f-1}$ denotes any set of real orthonormal sequences (each of length N_f). Evidently, the resultant MC codes are also orthogonal.

Similar to SC-UWB, the digital carriers (6) also give rise to multiband transmissions with multiple sinc main lobes within the frequency support of $P(f)$. This multiband feature of MC-UWB is illustrated by the discrete-time Fourier transform (DTFT) of $\bar{\boldsymbol{g}}_k$'s in Figure 11.2. Also similar to SC-UWB, each MC carrier has a distinct frequency support, which enables flexible NBI suppression by simply avoiding contaminated carriers.

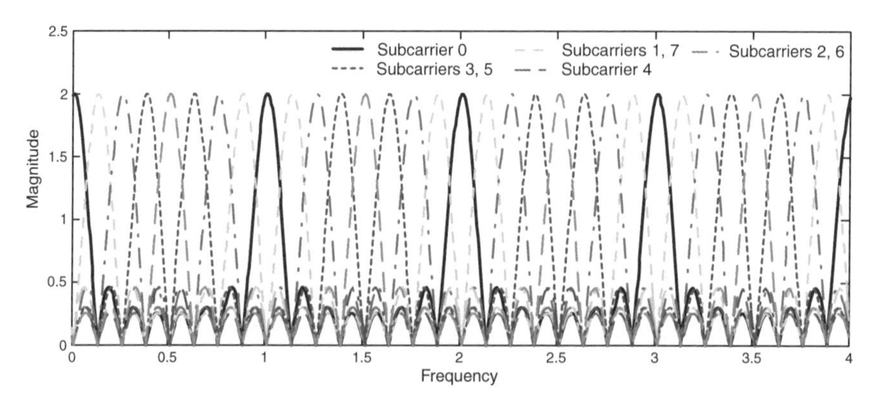

Figure 11.2 Subcarriers in baseband MC-UWB ($N_f = 8$, $N_p = 4$)

Though similar, the SC and MC codes are designed differently and have distinct merits. Comparing the digital SC and MC codes, we notice that the shift of $0.5/N_f$ in f_u is present in (2), but not in (6). The difference becomes evident when comparing Figures 11.1 and 11.2; chosen for SC-UWB as in (2), each digital carrier corresponds to the same number of sinc main lobes ($2N_p = 8$ in Figure 11.1); whereas chosen for MC-UWB as in (6), the 0th and the 4th carriers each contains only half as many sinc main lobes as other carriers (see Figure 11.2). Consequently, specializing (6) to SC (by setting $\boldsymbol{c}_u^{(o)} = \boldsymbol{e}_u$) transmissions will induce unbalanced user bandwidth, which we will show to imply user-dependent multipath diversity. However, since MC-UWB allows each user to utilize *all* carriers with the MC codes in (7), there is no need to equate the bandwidth of each carrier. This explains why the $0.5/N_f$ shift is not included in f_n for the MC spreading codes in (6).

Despite their differences, SC/MC codes share one attractive feature; they both facilitate low-complexity implementation using standard DCT circuits, thanks to their construction based on discrete cos/sin functions. This implementation advantage also distinguishes them from the analog SC-UWB codes introduced in [7] that aim to offer robustness against user asynchronism. Also, different from the WPAN multiband proposals that rely on analog carriers, these SC/MC codes achieve multiband transmission using baseband operations. Compared to analog multiband solutions that entail multiple local oscillators, the carrier-free multiband SC/MC-UWB not only enjoys low-complexity implementation but also are exempt from carrier frequency offsets that are known to degrade performance (e.g., OFDMA) severely.

In the discussions so far, we have introduced the construction of baseband real SC/MC codes, and the resultant multiband transmissions. Next, their performance in a variety of operating conditions will be compared with the conventional DS spreading codes.

11.3 Low Duty-Cycle Access in the Presence of NBI

Coexisting narrowband services introduce interference, which degrades performance of UWB radios that rely on spreading schemes to enable multiple access. In this section, performance comparisons will be carried out between the SC/MC codes and the widely adopted DS codes [1, 2]. Although the SC/MC codes provide improved flexibility to avoid NBI in the first place by excluding the affected digital carrier(s), their performance will be tested without invoking this flexibility. In other words, it will be assumed that no knowledge of the NBI is available. Comparisons in this section will be confined within the scope of low duty-cycle access, where the frame duration T_f is sufficiently large so that no interframe interference (IFI) is present[1].

11.3.1 General Rake Reception Model

In order to facilitate performance analysis and comparison, a general rake reception model is needed. Let us start with the signal received after multipath propagation. Let $\sum_{l=0}^{L} \alpha_l \delta(t - \tau_l)$ denote the multipath channel with $(L + 1)$ paths. The continuous-time

[1] A more general treatment that includes TH codes as well as IFI can be found in [8].

received waveform is then given by [cf. (1)]

$$\begin{aligned} r_u(t) &= \nu_u(t) \star h(t) + w(t) + j(t) \\ &= \sqrt{\mathcal{E}_u/N_f} \sum_{k=0}^{\infty} s_u(\lfloor k/N_f \rfloor) c_u(k) h(t - kT_f) + w(t) + j(t), \end{aligned} \tag{8}$$

where $h(t) := \sum_{l=0}^{L} \alpha_l p(t - \tau_l)$ is the composite pulse-multipath channel with delay spread $\tau_L + T_p$, $w(t)$ is the AWGN, and $j(t)$ denotes NBI. The frame duration will be chosen to satisfy $T_f > \tau_L + T_p$ to avoid IFI as well as intersymbol interference (ISI). This assumption simplifies the analysis considerably, but confines the scope to *low duty-cycle* UWB systems.

To collect multipath diversity, rake reception is often adopted by UWB systems [9]. Rake receivers with N fingers weight and sum, outputs from a bank of N correlators (diversity combining). During the kth frame of each symbol, the template for the nth correlator (rake finger with delay $\tilde{\tau}_n$)[2] is the pulse $p(t)$ delayed by $kT_f + \tilde{\tau}_n$. For fingers to capture *resolvable* multipath returns, their delay must satisfy $\tilde{\tau}_n - \tilde{\tau}_{n-1} \geq T_p$, which implies that at most $\bar{N} := \lfloor \tau_L / T_p \rfloor$ fingers, it is suffice to collect the available multipath diversity. In practice, $N \leq \bar{N}$ (all-rake, partial-rake, or selective-rake) is often chosen to trade off error performance with complexity [10].

During the kth frame of each symbol, the template for the nth finger's correlator is given by $p(t - kT_f - \tilde{\tau}_n)$. Accordingly, the nth finger's correlator output during the kth frame is

$$x_u(k; n) = \int_{kT_f + \tilde{\tau}_n}^{kT_f + \tilde{\tau}_n + T_p} r_u(t) p(t - kT_f - \tilde{\tau}_n)\, dt. \tag{9}$$

To establish the input-output (I/O) relationship in digital form, let us define $\tilde{\alpha}_n := \int_{\tilde{\tau}_n}^{\tilde{\tau}_n + T_p} h(t) p(t - \tilde{\tau}_n)\, dt$ to represent the 'effective channel amplitude' at the nth finger's correlator output. Using the low duty-cycle assumption $T_f > \tau_L + T_p$, it then follows that

$$\int_{kT_f + \tilde{\tau}_n}^{kT_f + \tilde{\tau}_n + T_p} h(t - mT_f) p(t - kT_f - \tilde{\tau}_n)\, dt = \tilde{\alpha}_n \delta_{km}. \tag{10}$$

Taking this into account when substituting (8) to (9), we obtain

$$x_u(k; n) = \sqrt{\mathcal{E}_u/N_f} s_u(\lfloor k/N_f \rfloor) c_u(k) \tilde{\alpha}_n + w(k; n) + j(k; n),$$

where $w(k; n)$ and $j(k; n)$ denote respectively the sampled additive noise and NBI at the correlator output of the nth finger, during the kth frame. Notice that the filtered and sampled AWGN $w(k; n)$ stays white, because the $\tilde{\tau}_n$'s are spaced sufficiently apart.

Stacking the correlator output samples from all N fingers during the kth frame, we form

$$\begin{aligned} \boldsymbol{x}_u(k) &:= [x_u(k; 0) \;\; x_u(k; 1) \cdots x_u(k; N-1)]^T \\ &= \sqrt{\mathcal{E}_u/N_f} s_u(\lfloor k/N_f \rfloor) c_u(k) \tilde{\boldsymbol{\alpha}} + \boldsymbol{w}(k) + \boldsymbol{j}(k), \end{aligned} \tag{11}$$

[2] The usage of tilde for rake finger delays ($\tilde{\tau}_n$) is to stress that they do not have to necessarily coincide with the multipath channel delays (τ_n).

where $\tilde{\boldsymbol{\alpha}}$, $\boldsymbol{w}(k)$, and $\boldsymbol{j}(k)$ are all $N \times 1$ vectors constructed by stacking $\tilde{\alpha}_n$, $w(k;n)$, and $j(k;n)$ for $n \in [0, N-1]$. Recalling that each symbol is conveyed by N_f pulses, a total of NN_f correlator output samples, N per frame, must be collected in order to decode one symbol. To this end, vectors $\{\boldsymbol{x}_u(k)\}_{k=n_s N_f}^{(n_s+1)N_f-1}$ are concatenated into a super vector of size $N_f N \times 1$ as [cf. (11)]:

$$\begin{aligned} \boldsymbol{y}_u(n_s) &:= [\boldsymbol{x}_u^T(n_s N_f) \cdots \boldsymbol{x}_u^T(n_s N_f + N_f - 1)]^T \\ &= \sqrt{\mathcal{E}_u / N_f}\, s_u(n_s)(\boldsymbol{c}_u \otimes \tilde{\boldsymbol{\alpha}}) + \boldsymbol{\eta}(n_s), \end{aligned} \tag{12}$$

where $\boldsymbol{c}_u := [c_u(0), \ldots, c_u(N_f-1)]^T$ is the spreading code vector and $\boldsymbol{\eta}(n_s) := [\boldsymbol{w}^T(n_s N_f) + \boldsymbol{j}^T(n_s N_f), \ldots, \boldsymbol{w}^T(n_s N_f + N_f - 1) + \boldsymbol{j}^T(n_s N_f + N_f - 1)]^T$ is the $N_f N \times 1$ noise vector that consists of AWGN, MUI, and NBI present over the n_sth symbol duration.

Notice that $\boldsymbol{y}_u(n_s)$ contains nothing but the correlator outputs collected from N fingers over N_f consecutive frames corresponding to the n_sth symbol. Relative to [11], which requires subpulse rate oversampling, (12) is formed by *frame-rate* samples per rake finger. To decode a symbol, diversity combining is required. With the $N_f N \times 1$ weight vector $\boldsymbol{\beta}$, diversity combining yields the following decision statistic for the n_sth symbol; $z_u(n_s) := \boldsymbol{\beta}^T \boldsymbol{y}_u(n_s)$.

In the *absence* of NBI, maximum ratio combining (MRC) is optimal since $\boldsymbol{\eta}$ in (12) becomes white Gaussian. MRC then corresponds to matched-filter (MF) weights $\boldsymbol{\beta}_{\text{mf}} := \boldsymbol{c}_u \otimes \tilde{\boldsymbol{\alpha}}$. However, in the *presence* of NBI, the color of $\boldsymbol{\eta}(n_s)$ renders MF reception suboptimal and motivates the usage of minimum mean-square error (MMSE) weights, which, for the sampled vector model in (12), are

$$\begin{aligned} \boldsymbol{\beta}_{\text{mmse}} &:= \frac{\mathcal{E}_u}{N_f}\left[\boldsymbol{R}_\eta + \frac{\mathcal{E}_u}{N_f}(\boldsymbol{c}_u \otimes \tilde{\boldsymbol{\alpha}})(\boldsymbol{c}_u \otimes \tilde{\boldsymbol{\alpha}})^T\right]^{-1}(\boldsymbol{c}_u \otimes \tilde{\boldsymbol{\alpha}}) \\ &= \frac{\mathcal{E}_u}{N_f}\boldsymbol{R}_\eta^{-1}(\boldsymbol{c}_u \otimes \tilde{\boldsymbol{\alpha}})\left[1 + \frac{\mathcal{E}_u}{N_f}(\boldsymbol{c}_u \otimes \tilde{\boldsymbol{\alpha}})^T \boldsymbol{R}_\eta^{-1}(\boldsymbol{c}_u \otimes \tilde{\boldsymbol{\alpha}})\right]^{-1}, \end{aligned} \tag{13}$$

where $\boldsymbol{R}_\eta := E\{\boldsymbol{\eta}(n_s)\boldsymbol{\eta}^T(n_s)\}$ and $\otimes$ denotes Kronecker product. As usual, if $\boldsymbol{\eta}$ is white, $\boldsymbol{\beta}_{\text{mmse}}$ boils down to $\boldsymbol{\beta}_{\text{mf}}$.

In order to quantify the error performance of MF and MMSE rake receivers, one needs to know the structure of $\boldsymbol{R}_\eta$. It is well known that uniformly sampling a wide sense stationary (WSS) process yields a WSS sequence. However, in order to encompass various selections of rake fingers, we generally allow the delays $\{kT_f + \tilde{\tau}_n\}$ to be nonuniformly spaced across k and n; that is, although for a given n, delays $kT_f + \tilde{\tau}_n$ are uniform across frames (k), for a given k, $\tilde{\tau}_n$s are often nonuniformly spaced within a frame, as for example, in selective-rake receivers. Consequently, the entries of $\boldsymbol{\eta}(n_s)$ are nonuniformly sampled noise variables. With N rake fingers per frame, nonuniformly sampled noise per frame turns out to be cyclostationary with period N. As a result, $\boldsymbol{R}_\eta$ has a block-Toeplitz structure, but each $N \times N$ block of $\boldsymbol{R}_\eta$ is not necessarily Toeplitz. It is useful to specify the form of $\boldsymbol{R}_\eta$ for various rake finger selections, which will be represented with a finger selection matrix $\boldsymbol{S}$, whose construction is detailed next.

Let T_f/T_p be an integer N_p without loss of generality. With rake finger delays being integer multiples of T_p, N_p is an upper bound on the number of rake fingers N; that is, $N \leq \bar{N} < N_p$ and the N rake fingers can be thought of as being *selected* from a total of N_p possible ones. This selection can be represented with a delay selection matrix $\boldsymbol{S}$. Let us sort the N (generally nonequispaced) finger delays in an increasing order and let $\tilde{\tau}_n = i_n \cdot T_p$, with integers $\{i_n\}_{n=0}^{N-1} \in [0, \bar{N} - 1]$. Using these i_n's, the $N_p \times N$ delay selection matrix $\boldsymbol{S}$ is constructed so that the nth column of $\boldsymbol{S}$ is $\boldsymbol{e}_{i_n+1}$; that is, the $(i_n + 1)$th column of $\boldsymbol{I}_{N_p}$. Let us take as an example the partial-rake reception, where the first $N < \bar{N}$ delays are chosen; that is, $\tilde{\tau}_n = n \cdot T_p$, $\forall n \in [0, N - 1]$. In this case, the delay selection matrix is given by $\boldsymbol{S} = [\boldsymbol{I}_N\ \boldsymbol{0}_{N\times(N_p-N)}]^T$. Since the latter matrix pads $(N_p - N)$ zeros to a $N \times 1$ vector, we term it a zero-padding (ZP) matrix and denote it from now on as $\boldsymbol{T}_{N_p,N}$.

Corresponding to the N_p possible uniformly spaced rake finger delays in a given frame, there are N_p uniform noise samples. Collecting these noise samples over N_f frames into a $N_f N_p \times 1$ vector $\bar{\boldsymbol{\eta}}(n_s)$ leads to a correlation matrix $\boldsymbol{R}_{\bar{\eta}} := \mathrm{E}\{\bar{\boldsymbol{\eta}}(n_s)\bar{\boldsymbol{\eta}}^T(n_s)\}$ that is Toeplitz, simply because the entries of $\bar{\boldsymbol{\eta}}(n_s)$ are uniformly sampled noise variables. As $\boldsymbol{S}$ selects the rake finger delays, it also selects from $\bar{\boldsymbol{\eta}}(n_s)$ the noise samples corresponding to the selected delays. With such a selection process being repeated in each frame comprising a symbol, we have $\boldsymbol{\eta} = (\boldsymbol{I}_{N_f} \otimes \boldsymbol{S}^T)\bar{\boldsymbol{\eta}}(n_s)$. As a result, its correlation matrix $\boldsymbol{R}_\eta = (\boldsymbol{I}_{N_f} \otimes \boldsymbol{S}^T)\boldsymbol{R}_{\bar{\eta}}(\boldsymbol{I}_{N_f} \otimes \boldsymbol{S})$ is a $N_f N \times N_f N$ block-Toeplitz matrix with submatrices of size $N \times N$.

An example of $\boldsymbol{R}_{\bar{\eta}}$ is illustrated in Figure 11.3(a), with $N_f = 4$ and $N_p = 16$. It's Toeplitz structure is evident. With the same parameters, Figure 11.3(b) illustrates an example of $\boldsymbol{R}_\eta$, using the selection matrix $\boldsymbol{S} = [\boldsymbol{e}_7\ \boldsymbol{e}_8\ \boldsymbol{e}_{10}\ \boldsymbol{e}_{11}\ \boldsymbol{e}_{14}]$ with $N = 5$ and i_n's being 6, 7, 9, 10 and 13. Evidently, $\boldsymbol{R}_\eta$ is not Toeplitz, but rather block-Toeplitz with identical submatrices sitting along any diagonal with direction from northwest to southeast. Also notice that because of the choice of $\boldsymbol{S}$, these 5×5 submatrices are not Toeplitz. They will become Toeplitz if and only if rake finger delays are equispaced. A couple of remarks are now in order.

Remark 1: Equation (12) constitutes a digital model for general rake reception; namely, all-rake, partial-rake, or selective-rake. Established on the basis of a two-step (namely,

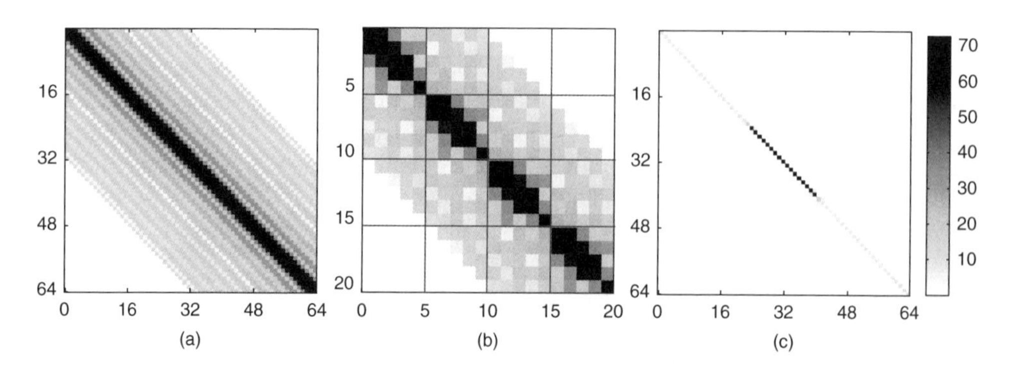

Figure 11.3 (a) Structure of $\boldsymbol{R}_{\bar{\eta}}$; (b) Structure of $\boldsymbol{R}_\eta$; and (c) Structure of $\boldsymbol{\Phi}$. $N_f = 4$, $N_p = 16$, $N = 5$

correlation followed by weighted combination) approach, (12) requires *frame-rate* sampling per finger. Interestingly, receiver processing can be implemented even with *symbol-rate* sampling. To prove this, recall first that the entries of $\boldsymbol{\beta}$ (that is, $[\boldsymbol{\beta}]_n \forall n$) are the diversity combining weights. Rake reception that yields the decision statistics z_u can be realized by correlating $r_u(t)$ with the *symbol-long* template $\bar{p}_s(t) = \sum_{k=0}^{N_f-1} \sum_{n=0}^{N-1} [\boldsymbol{\beta}]_{kN+n} p(t - kT_f - \tilde{\tau}_n)$, and sampling its output every $T_s = N_f T_f$ seconds. It can also be verified that, MF combining can be implemented using the template $\bar{p}_f(t) := \sum_{n=0}^{N-1} \tilde{\alpha}_n p(t - \tilde{\tau}_n)$, which has fixed combining weights that do not change across frames. The decision statistic is thus given by $z_u = \sum_{k=0}^{N_f-1} c_u(k) \int r_u(t) \bar{p}_f(t - kT_f - c_u^{th}(k)T_c)\, dt$. In other words, rake reception with $\boldsymbol{\beta}_{\mathrm{mf}}$ can be carried out frame-by-frame; whereas with $\boldsymbol{\beta}_{\mathrm{mmse}}$, rake can only be performed on a symbol-by-symbol basis. This also indicates that NBI renders frame-by-frame receiver processing suboptimal.

To generate the template $\bar{p}_s(t)$, multiple analog waveforms $p(t)$ have to be generated and delayed accordingly. The delay accuracy will affect decoding performance. However, differing from pulse-rate sampling that requires precise timing at each sampling instance, $\bar{p}_s(t)$ needs to be generated only once during the channel coherence time. The latter provides the timing circuits sufficient time to stabilize and thus reduces timing jitter effects.

Remark 2: Carried out for PAM modulation, the analysis and I/O model in (12) hold good for binary pulse position modulation (PPM) as well. In the latter case, one only needs to replace the correlator template $p(t)$ with $p(t) - p(t - \Delta)$, where Δ is the PPM modulation index.

11.3.2 SINR Analysis

In this section, performance of different transmission schemes will be compared on the basis of their corresponding SINR for different selections of $\boldsymbol{\beta}$.

From (12), it follows that the instantaneous SINR after diversity combining for the two receiver options is given by

$$\mathrm{SINR}_{\mathrm{mf}} := \frac{N_f \mathcal{E}_u |\tilde{\boldsymbol{\alpha}}^T \tilde{\boldsymbol{\alpha}}|^2}{(\boldsymbol{c}_u \otimes \tilde{\boldsymbol{\alpha}})^T \boldsymbol{R}_\eta (\boldsymbol{c}_u \otimes \tilde{\boldsymbol{\alpha}})}, \tag{14}$$

and

$$\mathrm{SINR}_{\mathrm{mmse}} := \frac{\mathcal{E}_u}{N_f} (\boldsymbol{c}_u \otimes \tilde{\boldsymbol{\alpha}})^T \boldsymbol{R}_\eta^{-1} (\boldsymbol{c}_u \otimes \tilde{\boldsymbol{\alpha}}), \tag{15}$$

respectively, where $\boldsymbol{R}_\eta = (\boldsymbol{I}_{N_f} \otimes \boldsymbol{S}^T) \boldsymbol{R}_{\bar{\eta}} (\boldsymbol{I}_{N_f} \otimes \boldsymbol{S})$. Recall that with $\bar{\boldsymbol{\eta}}$ being stationary Gaussian, its correlation matrix $\boldsymbol{R}_{\bar{\eta}}$ is a $N_f N_p \times N_f N_p$ Toeplitz matrix. Since $N_f N_p$ is large ($\gg$100), $\boldsymbol{R}_{\bar{\eta}}$ can be well approximated by a circulant matrix, which can be diagonalized by FFT matrices as in [12]; that is, $\boldsymbol{R}_{\bar{\eta}} \approx \boldsymbol{F}_{N_f N_p}^{\mathcal{H}} \boldsymbol{\Phi} \boldsymbol{F}_{N_f N_p}$, where $\boldsymbol{\Phi}$ is a diagonal matrix. In particular, if the NBI has a flat PSD on its nonzero frequency support, the nth diagonal entry of the diagonal matrix $\boldsymbol{\Phi}$ is given by

$$[\boldsymbol{\Phi}]_{n,n} = \begin{cases} \dfrac{J_0 + N_0}{2} & \text{if subband } n \text{ is hit} \\ \dfrac{N_0}{2} & \text{otherwise} \end{cases}, n \in (0, N_f N_c), \tag{16}$$

where $N_0/2$ is the AWGN power and $J_0/2$ is the NBI power. Figure 11.3(c) depicts an example of $\boldsymbol{F}_{N_fN_p}\boldsymbol{R}_{\bar{\eta}}\boldsymbol{F}^{\mathcal{H}}_{N_fN_p}$. The interferer has a flat PSD with power $J_0/N_0 = 8$ dB and $N_0/2 = 1$. Notice that with $N_fN_p = 4 \times 16 = 64$, $\boldsymbol{F}_{N_fN_p}\boldsymbol{R}_{\bar{\eta}}\boldsymbol{F}^{\mathcal{H}}_{N_fN_p}$ is approximately diagonal, as indicated in (16).

So far, we have developed closed-form expressions for the SINR at the rake outputs on the basis of the digital model (12). It follows from (14) and (15) that SINR is generally code dependent, which in turn implies that the BER generally depends on $\boldsymbol{c}_u$. However, it is possible to have code-independent SINR. For example, we know that when η is white, the MMSE weights boil down to MF weights and for all transmission schemes we have from (14) that $\mathrm{SINR}_{\mathrm{mmse}} = \mathrm{SINR}_{\mathrm{mf}} = (2\mathcal{E}_u/N_0)\tilde{\boldsymbol{\alpha}}^T\tilde{\boldsymbol{\alpha}}$, since $(\boldsymbol{c}_u \otimes \tilde{\boldsymbol{\alpha}})^T(\boldsymbol{c}_u \otimes \tilde{\boldsymbol{\alpha}}) = \boldsymbol{c}_u^T\boldsymbol{c}_u\tilde{\boldsymbol{\alpha}}^T\tilde{\boldsymbol{\alpha}} = N_f\tilde{\boldsymbol{\alpha}}^T\tilde{\boldsymbol{\alpha}}$. In *narrowband* systems, it has also been shown that the performance of MC-CDMA is code independent [12]. One may expect a similar result to hold for the real baseband MC-UWB with rake reception. In fact, it has been proved that the BER of real MC-UWB systems is indeed code independent, with any rake finger selection, even in the presence of NBI and dense multipath, as we summarize in the following proposition [8]:

Proposition 1 *In baseband MC-UWB with codes* $c_u(k) = \boldsymbol{g}_k^T\boldsymbol{c}_u^{(o)}$, *and* $\forall u, k \in [0, N_f - 1]$, *where no TH is employed, the SINR in* (14) *and* (15) *corresponding to MF and MMSE rake receivers are code/user independent even in the presence of NBI, regardless of the multipath channel and the rake finger delay selection, as long as* $\{|\bar{c}_u^{(o)}(k)|^2\}_{k=0}^{N_f-1}$ *are the same* $\forall u \in [0, N_f - 1]$, *where* $\{\bar{\boldsymbol{c}}_u^{(o)}\}_{u=0}^{N_f-1}$ *are new code vectors constructed from* $\boldsymbol{c}_u^{(o)}$

$$\left[\bar{\boldsymbol{c}}_u^{(o)}\right]_k = \begin{cases} \left[\boldsymbol{c}_u^{(o)}\right]_k, & \text{if } k = 0, \text{ or, } \dfrac{N_f}{2} \\ \dfrac{1}{\sqrt{2}}\left(\left[\boldsymbol{c}_u^{(o)}\right]_k + \jmath \cdot \left[\boldsymbol{c}_u^{(o)}\right]_{N_f-k}\right), & \text{if } k \in (1, N_f/2) \\ \dfrac{1}{\sqrt{2}}\left(\left[\boldsymbol{c}_u^{(o)}\right]_{N_f-k} - \jmath \cdot \left[\boldsymbol{c}_u^{(o)}\right]_k\right), & \text{if } k \in (N_f/2+1, N_f) \end{cases}.$$

Code-independent SINR implies that the corresponding BER is also code independent. As an example, consider $\boldsymbol{c}_u^{(o)}$ in (7) with binary $\{\pm 1\}$ entries. In this case, $|\bar{c}_u^{(o)}(k)|^2 = 1$, $\forall k, u \in [0, N_f - 1]$. The BER corresponding to such $\boldsymbol{c}_u^{(o)}$ will thus be code independent.

Unique to MC-UWB, Proposition 1 can also be justified intuitively. Recall that each MC-UWB user utilizes all N_f digital subcarriers, with each MC subcarrier occupying distinct frequency bands. When $\{|\bar{c}_u^{(o)}(k)|^2\}$ for each k is constant across all users $u \in [0, N_f - 1]$, all users have the same 'weight' on each subcarrier and its corresponding frequency bands. Therefore, NBI affects all users (codes) uniformly. In particular, when $|\bar{c}_u^{(o)}(k)|^2 = 1$, $\forall k, u \in [0, N_f - 1]$, each user's transmit PSD will be 'flat' (see also Figure 11.2).

11.3.3 Simulations and Numerical Results

In all examples, the number of frames per symbol is chosen to be $N_f = 32$. A Gaussian monocycle with duration $T_p = 0.7$ ns is used as the pulse shaper $p(t)$. Since the UWB transmission here is baseband real, the multipath channels are generated using the Saleh–Valenzuela (S-V) channel model [13] modified as in [14], where the channel taps are real. The pertinent parameters in [13, 14] to generate random channel realizations are chosen as $(1/\Lambda, 1/\lambda, \Gamma, \gamma) = (43, 0.4, 7.1, 4.3)$ ns, giving rise to maximum excess delays around 35 ns. The channel taps are truncated at $\tau_L = 34$ ns and choose $T_f = 35$ ns. As a result, we obtain $N_p := T_f/T_p = 50$. The TH codes are generated independently from a uniform distribution over $[0, N_c - 1]$. Two N_c values are used; $N_c = 10$ with $T_c = 3.5$ ns, and $N_c = 50$ with $T_c = 0.7$ ns. Walsh–Hadamard (W-H) codes are used both for DS-UWB and as the orthogonal codes $\boldsymbol{c}_u^{(o)}$ for the MC-UWB spreading in (7). Selective- and partial-rake are employed, where the N delays with strongest outputs are chosen for the former, and the first N delays are chosen for the latter. In AWGN channels, $N = 1$ is used. When multipath is also present, more fingers are used and will be specified in individual simulations. These results are generated using (14) and (15), averaged over 5000 channel realizations.

Three sets of parameters are used to generate NBI with a flat in-band power spectrum: NBI-1 with 1.2-GHz center frequency and 20-MHz bandwidth, NBI-2 with 900-MHz center frequency and 25-MHz bandwidth, and NBI-3 with 900-MHz center frequency and 80-MHz bandwidth. Notice that NBI-1 approximates the GPS band and NBI-2 approximates the GSM900 band. The power of NBI is $J_0/N_0 = 30$ dB relative to the AWGN power N_0.

In Figures 11.4(a) and 11.4(b), we plot BER corresponding to all 32 DS and MC spreading codes, in AWGN channels with a 1-finger MMSE rake receiver and in multipath channels with a 16-finger selective MF-rake receiver, both in the presence of NBI. Corroborating Proposition 1, MC exhibits identical performance across all codes/users; whereas DS exhibits code-dependent performance, in both AWGN and multipath channels with NBI. But as it is clear from Figures 11.5(a) and 11.5(b), which depict BER over multipath channels with MMSE rake, performance variation among different DS codes decreases as the number of rake fingers increases from 1 to 16. Also notice that though selective rake improves performance of all codes, it also increases the performance variation across codes, in comparison to partial-rake reception.

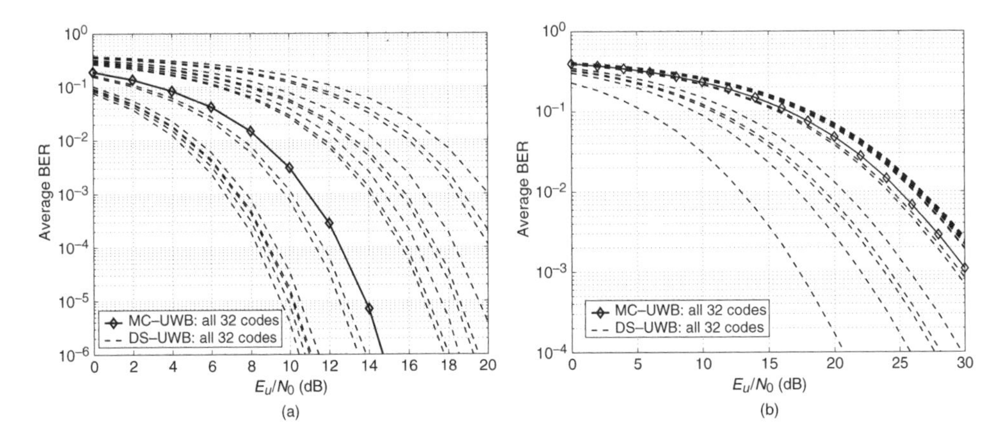

Figure 11.4 BER for DS- and MC-UWB in the presence of NBI-1 (center frequency 1.2 GHz, bandwidth 20 MHz). All codes' performance is plotted to illustrate code-(in)dependence. (a) AWGN channel with MMSE rake ($N = 1$); and (b) Multipath channel with Selective MF rake ($N = 16$)

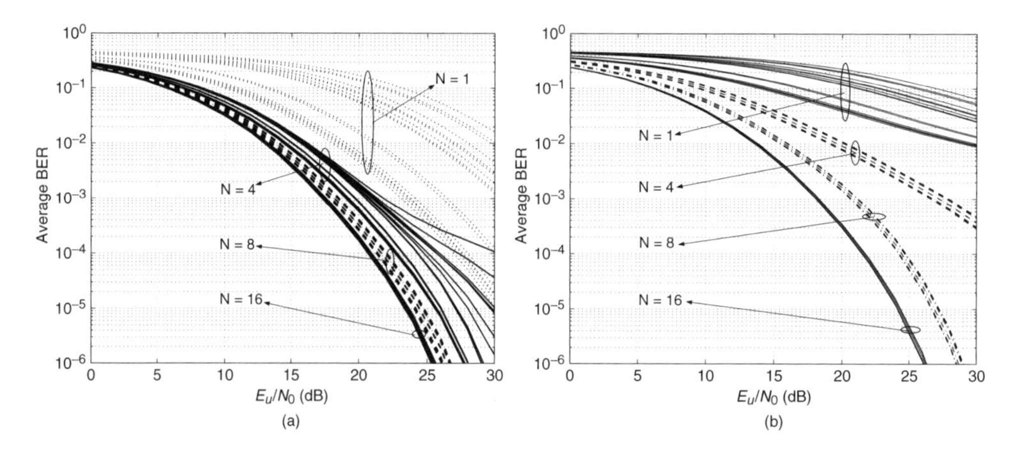

Figure 11.5 BER for DS-UWB over multipath channels with NBI-1 (center frequency 1.2 GHz, bandwidth 20 MHz). (a) Selective-rake; and (b) Partial-rake, with MMSE combining and various number of fingers: $N = 1, 4, 8, 16$. All codes' performance is plotted to illustrate code-dependence

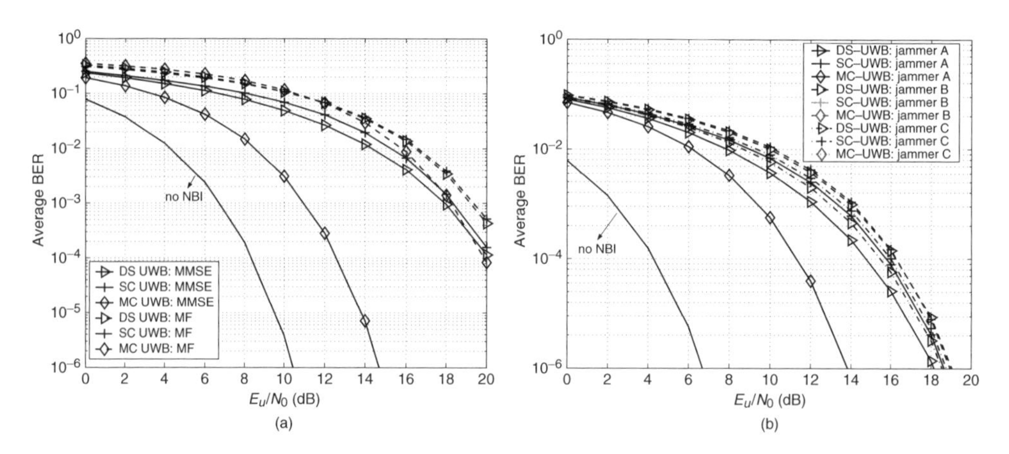

Figure 11.6 Average BER corresponding to DS-, SC-, and MC-UWB in AWGN channel. (a) In the presence of NBI-1 (center frequency 1.2 GHz, bandwidth 20 MHz); and (b) in the presence of three 25-MHz bandwidth jammers with different center frequencies: Jammer-A (300 MHz), Jammer-B (400 MHz) and Jammer-C (900 MHz)

When NBI is present but multipath is absent, the BER averaged over all DS, SC, and MC spreading codes is depicted in Figure 11.6(a) in the presence of NBI-1 (GPS) interferer. We observe that DS- and SC-UWB yield similar performance, while MC-UWB outperforms both. In Figure 11.6(b), we also plot the average BER of DS-, SC-, and MC-UWB in the presence of three different jammers. These jammers have the same bandwidth but different center frequencies. As shown in the figure, MC-UWB yields invariant performance, while DS- and SC-UWB exhibit variable BER, as the jammer's center frequency changes.

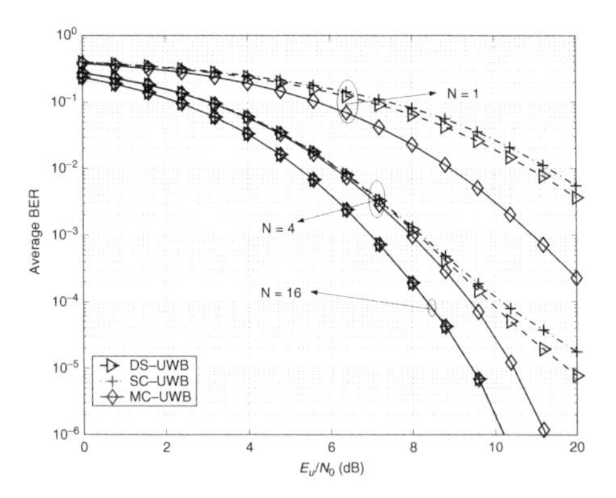

Figure 11.7 Average BER corresponding to DS-, SC-, and MC-UWB over dense multipath channels with AWGN, in the presence of NBI-1 (center frequency 1.2 GHz, bandwidth 20 MHz). Selective rake with $N = 1, 4, 16$ fingers is used

In the presence of both NBI and multipath effects, the average BER versus $\mathcal{E}_u/N_0$ is depicted in Figure 11.7 for DS/SC/MC-UWB. Notice that MC-UWB outperforms others, whereas the performance difference diminishes as more rake fingers are employed.

11.4 Improved Rate Access in the Presence of Multipath

As we have mentioned in Section 11.2, unlike OFDMA that has to resort to channel coding and/or frequency hopping to mitigate frequency-selective fading, UWB signaling even with SC spreading occupies multiple frequency bands, and the resulting multiband transmission enjoys multipath diversity gains. Next, performance of SC-, MC- and DS-UWB codes will be compared in terms of diversity order and coding gain.

To make such a comparison more general, we will relax the low duty-cycle constraint and allow for high data-rate. Since such relaxation induces IFI as well as ISI, the rake reception model we introduced in the preceding section needs some modifications, which will be detailed next.

11.4.1 Rake Reception Model with IFI

When N_u active users are present, the waveform arriving at the receiver becomes [cf. (8)]

$$r(t) = \sum_{u=0}^{N_u-1} \sqrt{\frac{\mathcal{E}_u}{N_f}} \sum_{k=0}^{\infty} s_u\left(\left\lfloor \frac{k}{N_f} \right\rfloor\right) c_u(k) h_u(t - kT_f) + \eta(t), \tag{17}$$

where $h_u(t) := \sum_{l=0}^{L_u} \alpha_{u,l}\, p(t - \tau_{u,l})$ is the composite 'pulse-multipath' channel corresponding to user u, and $\eta(t)$ is the aggregate noise including AWGN and possible NBI. Accordingly, the correlator output on the nth rake finger of a particular receiver is [cf. (9)]

$$x(k; n) = \int_{kT_f+\tilde{\tau}_n}^{kT_f+\tilde{\tau}_n+T_p} r(t) p(t - kT_f - \tilde{\tau}_n)\, dt. \tag{18}$$

In the presence of IFI, however, the condition in (10) does not hold anymore. To reach a reception model that accounts for IFI, let us introduce the notation $\rho_{u,h}(\tau) := \int_{\tau}^{\tau+T_p} p(t-\tau)h_u(t)\,dt$. It then follows from (18) that the correlator output for the nth rake finger during the kth frame is

$$x(k;n) = \sum_{u=0}^{N_u-1}\sqrt{\frac{\mathcal{E}_u}{N_f}}\sum_{l=0}^{\infty} s_u(\lfloor l/N_f\rfloor)c_u(l)\rho_{u,h}((k-l)T_f+\tilde{\tau}_n)+\eta(k;n)$$

$$= \sum_{u=0}^{N_u-1}\sqrt{\frac{\mathcal{E}_u}{N_f}}\sum_{l=0}^{+\infty}\tilde{\alpha}_{u,n}(l)c_u(k-l)s_u\left(\left\lfloor\frac{k-l}{N_f}\right\rfloor\right)+\eta(k;n), \tag{19}$$

where $\tilde{\alpha}_{u,n}(l) := \rho_{u,h}(lT_f+\tilde{\tau}_n)$. Eq. (19) represents the I/O relationship of the frame-sampled discrete-time equivalent pulse-multipath-rake system model, which accounts for the IFI effect. Using the definition of $\rho_{y,h}(\tau)$, it can be readily verified that (i) cascading the rake with the pulse-multipath channel yields a discrete-time equivalent channel with taps $\{\tilde{\alpha}_{u,n}(l)\}$ corresponding to the uth user per finger n and (ii) the summation over l captures the IFI.

Seemingly infinite, the number of IFI-inducing frames in (19) is actually finite. This is because the discrete-time equivalent channel is of finite length, as the underlying physical channel is (at least approximately). Indeed, for any u and n, we have $\tilde{\alpha}_{u,n}(l) = 0$, if $lT_f + \tilde{\tau}_n \geq \tau_{u,l} + T_p$. Therefore, the discrete-time equivalent channel $\{\tilde{\alpha}_{u,n}(l)\}_{l=0}^{M_{u,n}}$ corresponding to the uth user per finger n has order

$$M_{u,n} := \max\{l:\ \tilde{\tau}_n + lT_f < \tau_{u,l} + T_p\}. \tag{20}$$

Accordingly, (19) boils down to

$$x(k;n) = \sum_{u=0}^{N_u-1}\sqrt{\frac{\mathcal{E}_u}{N_f}}\sum_{l=0}^{M_{u,n}}\alpha_{u,n}(l)c_u(k-l)s_u\left(\left\lfloor\frac{k-l}{N_f}\right\rfloor\right)+\eta(k;n), \quad \forall n\in[1,N]. \tag{21}$$

Notice that IFI is present as long as the maximum channel order $\max_{u,n}\{M_{u,n}\} > 0$. If we select $T_f \geq \max_u\{\tau_{u,L_u}\} + T_p - \tilde{\tau}_1$, then $M_{u,n} = 0$, $\forall u, n$, and IFI vanishes. When IFI involves more than one symbol, ISI emerges on top of IFI. It can be verified though, that ISI is confined to two consecutive symbols, as long as $\max_u\{\tau_{u,L_u}\} + T_p - \tilde{\tau}_1 \leq T_s$. For notational simplicity, we assume that this condition is satisfied. However, the analysis hereafter can be generalized to cases where this condition is relaxed.

Let us now stack correlator outputs corresponding to the same finger n from the frames conveying the n_sth symbol, to form the block $\boldsymbol{x}(n_s;n) := [x(n_sN_f;n),\ldots,x(n_sN_f+N_f-1;n)]^T$. The following I/O relationship can then be obtained from (21)

$$\boldsymbol{x}(n_s;n) = \sum_{u=0}^{N_u-1}\sqrt{\frac{\mathcal{E}_u}{N_f}}\boldsymbol{H}_{u,n}^{(0)}\boldsymbol{c}_u s_u(n_s) + \sum_{u=0}^{N_u-1}\sqrt{\frac{\mathcal{E}_u}{N_f}}\boldsymbol{H}_{u,n}^{(1)}\boldsymbol{c}_u s_u(n_s-1) + \boldsymbol{\eta}(n_s;n), \tag{22}$$

where $\boldsymbol{\eta}(n_s;n) := [\eta(n_sN_f;n),\ldots,\eta(n_sN_f+N_f-1;n)]^T$, $\boldsymbol{H}_{u,n}^{(0)}$ is a $N_f \times N_f$ lower triangular Toeplitz matrix with first column $[\tilde{\alpha}_{u,n}(0),\ldots,\tilde{\alpha}_{u,n}(M_{u,n}),0,\ldots,0]$, and

$\boldsymbol{H}_{u,n}^{(1)}$ is a $N_f \times N_f$ upper triangular Toeplitz matrix with first row $[0, \ldots, 0, \tilde{\alpha}_{u,n}(M_{u,n}), \ldots, \tilde{\alpha}_{u,n}(1)]$.

To collect all the information related to the n_sth symbol, we concatenate vectors $\boldsymbol{x}(n_s; n)$ from all-rake fingers into a super vector $\boldsymbol{y}(n_s) := [\boldsymbol{x}^T(n_s; 1), \ldots, \boldsymbol{x}^T(n_s; N)]^T$ of size $N_f N \times 1$, which can be expressed as [cf. (22)]

$$\boldsymbol{y}(n_s) = \sum_{u=0}^{N_u-1} \sqrt{\frac{\mathcal{E}_u}{N_f}} \boldsymbol{H}_u^{(0)} \boldsymbol{v}_u(n_s) + \sum_{u=0}^{N_u-1} \sqrt{\frac{\mathcal{E}_u}{N_f}} \boldsymbol{H}_u^{(1)} \boldsymbol{v}_u(n_s - 1) + \boldsymbol{\eta}(n_s), \tag{23}$$

where the $N_f \times 1$ block $\boldsymbol{v}_u(n_s) := \boldsymbol{c}_u s_u(n_s)$ is the nth symbol spread over N_f frames, $\boldsymbol{\eta}(n_s) := [\boldsymbol{\eta}^T(n_s; 1), \ldots, \boldsymbol{\eta}^T(n_s; N)]^T$ is the $N_f N \times 1$ noise vector associated with the n_sth symbol, $\boldsymbol{H}_u^{(0)} := [\boldsymbol{H}_{u,1}^{(0)T}, \ldots, \boldsymbol{H}_{u,N}^{(0)T}]^T$ and $\boldsymbol{H}_u^{(1)} := [\boldsymbol{H}_{u,1}^{(1)T}, \ldots, \boldsymbol{H}_{u,N}^{(1)T}]^T$. Notice that the ISI (second sum in (22)) has given rise to an interblock interference (IBI) term (second summand in the right hand side of (23)).

Targeting block by block detection, IBI (and thus ISI) needs to be removed. From the definition of $M_{u,n}$ in (20), it follows that the maximum discrete-time equivalent channel order is $M_1 = \max_{u,n}\{M_{u,n}\}$. Consequently, padding each block $\boldsymbol{v}_u(n_s)$ with M_1 zero-guards allows the channel to settle down before the next block/symbol arrives, and thus eliminates the IBI terms of all users. After ZP, each block $\boldsymbol{v}_u(n_s)$ with M_1 trailing zeros prior to transmission, the I/O relationship in (23) simplifies to an IBI-free one

$$\boldsymbol{y}_{zp} = \sum_{u=0}^{N_u-1} \sqrt{\frac{\mathcal{E}_u}{N_f}} \bar{\boldsymbol{H}}_u \boldsymbol{v}_u + \boldsymbol{\eta}, \tag{24}$$

where the index n_s is dropped for notational simplicity, and $\bar{\boldsymbol{H}}_u := [\bar{\boldsymbol{H}}_{u,1}^T, \ldots, \bar{\boldsymbol{H}}_{u,N}^T]^T$ is the $NN_1 \times N_f$ channel matrix with $N_1 := N_f + M_1$. The lth block of the channel matrix $\bar{\boldsymbol{H}}_{u,n}^T$ is a $N_1 \times N_f$ lower triangular Toeplitz matrix with the first column given by $[\tilde{\alpha}_{u,n}(0), \ldots, \tilde{\alpha}_{u,n}(M_{u,n}), 0, \ldots, 0]^T$.

Instead of having M_1 zeros padded at the end of each block $\boldsymbol{v}_u$, an alternative way to eliminate IBI is by adding a cyclic prefix (CP) of length M_1 at the transmitter and removing it at the receiver, much like OFDMA. Intuitively, since only the first M_1 elements per block are contaminated by IBI, we can introduce redundancy at the transmission and discard it upon reception. In this case, the I/O relationship becomes

$$\boldsymbol{y}_{cp} = \sum_{u=0}^{N_u-1} \sqrt{\frac{\mathcal{E}_u}{N_f}} \tilde{\boldsymbol{H}}_u \boldsymbol{v}_u + \boldsymbol{\eta}, \tag{25}$$

where the channel matrix is now $\tilde{\boldsymbol{H}}_u := [\tilde{\boldsymbol{H}}_{u,1}^T, \ldots, \tilde{\boldsymbol{H}}_{u,N}^T]^T$. By inserting and removing CP, each block of the channel matrix $\tilde{\boldsymbol{H}}_{u,n}^T$ becomes a $N_f \times N_f$ column-wise circulant matrix with the first column given by $[\tilde{\alpha}_{u,n}(0), \ldots, \tilde{\alpha}_{u,n}(M_{u,n}), 0, \ldots, 0]^T$ [cf. Toeplitz in (24)].

Although *continuous-time* and *oversampled* rake receiver models are well documented in the UWB literature (e.g., [9–11]), (24) and (25) are novel and interesting because they describe in a *discrete-time frame-rate* sampled form, the aggregate pulse-multipath-rake model *in the presence of IFI*. Eqs. (24) and (25) also show that frame-by-frame rake correlator samples obey a matrix-vector I/O relationship free of IBI (ISI) even in dense multipath channels, provided that suitable guards (ZP or CP) are inserted in UWB transmissions. These expressions are surprisingly simple if one takes into account that they include IFI effects that are always present in high-rate UWB radios. On the basis of (24) and (25), we will benchmark the performance of UWB spreading codes in Section 11.4.2, and investigate their error performance in a multiuser scenario in Section 11.5.

11.4.2 Performance Comparisons

Next, we will assess the performance corresponding to different spreading codes by quantifying their diversity and coding gains in the presence of UWB multipath. This will allow comparisons of our novel spreading codes relative to existing DS ones in the presence of IFI. To quantify diversity and coding gains for a particular user, we set $N_u = 1$. We also suppose $\{\tilde{\boldsymbol{\alpha}}_{u,n}\}_{n=1}^{N} := [\tilde{\alpha}_{u,n}(0), \ldots, \tilde{\alpha}_{u,n}(M_{u,n})]^T$ are perfectly known at the receiver, to isolate these effects from channel estimation imperfections.

Let us consider the pairwise error probability (PEP) $P\left(\boldsymbol{v}_u \to \boldsymbol{v}'_u | \{\tilde{\boldsymbol{\alpha}}_{u,n}\}_{n=1}^{N}\right)$ of erroneously decoding $\boldsymbol{v}_u$ as $\boldsymbol{v}'_u \neq \boldsymbol{v}_u$, assuming a maximum-likelihood (ML) detector. The PEP can be upper bounded at high SNR using the Chernoff bound:

$$P\left(\boldsymbol{v}_u \longrightarrow \boldsymbol{v}'_u | \{\tilde{\boldsymbol{\alpha}}_{u,n}\}_{n=1}^{N}\right) \leq \exp\left(\frac{-d^2(\bar{\boldsymbol{y}}, \bar{\boldsymbol{y}}')}{4N_0}\right), \tag{26}$$

where $N_0/2$ is the noise variance, and $\bar{\boldsymbol{y}}(\bar{\boldsymbol{y}}')$ is the noise-free part of (24) [or (25)], corresponding to $\boldsymbol{v}_u(\boldsymbol{v}'_u)$, and $d(\bar{\boldsymbol{y}}, \bar{\boldsymbol{y}}') := \|\bar{\boldsymbol{y}} - \bar{\boldsymbol{y}}'\|$ is the Euclidean distance between them. Using (24) and (25), we have

$$d^2(\bar{\boldsymbol{y}}, \bar{\boldsymbol{y}}') = \frac{\mathcal{E}_u}{N_f} \epsilon^2 \boldsymbol{c}_u^T \boldsymbol{H}_u^T \boldsymbol{H}_u \boldsymbol{c}_u, \tag{27}$$

where $\epsilon := s_u - s'_u$, $\boldsymbol{H}_u = \bar{\boldsymbol{H}}_u$ with ZP guards, and $\boldsymbol{H}_u = \tilde{\boldsymbol{H}}_u$ with CP guards. Equation (27) implies that for each given channel realization, the PEP depends on the spreading code $\boldsymbol{c}_u$. We will show next how different spreading codes (DS, SC, or MC) affect the average PEP $P(\boldsymbol{v}_u \to \boldsymbol{v}'_u) := \mathrm{E}_{\tilde{\boldsymbol{\alpha}}}\left\{P\left(\boldsymbol{v}_u \to \boldsymbol{v}'_u | \{\tilde{\boldsymbol{\alpha}}_{u,n}\}_{n=1}^{N}\right)\right\}$.

Denoting the pairwise error as $\boldsymbol{\varepsilon}_u := \boldsymbol{v}_u - \boldsymbol{v}'_u$, it was shown in [15, Appendix I] that (26) can be reexpressed as

$$P\left(\boldsymbol{v}_u \longrightarrow \boldsymbol{v}'_u | \{\tilde{\boldsymbol{\alpha}}_{u,n}\}_{n=1}^{N}\right) \leq \exp\left(\frac{-\mathcal{E}_u \sum_{n=1}^{N} \tilde{\boldsymbol{\alpha}}_{u,n}^T \boldsymbol{\Theta}_{\varepsilon_u}^{(u,n)} \tilde{\boldsymbol{\alpha}}_{u,n}}{4N_f N_0}\right), \tag{28}$$

where $\boldsymbol{\Theta}_{\varepsilon_u}^{(u,n)} := \boldsymbol{E}_{u,n}^T \boldsymbol{E}_{u,n}$ in the ZP case, with $\boldsymbol{E}_{u,n}$ being a $N_f \times (M_{u,n} + 1)$ Toeplitz matrix with first column $[\varepsilon_u(0), \ldots, \varepsilon_u(N_f - 1), 0, \ldots, 0]^T$. With CP guards, $\boldsymbol{\Theta}_{\varepsilon_u}^{(u,n)} = N_f \boldsymbol{F}_{0:M_{u,n}}^{\mathcal{H}} \boldsymbol{D}_{\epsilon_u}^{\mathcal{H}} \boldsymbol{D}_{\epsilon_u} \boldsymbol{F}_{0:M_{u,n}}$, where $\boldsymbol{F}_{0:m}$ denotes the matrix formed by the first $(m + 1)$ columns of $\boldsymbol{F}_{N_f}$, and $\boldsymbol{D}_{\epsilon_u} := \mathrm{diag}\{\boldsymbol{\epsilon}_u\}$ with $\boldsymbol{\epsilon}_u := \boldsymbol{F}_{N_f} \boldsymbol{\varepsilon}_u$.

Let $r_{\varepsilon_u}^{(u,n)}$ denote the rank of $\boldsymbol{\Theta}_{\varepsilon_u}^{(u,n)}$, and $\{\lambda_{\varepsilon_u}^{(u,n)}(m)\}_{m=0}^{M_{u,n}}$ denote its nonincreasing eigenvalues. Supposing the entries of $\alpha_{u,n}(n)$ are zero-mean, uncorrelated Gaussian, with variance $\mathcal{E}_{u,n}(n) = \mathrm{E}\{\alpha_{u,n}^2(n)\}$, it has been shown that at high SNR ($\mathcal{E}_u/N_0 \gg 1$), the average PEP is upper bounded by [15, Appendix II]

$$P(\boldsymbol{v}_u \longrightarrow \boldsymbol{v}'_u) \leq \left(\frac{\mathcal{E}_u}{2N_0} G_{c,\varepsilon}\right)^{-G_{d,\varepsilon}}, \tag{29}$$

where $G_{d,\varepsilon} := (1/2)\sum_{n=1}^{N} r_{\varepsilon_u}^{(u,n)}$ denotes the diversity gain, and

$$G_{c,\varepsilon} := \left[\prod_{n=1}^{N}\prod_{m=0}^{r_{\varepsilon_u}^{(u,n)}-1} \frac{\mathcal{E}_{u,n}(m)\lambda_{\varepsilon_u}^{(u,n)}(m)}{N_f}\right]^{1/(2G_{d,\varepsilon})}$$

is the coding gain for the error vector $\boldsymbol{\varepsilon}_u$. Eq. (29) reveals that at high SNR, the average PEP is uniquely characterized by two parameters; the diversity gain $G_{d,\varepsilon}$ determining the PEP slope as a function of log-SNR, and the coding gain $G_{c,\varepsilon}$ determining its shift. As $\boldsymbol{\Theta}_{\varepsilon_u}^{(u,n)}$ is directly related to $\boldsymbol{c}_u$, the average PEP depends on the spreading code used. Interestingly, if $\alpha_{u,n}(m)$'s are uncorrelated *complex* Gaussian with variance $\mathcal{E}_{u,n}(m)/2$ per real dimension (see e.g., [16]), then $G_{d,\varepsilon}$ will be doubled, whereas $G_{c,\varepsilon}$ will be halved.

Both $G_{d,\varepsilon}$ and $G_{c,\varepsilon}$ depend on $\boldsymbol{\varepsilon}_u$. Accounting for all possible pairwise errors, we define the diversity order and coding gains as $G_d := \min_{\boldsymbol{\varepsilon}_u \neq \mathbf{0}}\{G_{d,\varepsilon}\}$, and $G_c := \min_{\boldsymbol{\varepsilon}_u \neq \mathbf{0}}\{G_{c,\varepsilon}\}$, respectively. Subject to the underlying physical channel and the UWB system parameters, we will subsequently upper bound these diversity and coding gains. To this end, we assume that the physical channel taps are zero-mean uncorrelated Gaussian, and establish first the following lemma[3] [15, Appendix III]:

Lemma 1 *In a UWB system with parameters N_f, T_f, and T_p, and N-finger rake reception with $N \leq L_u$ and delays $\{\tilde{\tau}_n\}_{n=1}^{N}$ spaced at least $2T_p$ apart, the following hold:*

1. *With the equivalent channel order $M_{u,n}$ as in (20), the maximum achievable diversity order is*

$$G_{d,max} = \frac{1}{2}\sum_{n=1}^{N}(M_{u,n}+1), \tag{30}$$

and can be guaranteed if and only if $\boldsymbol{\Theta}_{\varepsilon_u}^{(u,n)}$ is full rank, $\forall n \in [1, N]$, and $\forall \boldsymbol{\varepsilon}_u \neq \mathbf{0}$.

2. *With maximum diversity gain $G_{d,max}$ achieved, the maximum coding gain is*

$$G_{c,max} := d_{min}^2 \left[\prod_{n=1}^{N}\prod_{m=0}^{M_{u,n}} \mathcal{E}_{u,n}(m)\right]^{1/(2G_{d,max})}, \tag{31}$$

[3] In fact, the proof of Lemma 1 shows that full diversity only requires $\boldsymbol{\Theta}_{\varepsilon_u}^{(u,n)}$ to be full rank, $\forall l \in [1, L]$. Therefore, Lemma 1 can be generalized even to correlated channel vectors so long as the correlation matrix is of full rank. In addition, using the results of [17], it can be shown that our diversity and coding gain results apply to all non-Gaussian fading PDFs encountered in practice.

where $\mathcal{E}_{u,n}(m) := E\{\alpha_{u,n}^2(m)\}$, *and* d_{min} *is the minimum Euclidean distance of the* s_u *constellation. Further,* $G_{c,max}$ *can be achieved if and only if* $\boldsymbol{\Theta}_{\varepsilon_u}^{(u,n)}$ *is the scaled identity matrix* $N_f(s_u - s_u')^2 \boldsymbol{I}_{M_{u,n}+1}$.

Notice that Lemma 1 provides *upper bounds* on the achievable diversity order and coding gain. Although dependent on the underlying physical channel and system parameters, the results in (30) and (31) apply to all UWB systems with rake reception, irrespective of the underlying spreading code. Intuitively, Lemma 1 asserts that for each (say, the lth) rake finger, $(M_{u,n} + 1)$ taps of its corresponding discrete-time equivalent channel contribute a maximum diversity order of $(M_{u,n} + 1)/2$, where the factor $1/2$ appears because UWB transmissions are real. Summing up the diversity collected by all fingers $n \in [1, N]$, we obtain the maximum diversity order as in (30). With this in mind, the coding gain is nothing but the average SNR gain due to the energy collected from a total of $\sum_{n=1}^{N}(M_{u,n} + 1)$ multipath returns. Certainly, the concepts of diversity and coding gains in Lemma 1 are common to any fading channel. However, expressions (30) and (31) are tailored for UWB systems with rake reception, and can be specialized when IFI is absent as follows [15]:

Corollary 1 *If on top of the conditions in Lemma 1, one also selects* $T_f \geq \tau_{u,L_u} + T_p - \tau_{u,1}$ *to remove IFI, the resulting maximum diversity and coding gains are* $G_{d,max} = N/2$, *and* $G_{c,max} = d_{min}^2 [\prod_{n=1}^{N} \mathcal{E}_{u,n}(0)]^{1/(2G_{d,max})}$, *respectively.*

It is worth noticing that for a given number of fingers N, nonzero IFI works to our favor by boosting diversity gains [cf. (30)], which is intuitively appealing since we collect more energy. Of course, in the absence of IFI, one can increase the number of rake fingers N per frame, which will increase diversity at the expense of increasing complexity and reducing transmission rates (since T_f must be increased accordingly). As a special case, Corollary 1 quantifies this diversity in the absence of IFI, where, as expected, $G_{d,\max}$ is proportional to the number of rake fingers N. In the absence of MUI and NBI, these gains can be collected with MRC [18].

Having benchmarked the maximum possible diversity and coding gains, and having established conditions for achieving them, we are now ready to compare the gains enabled by DS, SC, and MC spreading codes, with ZP or CP guards.

11.4.2.1 Comparison with ZP Guards

When IBI is removed with ZP, we have $\boldsymbol{\Theta}_{\varepsilon_u}^{(u,n)} = \boldsymbol{E}_{u,n}^T \boldsymbol{E}_{u,n}$. Because $\{\boldsymbol{E}_{u,n}\}_{n=1}^{N}$ are Toeplitz matrices that guarantee full column rank $\forall \boldsymbol{\varepsilon}_u \neq \boldsymbol{0}$, matrix $\boldsymbol{\Theta}_{\varepsilon_u}^{(u,n)}$ has full rank $r_{\varepsilon_u}^{(u,n)} = M_{u,n} + 1, \forall n \in [1, N]$. Therefore, *regardless of the spreading code* used, the maximum achievable diversity order is always guaranteed: $G_{d,zp} = G_{d,\max}$, that is,

Proposition 2 *In a UWB system with parameters* N_f, T_f, *and* T_p, *and* N*-finger rake reception with delays* $\{\tilde{\tau}_n\}_{n=1}^{N}$ *spaced at least* $2T_p$ *apart, IBI removal with ZP enables the maximum achievable diversity order* $G_{d,zp} = G_{d,max}$, *regardless of the spreading codes used.*

Having the same diversity order, the relative error performance of DS-, SC-, and MC-UWB is dictated by their corresponding coding gains. To achieve maximum coding gain, however, requires the spreading codes to have perfect correlation; that is,

$\sum_{k=0}^{N_f-1} c_u(k)c_u(k+l) = N_f\delta(l)$. Generally, this is not guaranteed. But interestingly, simulations will illustrate that MC-UWB signaling with W–H $\boldsymbol{c}_u^{(o)}$ approaches $G_{c,\max}$, $\forall u$.

11.4.2.2 Comparison with CP Guards

In the case of CP Guards, we have $\boldsymbol{\Theta}_{\mathcal{E}_u}^{(u,n)} = N_f \boldsymbol{F}_{0:M_{u,n}}^{\mathcal{H}} \boldsymbol{D}_{\epsilon_u}^{\mathcal{H}} \boldsymbol{D}_{\epsilon_u} \boldsymbol{F}_{0:M_{u,n}}$; and with $\boldsymbol{F}_{0:M_{u,n}}$ having full column rank, the rank of $\boldsymbol{\Theta}_{\mathcal{E}_u}^{(u,n)}$ is $r_{\mathcal{E}_u}^{(u,n)} = \min\{\mathcal{D}_u, M_{u,n}+1\}$, where $\mathcal{D}_u$ is the number of nonzero entries of $\boldsymbol{\epsilon}_u$. To guarantee $G_{d,\max}$ for any $M_{u,n}$, all entries of $\boldsymbol{\epsilon}_u$ must be nonzero. It can also be verified that to further guarantee $G_{c,\max}$, the entries of $\boldsymbol{\epsilon}_u$ must have identical magnitudes.

Now let us consider whether and under what conditions our SC/MC-UWB codes with CP guards can enable the benchmark diversity and coding gains established by Lemma 1.

For DS-UWB, we have $\boldsymbol{\epsilon}_u = (s_u - s_u')\boldsymbol{F}_{N_f}\boldsymbol{c}_u$ by definition. With $c_u(k) \in \{\pm 1\}$, it cannot be guaranteed that $\boldsymbol{\epsilon}_u$ contains no zero. Enabling $G_{d,\max}$ is therefore not guaranteed either. In fact, if Walsh–Hadamard codes are used, it can be shown that the *diversity order is user dependent*:

$$G_{d,\text{cp}}^{(ds)}(u) = \begin{cases} \dfrac{1}{2}\displaystyle\sum_{n=1}^{N} \min\{1, M_{u,n}+1\} = N/2, & \text{if } u = 0, 1, \\ \dfrac{1}{2}\displaystyle\sum_{n=1}^{N} \min\{2^{\lfloor \log_2 u \rfloor}, M_{u,n}+1\} \le G_{d,\max}, & \text{otherwise.} \end{cases} \tag{32}$$

For SC-UWB with carriers as in (2), the entries of $\boldsymbol{\epsilon}_u$ can be found $\forall n_f \in [0, N_f - 1]$ to be

$$[\boldsymbol{\epsilon}_u]_{n_f} = \sqrt{\frac{2}{N_f}}(s_u - s_u') \times \begin{cases} \jmath \dfrac{\sin(2\pi n_f/N_f)}{\cos(2\pi n_f/N_f) - \cos(2\pi(u+0.5)/N_f)} + 1, \\ \qquad \text{if } u \in [0, N_f/2 - 1], \\ \dfrac{\sin(2\pi(u+0.5)/N_f)}{\cos(2\pi n_f/N_f) - \cos(2\pi(u+0.5)/N_f)}, \\ \qquad \text{if } u \in [N_f/2, N_f - 1]. \end{cases} \tag{33}$$

It is clear from (33) that $\boldsymbol{\epsilon}_u$ will have nonzero entries, for all nonzero errors. As a result, SC-UWB guarantees $G_{d,\text{cp}}^{(sc)} = G_{d,\max}$ even with CP guards.

For MC-UWB with carriers as in (6), we have $\boldsymbol{\epsilon}_u = (s_u - s_u')\boldsymbol{F}_{N_f}\boldsymbol{G}_{mc}\boldsymbol{c}_u^{(o)}$. And its n_fth element is

$$[\boldsymbol{\epsilon}_u]_{n_f} = (s_u - s_u') \times \begin{cases} \sqrt{N_f}\left[\boldsymbol{c}_u^{(o)}\right]_{n_f}, & n_f = 0 \text{ or } \dfrac{N_f}{2} \\ \sqrt{\dfrac{N_f}{2}}\left(\left[\boldsymbol{c}_u^{(o)}\right]_{n_f} + \jmath\left[\boldsymbol{c}_u^{(o)}\right]_{N_f-n_f}\right), & n_f \in \left[1, \dfrac{N_f}{2} - 1\right] \\ \sqrt{\dfrac{N_f}{2}}\left(\left[\boldsymbol{c}_u^{(o)}\right]_{N_f-n_f} - \jmath\left[\boldsymbol{c}_u^{(o)}\right]_{n_f}\right), & n_f \in \left[\dfrac{N_f}{2} + 1, N_f - 1\right]. \end{cases} \tag{34}$$

Equation (34) implies that $\boldsymbol{\epsilon}_u$ does not contain any zero entry, for all nonzero errors, if and only if the real orthonormal codes $\boldsymbol{c}_u^{(o)}$ satisfy the following conditions:

$$\left[\boldsymbol{c}_u^{(o)}\right]_{n_f} + \jmath \left[\boldsymbol{c}_u^{(o)}\right]_{\mathrm{mod}(N_f - n_f, N_f)} \neq 0, \ \forall n_f \in [0, N_f - 1], \tag{35}$$

or equivalently, at least one of the two real numbers $\left[\boldsymbol{c}_u^{(o)}\right]_{n_f}$ or $\left[\boldsymbol{c}_u^{(o)}\right]_{\mathrm{mod}(N_f - n_f, N_f)}$ is nonzero. This implies that MC-UWB achieves maximum diversity order $G_{d,\mathrm{cp}}^{(mc)} = G_{d,\max}$, even with CP, if and only if (35) is satisfied. Intuitively speaking, when (35) is satisfied, the resultant $\boldsymbol{c}_u$ will give rise to a transmit spectrum that covers the entire bandwidth, and thus enables the maximum multipath diversity provided by the UWB channel. To gain more insight, let us consider two special choices of $\boldsymbol{c}_u^{(o)}$.

Special Case 1 (MC-I): Suppose that the entries of $\boldsymbol{c}_u^{(o)}$ take binary values $\{\pm 1/\sqrt{N_f}\}$, where the factor $\sqrt{N_f}$ is used for normalization. In fact, such a choice of $\boldsymbol{c}_u^{(o)}$ also guarantees the maximum coding gain. From (34), we deduce that entries of $\boldsymbol{\epsilon}_u$ are either $\pm(s_u - s_u')$, or $(\pm 1 \pm \jmath)(s_u - s_u')/\sqrt{2}$, both of which have the same magnitude. It then follows that $\boldsymbol{\Theta}_{\varepsilon_u}^{(u,n)} = N_f (s_u - s_u')^2 \boldsymbol{I}_{M_{u,n}+1}$. According to Lemma 1, MC-UWB with $\boldsymbol{c}_u^{(o)}$ having $\{\pm 1/\sqrt{N_f}\}$ entries not only guarantees full diversity but also enjoys maximum coding gain.

Special Case 2 (MC-II): Consider here that the orthonormal codes are chosen as $\boldsymbol{c}_u^{(o)} = \boldsymbol{e}_u/\sqrt{N_f}$, and $\forall u \in [0, N_f - 1]$. Evidently, such a choice of $\boldsymbol{c}_u^{(o)}$ does not satisfy condition (35). In fact, it can be shown that the resulting diversity order for user u now becomes

$$G_{d,\mathrm{cp}}^{(mc2)}(u) = \begin{cases} \dfrac{1}{2}\displaystyle\sum_{n=1}^{N} \min\{1, M_{u,n} + 1\} = N/2, & \text{if } u = 0 \text{ or } \dfrac{N_f}{2}, \\ \dfrac{1}{2}\displaystyle\sum_{n=1}^{N} \min\{\mathcal{D}_{u,n}, M_{u,n} + 1\} \in [N/2, N], & \text{otherwise.} \end{cases} \tag{36}$$

As with W–H DS codes, the diversity order in this MC-II case becomes user (or code) dependent, and full diversity cannot be guaranteed, because $G_{d,\mathrm{cp}}^{(mc2)}(u) \leq G_{d,\max}$.

We summarize these results in the following proposition.

Proposition 3 *In a UWB system with parameters N_f, T_f, and T_p, and N-finger rake with delays $\{\tilde{\tau}_n\}_{n=1}^{N}$ spaced at least $2T_p$ apart, the achievable diversity order depends on the spreading code, when IBI is removed with CP. Specifically,*

1. *DS-UWB generally does not guarantee $G_{d,max}$; for example, with W–H codes, $G_{d,cp}^{(ds)}(u) \leq G_{d,max}$ as in (32) is user/code dependent.*
2. *SC-UWB using $\boldsymbol{G}_{sc}$ with carriers as in (2) achieves maximum diversity order $G_{d,cp}^{(sc)} = G_{d,max}$.*
3. *MC-UWB guarantees $G_{d,max}$, if and only if (35) is satisfied.*

4. *In MC-UWB, if* (35) *is satisfied and* $|[\boldsymbol{c}_u^{(o)}]_{n_f} + \jmath[\boldsymbol{c}_u^{(o)}]_{N_f - n_f}|$ *remains invariant* $\forall n_f$, *then MC-UWB achieves both the maximum diversity order* $G_{d,max}$, *and the maximum coding gain* $G_{c,max}$.

To corroborate these results, let us first recall that the diversity gain relies on the number of nonzero entries of $\boldsymbol{F}_{N_f}\boldsymbol{c}_u$. Since $\boldsymbol{F}_{N_f}\boldsymbol{c}_u$ is nothing but the frequency response of the spreading code $\boldsymbol{c}_u$, it is not surprising that MC-I codes guarantee $G_{d,\max}$, whereas MC-II codes do not. Indeed, under MC-I, each user utilizes all carriers; whereas under MC-II, each user actually utilizes a single carrier. Interestingly however, with carriers as in (2), SC-UWB enables $G_{d,\max}$ even with a single carrier, thanks to its multiband signaling characteristic and the critical $0.5/N_f$ shift in its digital frequencies $\{f_u\}_{u=0}^{N_u-1}$. Insofar as coding gains are concerned, $G_{c,\max}$ can be achieved if and only if entries of $\boldsymbol{F}_{N_f}\boldsymbol{c}_u$ share the same magnitude.

In a nutshell, for a given spreading gain N_f, single-user performance heavily depends on the UWB spreading code selection. Existing DS codes do not guarantee $G_{d,\max}$ when CP guards are employed. Even with ZP guards, the error performance with DS codes is suboptimum as $G_{c,\max}$ is not guaranteed. On the other hand, the SC/MC codes in [15] enable maximum diversity order with ZP or CP guards. In particular, MC-UWB can also achieve (approach) $G_{c,\max}$ with CP (ZP) guards.

11.4.3 Simulated Examples

In this section, simulations and comparisons will be performed to validate the analysis of the previous subsection. The pulse shaper $p(t)$ is chosen as the 2nd derivative of the Gaussian function with unit energy, and $T_p \approx 1.0$ ns. Each symbol contains $N_f = 32$ frames, each with $T_f = 24$ ns. The random channels are generated according to [13], with parameters $(1/\Lambda, 1/\lambda, \Gamma, \gamma) = (2, 0.5, 30, 5)$ ns. The resulting maximum delay spread of the multipath channel is 90 ns. The rake receiver uses $N = 2$ fingers per frame, selected randomly but kept fixed for all testing scenarios. Consequently, we have $M_{\mu,1} = 3$, and $M_{\mu,2} = 2$, where μ is the index of the desired user. Accounting for the ZP or CP guard of length $M_1 = 3$, the transmission rate is about 1.2 Mbps for binary PAM.

According to Lemma 1, the maximum achievable diversity order is $G_{d,\max} = \frac{1}{2}\sum_{n=1}^{N}(M_{\mu,n} + 1) = 7/2$, which is the same as that of a system with $N = 7$ fingers but free from IFI. In the presence of IFI, DS- and SC-UWB may result in diversity order as low as 1, which coincides with that of a system with $N = 2$ fingers in the absence of IFI. Therefore, BER curves corresponding to these two IFI-free systems are also plotted as benchmarks. According to Corollary 1, these benchmark curves exhibit both $G_{d,\max}$ and $G_{c,\max}$.

ML detection is applied to decode individual users with both ZP and CP guards. Average BER versus $\mathcal{E}_u/N_0$ with ZP is shown in Figures 11.8(a), and 11.8(b), for DS, SC, and MC codes. W–H codes are used for DS-UWB spreading and also for the $\boldsymbol{c}_u^{(o)}$ part of MC-UWB codes. Although all spreading codes can enable $G_{d,\max} = 7/2$, the BER curves corresponding to all 32 MC spreading codes are almost identical to this $N = 7$ benchmark performance (see Figure 11.8(b)); whereas those of the other two are distributed over a rather wide range (see Figures 11.8(a) and 11.8(b)). The performance difference between them comes from the discrepancy in their corresponding coding gains. Although the coding gain corresponding to MC-UWB is not maximum, it comes very close to $G_{c,\max}$.

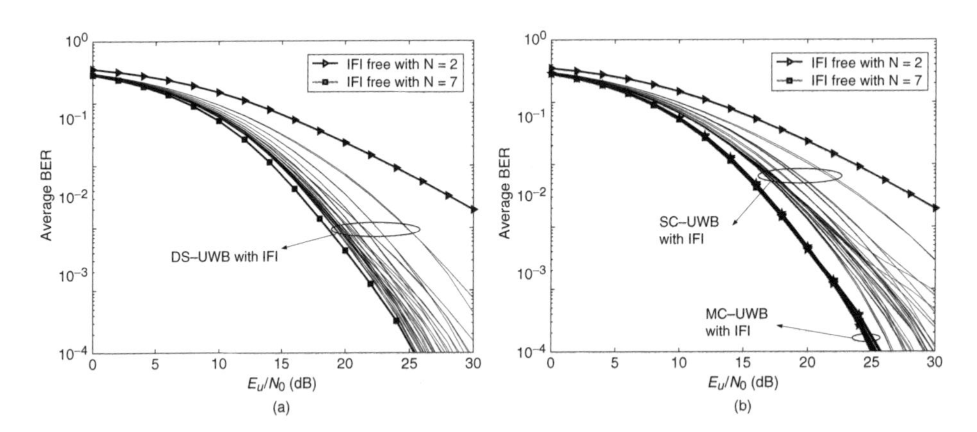

Figure 11.8 Code-dependent performance with ZP guards and IFI; average BER corresponding to 32 individual codes is shown; $\tau_{u,L_u} = 90$ ns, $T_f = 24$ ns, $N = 2$ with IFI, $M_{u,1} = 3$, and $M_{u,2} = 2$. (a) DS-UWB; (b) SC- and MC-UWB

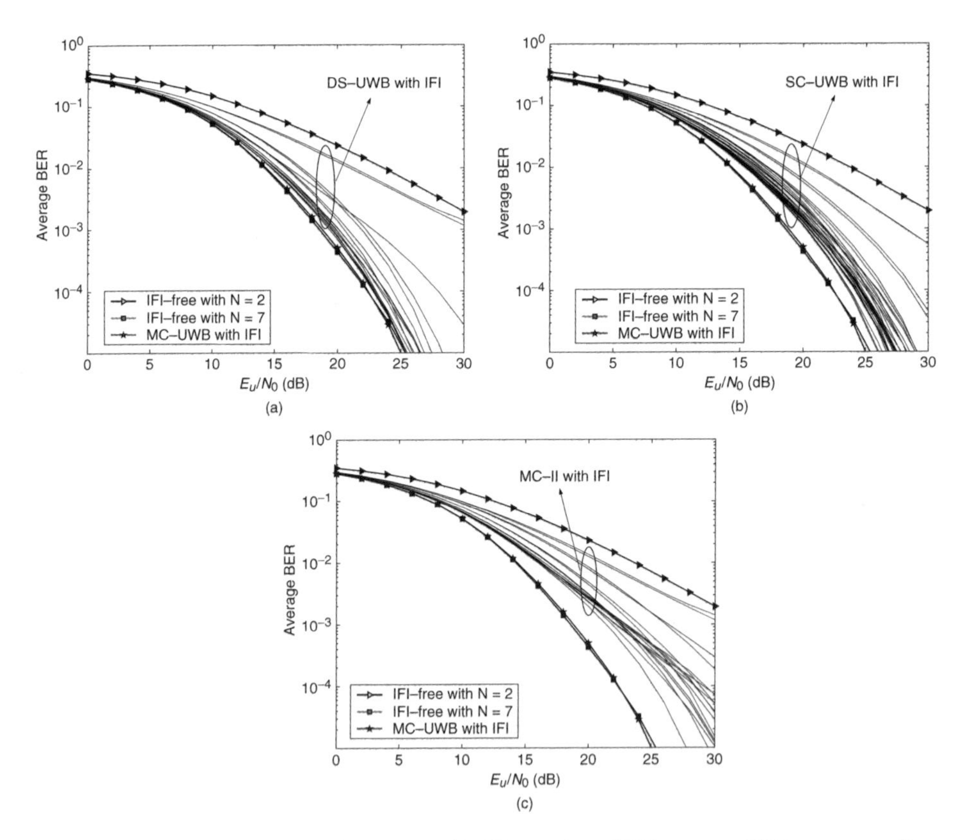

Figure 11.9 Code-dependent performance with CP guards and IFI; average BER corresponding to 32 individual codes is shown; $\tau_{u,L_u} = 90$ ns, $T_f = 24$ ns, $N = 2$ with IFI, $M_{u,1} = 3$, and $M_{u,2} = 2$. (a) DS-UWB; (b) SC-UWB; and (c) MC-II

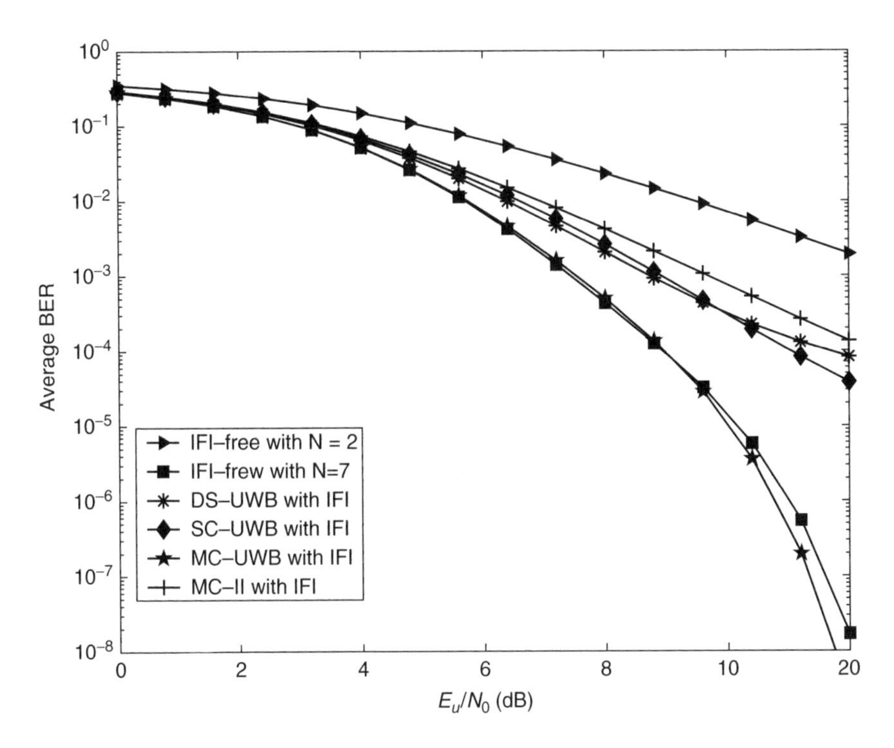

Figure 11.10 CP for IBI removal; $\tau_{u,L_u} = 90$ ns, $T_f = 24$ ns, $N = 2$, $M_{u,1} = 3$, and $M_{u,2} = 2$

With CP, the average BER is shown in Figures 11.9(a), 11.9(b), and 11.9(c), for DS-, SC-, and MC-UWB (the special case of MC-II codes is also plotted). We observe that MC-UWB with $N = 2$ rake fingers in the presence of IFI yields BER curves identical to the $N = 7$ benchmark without IFI; SC-UWB guarantees full diversity, but not maximum coding gain; DS-UWB enjoys full diversity for most users, but for some users it exhibits diversity order of only 1, as we predicted in (32); and MC-II exhibits diversity orders ranging from 1 to 2, as we predicted in (36). In Figure 11.10, the BER averaged over all the codes is also plotted, where MC-UWB is confirmed to outperform all other spreading codes.

Next, we will investigate usage of the MC codes in a multiaccess setup. Relying on low-complexity single-user rake reception, we will show how MC-UWB affects MUI mitigation.

11.5 Multiuser Interference Mitigation

In multiaccess scenarios, employment of multiuser detection (MUD) approaches (such as the ML ones) generally require knowledge of all users' channels and spreading codes, which is often unrealistic. Moreover, the computational complexity may be prohibitive for the stringent size and power limitations of UWB radios. Relying on simple receiver processing with rake reception, we will focus on *single-user* MF-rake detection using MRC.

Collecting outputs of the rake correlators, N per frame, the I/O relationship is given by [cf. (24) and (25)]

$$\begin{aligned} \mathbf{y} &= \sqrt{\frac{\mathcal{E}_\mu}{N_f}} \mathbf{H}_\mu \mathbf{v}_\mu + \sum_{u \neq \mu} \sqrt{\frac{\mathcal{E}_u}{N_f}} \mathbf{H}_u \mathbf{v}_u + \boldsymbol{\eta} \\ &= \sqrt{\frac{\mathcal{E}_\mu}{N_f}} \mathbf{H}_\mu \mathbf{c}_\mu s_\mu + \sum_{u \neq \mu} \sqrt{\frac{\mathcal{E}_u}{N_f}} \mathbf{H}_u \mathbf{c}_u s_u + \boldsymbol{\eta}, \end{aligned} \quad (37)$$

where $\mathbf{y} = \mathbf{y}_{zp}$ and $\mathbf{H}_u = \bar{\mathbf{H}}_u$ with ZP, and $\mathbf{y} = \mathbf{y}_{cp}$ and $\mathbf{H}_u = \tilde{\mathbf{H}}_u$ with CP. With combining weights $\boldsymbol{\beta}_\mu$, the decision statistic for detecting the symbol s_μ is $z_\mu := \boldsymbol{\beta}_\mu^T \mathbf{y}$.

Corresponding to MRC, MF combining weights for rake reception are given by $\boldsymbol{\beta}_\mu^{mf} := \sqrt{1/N_f} \mathbf{H}_\mu \mathbf{c}_\mu$. Incorporating both the desired and interfering user terms, the decision statistic after MF combining in the absence of noise becomes

$$(\boldsymbol{\beta}_\mu^{mf})^T \mathbf{y} = \frac{\sqrt{\mathcal{E}_\mu}}{N_f} \sum_{u=0}^{N_u-1} \mathbf{c}_\mu^T \mathbf{H}_\mu^T \mathbf{H}_u \mathbf{c}_u s_u = \frac{\sqrt{\mathcal{E}_\mu}}{N_f} \sum_{u=0}^{N_u-1} \sum_{n=1}^{N} \mathbf{c}_\mu^T \mathbf{H}_{\mu,n}^T \mathbf{H}_{u,n} \mathbf{c}_u s_u. \quad (38)$$

In general, MF rake does not guarantee MUI elimination. However, if CP is coupled with MC spreading codes, it becomes possible to mitigate MUI even with low-complexity MF rake. In fact, we will see that special choices of $\{\mathbf{c}_u^{(o)}\}_{u=0}^{N_f-1}$ can suppress MUI significantly, while maintaining the capability of simultaneously accommodating $N_u = N_f$ active users.

To confirm this, we prove in [15, Appendix IV], that when $\mathbf{c}_u^{(o)} = \mathbf{e}_u$, $\forall u \in [0, N_f - 1]$, the decision statistic in (38) boils down to

$$(\boldsymbol{\beta}_\mu^{mf})^T \mathbf{y} = \begin{cases} \sqrt{\mathcal{E}_\mu} \sum_{n=1}^{N} [\mathbf{d}_{\mu,\mu,n}]_\mu \; s_\mu, & \mu = 0 \text{ or } \dfrac{N_f}{2}, \\ \sqrt{\mathcal{E}_\mu} \sum_{n=1}^{N} \left[\mathcal{R}([\mathbf{d}_{\mu,\mu,n}]_\mu) \; s_\mu + \mathcal{I}([\mathbf{d}_{\mu,N_f-\mu,n}]_\mu) \; s_{N_f-\mu} \right], & \text{otherwise.} \end{cases} \quad (39)$$

where $\mathbf{d}_{\mu,u,l}$ consists of the element-by-element product of vectors $\sqrt{N_f} \mathbf{F}_{0:M_{\mu,l}}^* \boldsymbol{\alpha}_{\mu,l}$ and $\sqrt{N_f} \mathbf{F}_{0:M_{u,n}} \boldsymbol{\alpha}_{u,n}$.

Surprisingly, (39) implies that even with simple MF rake, the number of interfering users is reduced to one at the most, as opposed to $N_f - 1$. As a result, with low-cost UWB receivers equipped with MF rake, the MC-UWB can accommodate $(N_f/2 + 1)$ users while still achieving *single-user* performance; whereas with DS-UWB, single-user performance can be achieved only when 1 user is active. In typical UWB systems with large N_f, this translates to a significant user capacity increase by $N_f/2$.

Reducing the number of interfering users also reduces the complexity of ML detection considerably, and renders it feasible for UWB applications. Notice that this selection corresponds to our MC-II codes in Section 11.4.2, where we assign to each user a single real digital carrier. As mentioned earlier, full diversity is not guaranteed in this case. With each user employing more than one digital carrier, the diversity order can be increased at the price of reduced user capacity, or increased MUD complexity. However, clearly distinct from narrowband OFDMA that has diversity order 1, even with a single carrier (from the set of digital carriers in (6)), the minimum achievable diversity order is already $N/2$, as shown in (36).

Remark 3: We have derived a MF-rake model for digital receiver processing of UWB transmissions. Recall that in the development of our digital model, we did not impose any constraints on the number, or the placement of rake fingers. MF weights were developed also in [11, 19], by oversampling the received waveform. We have shown here that frame-rate rake reception can be implemented with analog waveforms, without oversampling. Specifically, the decision statistic z_μ can be generated by correlating $r(t)$ with the template $\bar{p}_{\mu,k}(t)$ and sampling its output once per frame. The template is given by

$$\bar{p}_{\mu,k}(t) = \sum_{n=1}^{N} [\boldsymbol{\beta}_\mu]_{kN+n-1} p(t - \tilde{\tau}_n)$$

during the kth frame, where $k \in [0, N_1 - 1]$ with ZP, and $k \in [M_1, N_1 - 1]$ with CP.

In Figure 11.11, BER of MF-rake detectors is depicted in multiaccess UWB operation under variable user loads: fully loaded with $N_u = 32$ users (dashed curves), medium

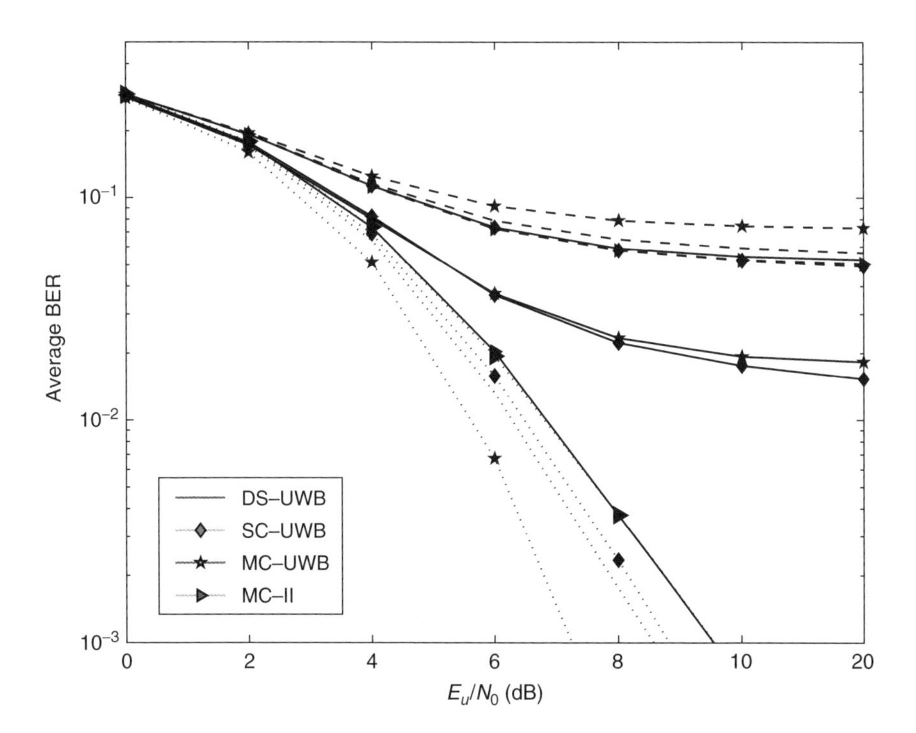

Figure 11.11 BER performance of MUD, with MF and CP used for IBI removal. Dashed, solid, and dotted curves correspond to cases with $N_u = 32$, $N_u = 17$, and $N_u = 1$, respectively; $\tau_{u,L_u} =$ 90 ns, $T_f = 24$ ns, $N = 2$ with IFI

loaded with $N_u = 17$ users (solid curves), and lightly loaded with $N_u = 1$ (dotted curves). Under light user loads MC-UWB outperforms all others. In the medium loaded system, the performance of MC-II codes is identical to the single-user case, as expected. In the fully loaded case, all of them exhibit error floor. Among all spreading codes, MC-UWB exhibits highest sensitivity to MUI, possibly because all users have identically flat transmit spectra. For these (even close to) fully loaded systems, the single-user rake receivers selected due to their low complexity are not sufficient to cope with the near-far effects that cause the error floors in Figure 11.11. For these fully loaded cases, the spreading codes in [5] are more suitable because they are capable of eliminating MUI by design, while being able to afford single-user complexity.

11.6 Summary

In this chapter, we introduced two real-valued spreading codes (SC and MC) for baseband UWB multiple access. Constructed using discrete cos/sin functions, our SC/MC codes can be implemented with low-complexity DCT circuits. Relative to DS-UWB, SC-UWB and MC-UWB offer flexibility in handling NBI, simply by avoiding the corresponding digital carriers. Moreover, utilizing single or multiple digital carriers, these codes have multiband UWB spectra. Consequently, differing from narrowband OFDMA systems, each SC-UWB user occupies multiple frequency bands, and enjoys full multipath diversity, even with a single digital carrier. In addition, MC-UWB achieves maximum coding gain. We also introduced a general rake reception model for UWB multiple access encompassing various spreading codes. The model leads to closed-form SINR performance analysis, which is used to assess the relative merits of several UWB systems in the presence of NBI, multipath, and AWGN, for partial-and selective-rake receivers with MF of MMSE combining weights. Finally, even with frame-rate samples and simple matched filtering operations, SC- and MC-UWB are capable of reducing multiuser interference, which in turn reduces receiver complexity.

References

[1] J. R. Foerster, "The Performance of a Direct-Sequence Spread Ultra Wideband System in the Presence of Multipath, Narrowband Interference, and Multiuser Interference," *Proc. Conference on Ultra-Wideband Systems and Technologies*, Baltimore, MD, pp. 87–92, May 20–23, 2002.

[2] B. M. Sadler and A. Swami, "On the Performance of UWB and DS-Spread Spectrum Communication Systems," *Proc. Conference on Ultra-Wideband Systems and Technologies*, Baltimore, MD, pp. 289–292, May 20–23, 2002.

[3] R. A. Scholtz, "Multiple Access with Time-Hopping Impulse Modulation," *Proc. MILCOM Conference*, Boston, MA, pp. 447–450, Oct. 11–14, 1993.

[4] H. Sari and G. Karam, "Orthogonal Frequency-Division Multiple Access and its Application to CATV Network," *European Trans. Telecommun.*, vol. 9, pp. 507–516, Nov./Dec. 1998.

[5] L. Yang and G. B. Giannakis, "Multi-Stage Block-Spreading for Impulse Radio Multiple Access Through ISI Channels," *IEEE J. Select. Areas Commun.*, vol. 20, no. 9, pp. 1767–1777, Dec. 2002.

[6] J. Proakis, *Digital Communications*, 4th edition, McGraw-Hill, New York, Feb. 2001.

[7] Z. Wang, "Multi-Carrier Ultra-Wideband Multiple-Access with Good Resilience Against Multiuser Interference," *Proc. Conference on Information Sciences & Systems*, Baltimore, MD, Mar. 12–14, 2003.

[8] L. Yang and G. B. Giannakis, "A General Model and SINR Analysis of Low Duty-Cycle UWB Access Through Multipath with Narrowband Interference and Rake Reception," *IEEE Trans. Wireless Commun.*, vol. 4, no. 4, pp. 1818–1833, 2005.

[9] M. Z. Win and R. A. Scholtz, "On the Energy Capture of Ultrawide Bandwidth Signals in Dense Multipath Environments," *IEEE Commun. Lett.*, vol. 2, no. 9, pp. 245–247, Sept. 1998.

[10] D. Cassioli, M. Z. Win, F. Vatalaro, and A. F. Molisch, "Performance of Low-Complexity Rake Reception in a Realistic UWB Channel," *Proc. International Conference on Communications*, New York, pp. 763–767, Apr. 28/May 2, 2002.

[11] G. Durisi, A. Tarable, J. Romme, and S. Benedetto, "A General Method for Error Probability Computation of UWB Systems for Indoor Multiuser Communications," *J. Commun. Netw.*, vol. 5, no. 4, pp. 354–364, Dec. 2003.

[12] S. Zhou, G. B. Giannakis, and A. Swami, "Digital Multi-Carrier Spread-Spectrum Versus Direct-Sequence Spread-Spectrum for Resistance to Jamming and Multipath," *IEEE Trans. Commun.*, vol. 50, no. 4, pp. 643–655, Apr. 2002.

[13] A. A. M. Saleh and R. A. Valenzuela, "A Statistical Model for Indoor Multipath Propagation," *IEEE J. Select. Areas Commun.*, vol. 5, no. 2, pp. 128–137, Feb. 1987.

[14] J. R. Foerster, *Channel Modeling Sub-Committee Report Final*, IEEE P802.15-02/368r5-SG3a, IEEE P802.15 Working Group for WPAN, Nov. 2002.

[15] L. Yang and G. B. Giannakis, "Digital-Carrier Multi-Band User Codes for Baseband UWB Multiple Access," *J. Commun. Netw.*, vol. 5, no. 4, pp. 374–385, Dec. 2003.

[16] Z. Wang and G. B. Giannakis, "Complex-Field Coding for OFDM Over Fading Wireless Channels," *IEEE Trans. Inf. Theory*, vol. 49, no. 3, pp. 707–720, Mar. 2003.

[17] Z. Wang and G. B. Giannakis, "A Simple and General Parameterization Quantifying Performance in Fading Channels," *IEEE Trans. Commun.*, vol. 51, no. 8, pp. 1389–1398, Aug. 2003.

[18] L. Yang and G. B. Giannakis, "Analog Space-Time Coding for Multi-Antenna Ultra-Wideband Transmissions," *IEEE Trans. Commun.*, vol. 52, no. 3, pp. 507–517, Mar. 2004.

[19] C. L. Martret and G. B. Giannakis, "All-Digital Impulse Radio for Wireless Cellular Systems," *IEEE Trans. Commun.*, vol. 50, no. 9, pp. 1440–1450, Sept. 2002.

12

Localization

Kegen Yu, Harri Saarnisaari, Jean-Philippe Montillet, Alberto Rabbachin, Ian Oppermann and Giuseppe Thadeu Freitas de Abreu

12.1 Introduction

Precision localization has been one of the fascinating application areas for impulse radio ultra-wideband (IR-UWB) technology. These applications exploit the fine time resolution of UWB signals. The ultrashort pulse waveform enables UWB receivers to accurately determine the time of arrival (TOA) of the signal transmitted from another UWB transmitter. For example, the accuracy of TOA measurements better than 40 ps [1] has been achieved, which corresponds to 1.2 cm spatial uncertainty.

There are various existing/potential applications of precision localization by making use of UWB technology. A UWB precision localization system [2] can be used to identify and locate valuable assets in hospitals, industrial fields, government offices, and so on. A UWB ASIC (application-specific integrated circuit) device (low cost and small size) can be employed for recreational activities [3, 4]. Another important/potential application is in the sensor networks. Awareness of sensor positions may effectively improve network performance. For instance, location-aware routing protocols can reduce routing overhead and save energy by avoiding a route search [5]. UWB technology is particularly well suited for sensor network applications due to its low power consumption.

This chapter aims to provide a comprehensive and detailed view of the localization techniques. It also provides an opportunity for the authors to present the results of their research.

12.2 Time-of-Arrival Estimation

Positioning techniques exploit one or more characteristics of radio signals to estimate the position of their sources. Some of the parameters that have been traditionally used for positioning are the received signal strength intensity (RSSI), the angle of arrival (AOA) and time of arrival (TOA). Amongst these positioning parameters, RSSI is the

Ultra-wideband Wireless Communications and Networks
Edited by Xuemin (Sherman) Shen, Mohsen Guizani, Robert Caiming Qiu and Tho Le-Ngoc

least adequate in the case of UWB, because it does not exploit from the fine space-time resolution of impulsive signals and requires a site-specific path loss model [6]. The estimation of AOA, on the other hand, requires multiple antennas (or at least an antenna capable of beamforming) at the receiver. This requirement implies size and complexity demands that are often not compatible with the low-cost, small-size constraints associated with applications such as wireless sensor networks, which UWB technology is particularly suited for.

One of the most attractive characteristics of IR-UWB signals is the fine time resolution, which makes impulse-based UWB a prominent candidate technology for indoor positioning. Therefore, TOA stands out as the most suitable signal parameter to be used for positioning with UWB devices. However, because of the ultrashort (usually subnanosecond) pulses, it poses challenges for synchronization in UWB systems. Some techniques have been proposed to estimate the TOA of UWB signals, for instance, correlation in conjunction with serial search [7], special code design [8], and frequency-domain processing [9]. However, all these solutions seem to be in conflict with the strict requirements of low cost and low complexity imposed on some UWB applications and may not provide satisfactory TOA estimate. Another TOA estimation scheme for UWB signals is the generalized likelihood ratio test [10] that seems to require high hardware complexity.

In order to further reduce the complexity of UWB systems, noncoherent receivers using energy collection [11–13] and transmitted reference [14–17] have been recently proposed.

In the following subsections, we first briefly discuss the TOA estimation accuracy. Then, a practical energy-collection–based approach is given a detailed description. Finally, a two-stage TOA estimation scheme is presented.

12.2.1 Estimation Accuracy

The accuracy of the delay estimator can be estimated by the Cramér-Rao bound (CRB). Assuming a half sinusoidal pulse waveform, the standard deviation of the delay estimator according to the CRB becomes

$$\sigma_\tau = \frac{1}{\sqrt{2N\pi^2\gamma}} T_p, \tag{1}$$

where T_p is the pulse duration, N is the number of used pulses, and γ is the signal-to-noise ratio (SNR) of a pulse defined as $\gamma = E/N_0$, where E is the pulse energy and N_0 is the spectral density of the noise. Since the bound is proportional to the pulse width, or equivalently, antiproportional to the bandwidth of the signal, IR-UWB signals offer a significant advantage in TOA/TDOA (time difference of arrival) (and therefore distance) estimation due to the ultrashort pulse/ultra-wide bandwidth.

A typical delay estimator includes a correlator and a delay lock loop. In packet-based systems, such an arrangement may be inadequate and, preferably, the maximum output of the correlator should be searched. Therein, the interpolation techniques may increase the accuracy. In an interpolation technique, a parabola is fitted to the maximum and its neighboring samples [18–20]. Let k denote the sample index of the maximum and let $c(k)$, $c(k-1)$, $c(k+1)$ be the maximum value of the envelope of the correlator's output and its neighbors. Then, the corrected delay estimate is

$$\hat{\tau} = \left(k + \frac{1}{2} \frac{c(k-1) - c(k+1)}{c(k-1) - 2c(k) + c(k+1)} \right) \frac{T_p}{q}. \tag{2}$$

More details about interpolation effects may be found from [21].

12.2.2 Energy-Collection–Based TOA Estimation

Figure 12.1 shows the block diagram of the energy detection receiver. The received signal is first passed through a band-pass filter (BPF) to reduce the noise power. After low noise amplification, the signal is squared and then passed to a block of integrators that integrate the received signal in different time slots. The advantage of this receiver scheme is that it is relatively easy in implementation when compared with the correlation-based scheme. The drawback of the noncoherent approach is noise enhancement due to the squaring and the degradation in time resolution, which is proportional to the length of the integration.

In order to reuse the same receiver architecture, TOA estimation can also be performed on the basis of energy detection. After initial synchronization is completed, the TOA estimation is performed by dividing the uncertainty region (T_u) around the synchronization point into N integration windows, where N represents the number of integrators available in the receiver. Intuitively, the estimation accuracy is dependent on the size of the uncertainty region and on the number of integrators. Differing from synchronization, which provides the time reference that ensures maximum signal energy detection, the TOA estimation can be seen as fine synchronization, searching for the arrival time of the received signal.

On the basis of the energy measurements, a decision is made according to a chosen criterion. For example, a threshold crossing (TC) criterion can be used. With TC, the search is performed serially and is stopped once a measurement value crosses the threshold. The corresponding window is then chosen and its starting point provides the TOA information. If necessary, a verification process may be pursued. For example, new measurements are taken from the chosen window and are tested against the threshold. If the threshold is crossed B out of A ($B \leq A$) tests, the chosen window is finally accepted. Otherwise, the search resumes. In the event that no measurement crosses the threshold, new measurements are taken and the search resumes.

The other approach is the maximum (MAX) selection criterion. With this criterion, measurements at all windows are first compared. Then, the maximal measurement is produced and the relevant window is selected. In the event that no appropriate thresholds can be readily obtained, the MAX criterion would be desirable. Another criterion is

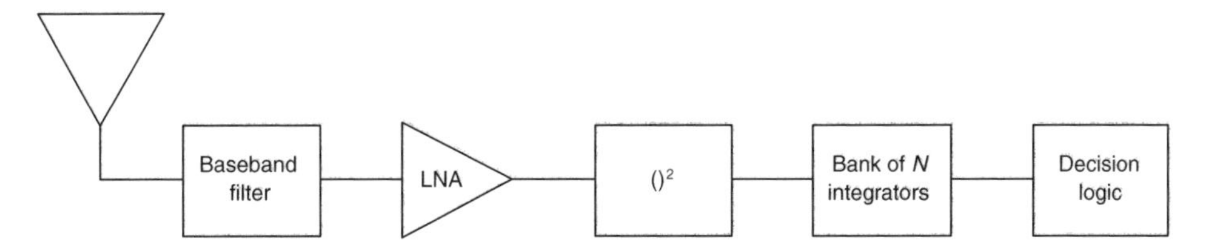

Figure 12.1 Block diagram of energy detection receiver

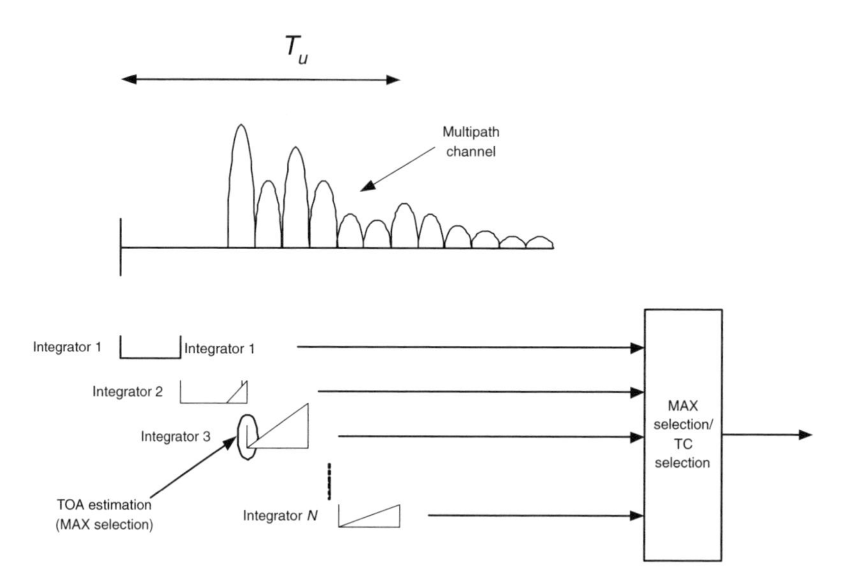

Figure 12.2 Illustration of TOA estimation

the hybrid of MAX and TC. In this hybrid criterion, the maximal measurement is first obtained. Then, the maximum is examined against the threshold. If the threshold is crossed, the related integration window is selected. If the maximum does not cross the threshold, the search resumes. Figure 12.2 shows the basic principle of this approach. When using the MAX selection criterion, the switch-on time of integrator 3 is chosen to be the TOA estimate.

The TC algorithm and the hybrid approach require the setting of a threshold. The threshold can be determined on the basis of a given probability of false alarm and the noise distribution. This will be further studied in Section 12.2.3.1. The threshold can also be determined adaptively on the basis of the received signal strength (RSS).

Note that there exist other simple methods such as the leading-edge detection and first-peak detection approaches, however, the noncoherent energy-collection–based method is more suited in rich multipath UWB indoor communications.

12.2.3 Two-Stage TOA Estimation

The previously discussed energy-collection–based approach has low complexity and is easy to implement, however, it may not provide satisfactory TOA estimation performance for systems demanding high accuracy. For low data rate communications, it is desirable to achieve fast synchronization or TOA estimation. In this section, we study a two-stage TOA estimation scheme to achieve the goal of rapid acquisition and high accuracy.

12.2.3.1 First Stage Processing

As shown in Figure 12.3, in the preamble of the packet, there is a sequence of (N_{pr}) symbols assigned for synchronization, each of which consists of a sequence of coded

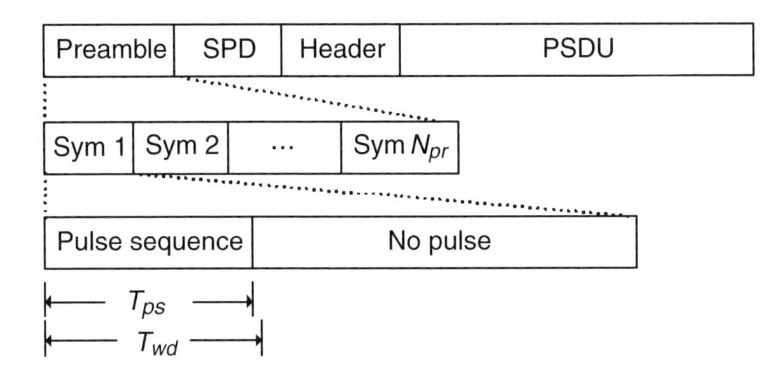

Figure 12.3 Illustration of timing diagram of the preamble

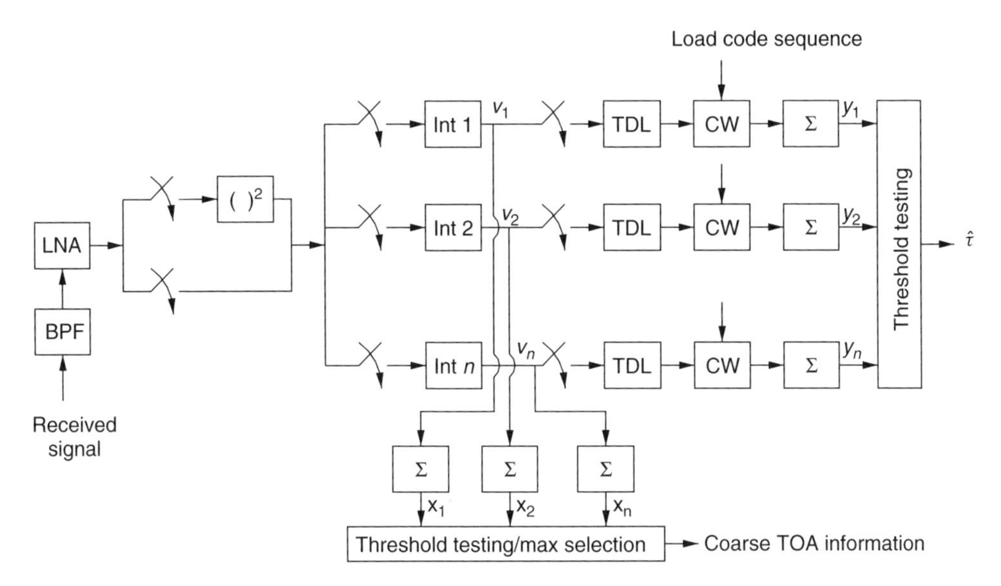

Figure 12.4 Acquisition system structure. BPF: band-pass filter. LNA: low noise amplifier. Int: integrator. TDL: tapped delay line. CW: code weighting

pulses of length T_{ps}, which is a small fraction of the symbol period. During the first stage, a search is performed over a duration of one or more symbols. The symbol duration is divided into K windows of equal size $T_{wd} = T/K$. The aim is to determine which window will contain the pulses. Then, code acquisition can be rapidly achieved at the second stage. Figure 12.4 shows the block diagram of the proposed acquisition system.

After passing through a BPF and a low noise amplifier (LNA), the received signal is squared and then forwarded to a bank of n integrators as shown in Figure 12.4. The first integrator starts integration at a chosen instant of time. Each of the other integrators switches on once its preceding integrator switches off. The integration interval of each integrator is the same and equals the window size T_{wd}. The process is similar to the technique employed in the previous section. It may go over κ symbols and the energy measurements at the same window in each symbol are accumulated.

Once the energy measurements are available, a decision is made according to one of the three criteria mentioned in Section 2.2. When the TC is applied, a decision is made by choosing one of the two hypotheses:

- $\mathcal{H}_{1,0}$: No signal pulses appear in the window examined at the first stage.
- $\mathcal{H}_{1,1}$: At least one signal pulse appears in the window examined at the first stage.

Note that $\mathcal{H}_{m,n}$ means a hypothesis at the mth stage with $n = 0$ indicating the absence of pulses and $n = 1$ indicating the presence of pulses. Then, two types of errors can be made when making a decision:

- Decide $\mathcal{H}_{1,1}$ when $\mathcal{H}_{1,0}$ is true and the corresponding probability is denoted by $P(\mathcal{H}_{1,1}; \mathcal{H}_{1,0})$.
- Decide $\mathcal{H}_{1,0}$ when $\mathcal{H}_{1,1}$ is true and the probability is denoted by $P(\mathcal{H}_{1,0}; \mathcal{H}_{1,1})$.

Similar to the terminologies used in code acquisition, $P(\mathcal{H}_{1,1}; \mathcal{H}_{1,0})$ is termed the *probability of false alarm* and is denoted by P_{F1}. The probability, $P(\mathcal{H}_{1,1}; \mathcal{H}_{1,1})$, is termed the *probability of detection*, which is denoted by P_{D1}. The probability of missing is the probability that all measurements are less than the threshold.

According to the Neyman–Pearson theorem [22], for a given $P_{F1} = \alpha_1$, P_{D1} is maximized by deciding $\mathcal{H}_{1,1}$ if

$$L(x_i) = p(x_i; \mathcal{H}_{1,1})/p(x_i; \mathcal{H}_{1,0}) > \gamma_0, \tag{3}$$

where $p(x_i; \mathcal{H}_{1,j})$, $j = 1$ and 2, is the probability density function (pdf) and the threshold γ_0 is found from

$$P_{F1} = \int_{x_i; L(x_i)>\gamma_0} p(x_i; \mathcal{H}_{1,0})\, dx_i = \alpha_1. \tag{4}$$

Here x_i is the energy measurement. Suppose that there are L_{sect} samples in each window and κ measurements are accumulated. When no signal pulses appear, $x_i = \sum_{j=1}^{\kappa L_{sect}} z_j^2$, where z_j is a Gaussian random variable of mean zero and variance σ^2, and has a chi-square (or gamma) distribution [23] with pdf

$$p(x) = \frac{1}{(2\sigma^2)^{m_1/2}\Gamma(m_1/2)} x^{m_1/2-1} \exp\left(-\frac{x}{2\sigma^2}\right),$$

where $m_1 = \kappa L_{sect}$. Without loss of generality, let m_1 be even: $m_1 = 2m$ and the solution of $L(x_i) > \gamma_0$ is $x_i > \gamma_1$. Then, the false alarm can be expressed by

$$P_{F1} = \int_{\gamma_1}^{\infty} p(x)\, dx = \exp\left(-\frac{\gamma_1}{2\sigma^2}\right) \sum_{k=0}^{m-1} \frac{1}{k!} \left(\frac{\gamma_1}{2\sigma^2}\right)^k. \tag{5}$$

Since there are many cases when at least one signal pulse appears in a window of interest, it would be impractical to derive a tractable expression for the probability of detection (as well as the probability of missing) even when the multipath channel information is available. This also happens to the probabilities when either the MAX or the hybrid criterion is employed.

For clarity, we assume that the first symbol of the received signal is among the symbol(s) examined. We also assume that the ith window is chosen and it was switched on at t_i. In a probability dependent on the SNR, some pulses of the received signal will be locked in the ith window and the TOA would be at some point around t_i. Therefore, at the second (code acquisition) stage, we may constrain the TOA to

$$t_i - (T_{ps} + T_m) < t < t_i + T_{wd}, \tag{6}$$

where T_m is the multipath spread, which may be unknown; however, it may be roughly predicted on the basis of the channel condition, if available. The goal of the first stage is to lock the pulse sequence in the region given by (6).

12.2.3.2 Second Stage Processing

In the second stage, that is, the code acquisition stage, the received signal is passed directly to the integrators, which now become part of the matched filter correlator (consisting of the integrator, the tapped delay line (TDL), the code weighting (CW), and the summation) [24, 25]. The located region given by (6) is divided into a group of code offsets. As shown in Figure 12.5, the uncertain region, $t_0 \le t \le t_8$, is divided into eight offsets with the ith offset located between t_{i-1} and t_i.

The MAX and the hybrid criterion may not be appropriate for the second stage due to the multipath channels, so the threshold crossing criterion is exploited. The correlations (y_1 to y_n in Figure 12.4) are examined serially. Once the correlation of an offset crosses the threshold, it is selected as the correct offset whose time instant becomes the TOA estimate. Similar to the technique at the first stage, correlations of the same code offset can be produced over several symbols and then be accumulated (averaged) to increase the accuracy.

For an L-path channel, there are $(L + 1)$ hypotheses denoted by $\mathcal{H}_{2,0}$: no signal appears, and $\mathcal{H}_{2,i}$, $i = 1, 2, \ldots, L$: ith path signal appears. Since the signal of the first (shortest) path is of interest, only two decisions will be made, that is, $\mathcal{H}_{2,1}$ is true or not. If $\mathcal{H}_{2,0}$ is true and it is assumed that κ_2 measurements are accumulated, the correlation has a Gaussian distribution of mean zero and variance $\sigma_2^2 = \kappa_2 m_s \sigma^2$, where m_s is the code length. Let P_{F2} denote the probability of false alarm at the second stage, which is the probability of deciding $\mathcal{H}_{2,1}$ when $\mathcal{H}_{2,0}$ is true. Then,

$$P_{F2} = \int_{\gamma_2}^{\infty} \frac{1}{\sqrt{2\pi}\sigma_2} \exp\left[-\frac{x^2}{2\sigma_2^2}\right] dx, \tag{7}$$

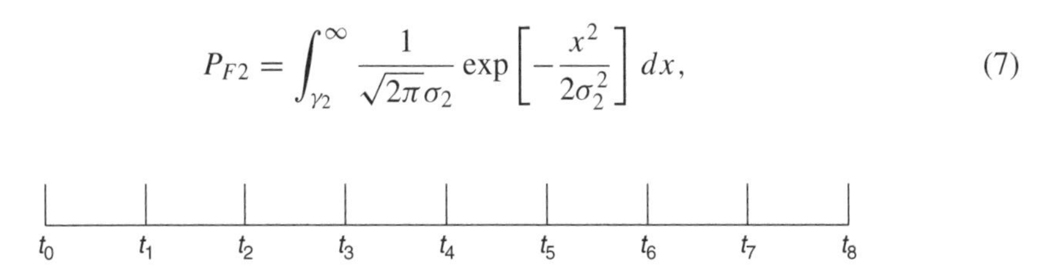

Figure 12.5 Illustration of uncertain region divided into code offsets

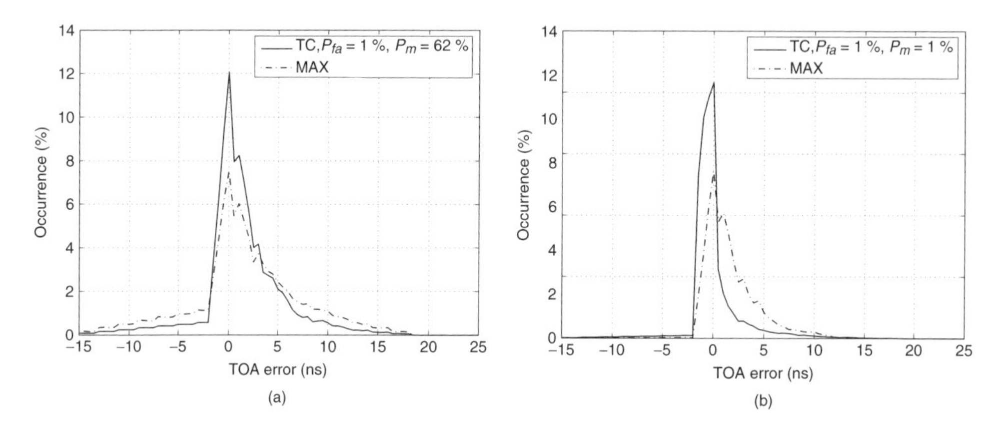

Figure 12.6 Comparison between MAX and TC algorithm. $T_u = 15$ ns, $N = 10$, window size = 2 ns and $E_b/N_0 = 10$ dB (a) and $E_b/N_0 = 20$ dB (b)

where γ_2 is the threshold at the second stage. Denote P_{D2} as the probability of detection at the second stage, which is also channel-dependent. Then, when $\mathcal{H}_{2,1}$ is true, we have

$$P_{D2} = \int_{\gamma_2}^{\infty} \frac{1}{\sqrt{2\pi}\sigma_2} \exp\left(-\frac{(x - \kappa_2 m_s a_0)^2}{2\sigma_2^2}\right) dx,$$

where a_0 is the constant amplitude of the first path signal.

12.2.4 Simulation Results

This subsection presents some simulation results of the energy-collection–based approach. In the case of line-of-sight (LOS) or nonsevere nonline-of-sight (NLOS) propagation, the channel is characterized by the presence of high cluster of energy in the first nanoseconds of the received signal. In this case, the MAX algorithm can achieve the same performance of the TC but with simpler implementation and negligible probability of missed detection. Figure 12.6 shows the performance of the two proposed algorithms. These results are obtained by employing the channel model 3 (CM3) as defined by the 802.15.4a standardization group [26] using 1000 channel realizations. The TC algorithm tends to outperform the MAX algorithm, however, at low SNR, the probability of missed detection (P_m) can considerably increase. In these specific cases, the threshold is set on the basis of the probability of false alarm (P_{fa}) that is estimated over a number of integrated noise signal values. An adaptive threshold based on the RSS might reduce the probability of missed detection; however, SNR estimation of a UWB signal is not an easy task and it may not be suited for low-complexity devices. Another option to reduce the probability of missed detection is to increase the integration window to increase the received signal energy.

12.3 Location and Tracking

In this section, an overview of both noniterative and iterative position estimation algorithms is provided, and details of several typical algorithms are presented. Also presented

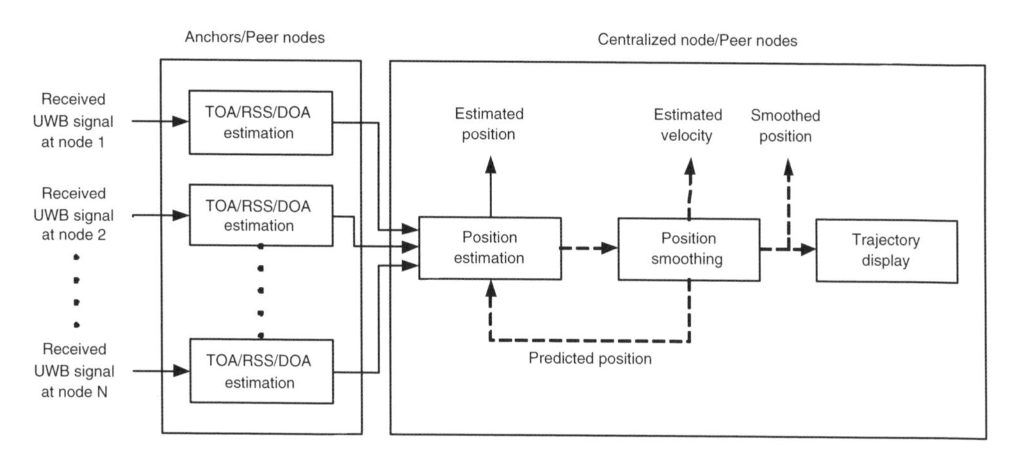

Figure 12.7 Block diagram of position location and tracking

are mobile node/station tracking and mobile track smoothing algorithms. The block diagram of the location and tracking system is illustrated in Figure 12.7.

12.3.1 Position Estimation

Position estimation algorithms using radio signals[1], including UWB, may be broken into two broad categories: iterative and noniterative methods, for both cellular systems and sensor networks.

12.3.1.1 Noniterative Algorithms

A variety of noniterative algorithms have been developed for position estimation. The most straightforward one is the direct method (DM) [27, 28], which directly solves a set of simultaneous equations with four anchors for 3-D positioning using TDOA measurements. This method, however, may not effectively exploit extra measurements from extra sensors to improve position accuracy. The spherical-interpolation (SI) method [29] and related approaches [30, 31] were developed to exploit extra measurements. To approach optimal estimation, the two-stage maximum-likelihood approach was considered in [32] and the linear-correction least-square approach was considered in [33]. When using range measurements, the standard least-squares (LS) approach is commonly considered [34–36]. For quick reference, the DM, the LS, and the SI approaches are described as follows. For convenience, we assume that the nodes with known positions, $(x_i, y_i, z_i), i = 1, 2, \ldots, N$, are anchors while the nodes with unknown positions, (x, y, z), are sensors.

Let t_i be the TOA of the signal at anchor i, t_0 be the transmit time at the sensor, c be the speed of light, and $d_i = c(t_i - t_0)$ be the range between the sensor and the ith anchor.

[1] Several parameters of the received signal have been commonly exploited: TOA (and thus time-difference-of-arrival (TDOA) and round-trip time (RTT)), received signal strength (RSS), and angle of arrival (AOA).

Then, we obtain

$$d_i = \sqrt{(x_i - x)^2 + (y_i - y)^2 + (z_i - z)^2}, \quad i = 1, 2, \ldots, N. \tag{8}$$

Squaring both sides of (8) yields

$$(x_i - x)^2 + (y_i - y)^2 + (z_i - z)^2 = d_i^2. \tag{9}$$

Subtracting (9) for $i = 2, 3, \ldots, N$ by (9) for $i = 1$ produces

$$x_{i,1}x + y_{i,1}y + z_{i,1}z = g_{i,1}, \quad i = 2, \ldots, N, \tag{10}$$

where

$$x_{i,1} = x_i - x_1, \quad y_{i,1} = y_i - y_1, \quad z_{i,1} = z_i - z_1,$$
$$g_{i,1} = 0.5[(x_i^2 + y_i^2 + z_i^2) - (x_1^2 + y_1^2 + z_1^2) + d_1^2 - d_i^2].$$

In the DM, only three anchors are needed (say $i = 1, 2, 3$). Define

$$A_1 = \frac{x_{2,1}z_{3,1} - x_{3,1}z_{2,1}}{x_{3,1}y_{2,1} - x_{2,1}y_{3,1}}, \qquad B_1 = \frac{x_{3,1}g_{2,1} - x_{2,1}g_{3,1}}{x_{3,1}y_{2,1} - x_{2,1}y_{3,1}}, \tag{11}$$

$$C_1 = \frac{y_{3,1}z_{2,1} - y_{2,1}z_{3,1}}{x_{3,1}y_{2,1} - x_{2,1}y_{3,1}}, \qquad D_1 = \frac{y_{2,1}g_{3,1} - y_{3,1}g_{2,1}}{x_{3,1}y_{2,1} - x_{2,1}y_{3,1}}. \tag{12}$$

Then we have

$$\hat{z} = \frac{F_1}{E_1} \pm \sqrt{\left(\frac{F_1}{E_1}\right)^2 - \frac{G_1}{E_1}}, \tag{13}$$

where

$$E_1 = A_1^2 + C_1^2 + 1, \qquad F_1 = A_1(y_1 - B_1) + C_1(x_1 - D_1) + z_1, \tag{14}$$

$$G_1 = (x_1 - D_1)^2 + (y_1 - B_1)^2 + z_1^2 - d_1^2, \tag{15}$$

and

$$\hat{x} = C_1\hat{z} + D_1, \qquad \hat{y} = A_1\hat{z} + B_1. \tag{16}$$

Clearly, there are two solutions, however, only one solution is desirable. We remove the one either with no physical meaning or which is beyond the monitored area. If both solutions are reasonable and they are very close, we may choose the average as the position estimate. Otherwise, an ambiguity occurs.

When considering the LS method, there are at least four anchors. Write (10) in compact form as

$$\mathbf{Ap} = \mathbf{b}, \tag{17}$$

where

$$\mathbf{A} = \begin{bmatrix} x_{2,1} & y_{2,1} & z_{2,1} \\ x_{3,1} & y_{3,1} & z_{3,1} \\ \vdots & \vdots & \vdots \\ x_{N,1} & y_{N,1} & z_{N,1} \end{bmatrix}, \qquad \mathbf{p} = \begin{bmatrix} x \\ y \\ z \end{bmatrix}, \qquad \mathbf{b} = \begin{bmatrix} g_{2,1} \\ g_{3,1} \\ \vdots \\ g_{N,1} \end{bmatrix}. \tag{18}$$

The standard LS solution to (17) is given by

$$\hat{\mathbf{p}} = \left(\mathbf{A}^T\mathbf{A}\right)^{-1}\mathbf{A}^T\mathbf{b}, \tag{19}$$

where the matrix inverse is supposed to exist and d_i are replaced by their corresponding estimates.

When the transmit time t_0 is unknown, TDOA measurements may be employed as for centralized networks. In this case, the SI method would be a suitable candidate. First, map the spatial origin to one of the anchors, say anchor 1, as shown in Figure 12.8. Define

$$R_i = \sqrt{x_i^2 + y_i^2 + z_i^2}, \quad \mathbf{p}_i = [x_i,\ y_i,\ z_i]^T, \quad R = \sqrt{x^2 + y^2 + z^2}, \quad d_{i,j} = d_i - d_j.$$

Then, from the Pythagorean theorem, we have

$$(R + d_{i,1})^2 = R_i^2 - 2\mathbf{p}_i^T\mathbf{p} + R^2, \quad i = 2, 3, \ldots, N, \tag{20}$$

where we used

$$R + d_{i,1} = d_i.$$

In the presence of measurement errors, (20) usually does not hold. Introduce the equation error [29] ϵ_i and add it to the left-hand side of (20) to make the equation valid. Then, we obtain

$$\epsilon_i = R_i^2 - d_{i,1}^2 - 2Rd_{i,1} - 2\mathbf{p}_i^T\mathbf{p}. \tag{21}$$

Equation (21) can be written in compact form as

$$\boldsymbol{\epsilon} = \boldsymbol{\delta} - 2R\mathbf{d} - 2\mathbf{A}\mathbf{p}, \tag{22}$$

where $\mathbf{A}$ is given by (18), $\boldsymbol{\epsilon}$ is the equation error vector, and

$$\boldsymbol{\delta} = \left[R_2^2 - d_{2,1}^2, R_3^2 - d_{3,1}^2, \ldots, R_N^2 - d_{N,1}^2\right]^T, \quad \mathbf{d} = [d_{2,1}, d_{3,1}, \ldots, d_{N,1}]^T. \tag{23}$$

The standard LS solution for $\mathbf{p}$, given R, is

$$\mathbf{p} = \tfrac{1}{2}\left(\mathbf{A}^T\mathbf{A}\right)^{-1}\mathbf{A}^T(\boldsymbol{\delta} - 2R\mathbf{d}). \tag{24}$$

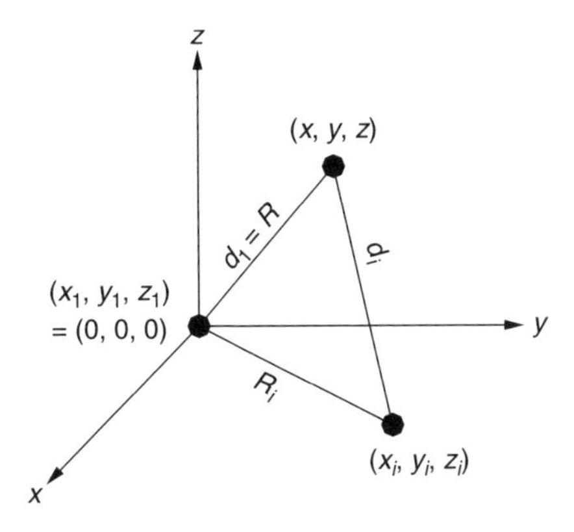

Figure 12.8 Illustration for spherical-interpolation approach

The key idea of the SI approach is to substitute (24) into (22) and minimize the equation error again, but with respect to R. The source location estimate is then obtained as

$$\hat{\mathbf{p}} = \frac{1}{2}\left(\mathbf{A}^T\mathbf{W}\mathbf{A}\right)^{-1}\mathbf{A}^T\mathbf{W}\left(\mathbf{I} - \frac{\mathbf{d}\mathbf{d}^T\mathbf{B}\mathbf{V}\mathbf{B}}{\mathbf{d}^T\mathbf{B}\mathbf{V}\mathbf{B}\mathbf{d}}\right), \tag{25}$$

where $\mathbf{W}$ and $\mathbf{V}$ are weighting matrixes and

$$\mathbf{B} = \mathbf{I} - \mathbf{A}\left(\mathbf{A}^T\mathbf{W}\mathbf{A}\right)^{-1}\mathbf{A}^T\mathbf{W}.$$

The purpose of introducing the weighting matrices, $\mathbf{W}$ and $\mathbf{V}$, is to emphasize the contributions of the measurements that are more reliable [37]. In the absence of *a priori* confidences, the weighting matrices may be simply chosen to be an identity matrix.

12.3.1.2 Iterative Algorithms

For both cellular networks and sensor networks, two iterative methods are often considered. One is the Taylor-series method [32, 38–42] and the other is the optimization-based method [43–45].

In the Taylor-series method, a set of nonlinear equations is linearized by expanding it in a Taylor series around a point (initially an estimate of the actual position) and keeping only terms below second order. The set of linearized equations is solved to produce a new approximate position, and the process continues until a prespecified convergence criterion is satisfied (e.g., the position estimate changes by less than some threshold amount). When TDOA measurements are employed, this method may be described as follows:

Subtracting (8) for $i = 1$ from (8) for $i = 2, 3, \ldots, N$ produces

$$\sqrt{(x - x_i)^2 + (y - y_i)^2 + (z - z_i)^2} - \sqrt{(x - x_1)^2 + (y - y_1)^2 + (z - z_1)^2} = c(t_i - t_1), \quad i = 2, 3, \ldots, N. \tag{26}$$

Define

$$f_i(x, y, z) = \sqrt{(x - x_{i+1})^2 + (y - y_{i+1})^2 + (z - z_{i+1})^2} - \sqrt{(x - x_1)^2 + (y - y_1)^2 + (z - z_1)^2}, \quad i = 1, 2, \ldots, N - 1, \tag{27}$$

and let $\hat{t}_i$ be the TOA estimate at anchor i. Then,

$$f_i(x, y, z) = \hat{d}_{i+1,1} + \epsilon_{i+1,1}, \quad i = 1, 2, \ldots, N - 1, \tag{28}$$

where $\hat{d}_{i,1} = c(\hat{t}_i - \hat{t}_1)$ and $\epsilon_{i,1}$ is the corresponding range difference estimation error with covariance $\mathbf{R}$. If x_v, y_v, and z_v are guesses of the actual sensor position, then,

$$x = x_v + \delta_x, \; y = y_v + \delta_y, \; z = z_v + \delta_z,$$

where δ_x, δ_y, and δ_z are the position errors to be determined. Expanding f_i in a Taylor series and retaining the first two terms produces

$$f_{i,v} + a_{i,1}\delta_x + a_{i,2}\delta_y + a_{i,3}\delta_z \approx \hat{d}_{i+1,1} + \epsilon_{i+1,1}, \quad i = 1, 2, \ldots, N - 1, \tag{29}$$

where

$$f_{i,v} = f_i(x_v, y_v, z_v)$$

$$a_{i,1} = \frac{\partial f_i}{\partial x}|_{x_v,y_v,z_v} = \frac{x_1 - x_v}{\hat{d}_1} - \frac{x_{i+1} - x_v}{\hat{d}_{i+1}}, \quad \hat{d}_i = \sqrt{(x_v - x_i)^2 + (y_v - y_i)^2 + (z_v - z_i)^2}$$

$$a_{i\,2} = \frac{\partial f_i}{\partial y}|_{x_v,y_v,z_v} = \frac{y_1 - y_v}{\hat{d}_1} - \frac{y_{i+1} - y_v}{\hat{d}_{i+1}}, \quad a_{i,3} = \frac{\partial f_i}{\partial z}|_{x_v,y_v,z_v} = \frac{z_1 - z_v}{\hat{d}_1} - \frac{z_{i+1} - z_v}{\hat{d}_{i+1}}.$$

Equation (29) can be rewritten as

$$\mathbf{A}\boldsymbol{\delta} = \mathbf{D} + \mathbf{e}, \tag{30}$$

where

$$\mathbf{A} = \begin{bmatrix} a_{1,1} & a_{1,2} & a_{1,3} \\ a_{2,1} & a_{2,2} & a_{2,3} \\ \vdots & \vdots & \vdots \\ a_{N-1,1} & a_{N-1,2} & a_{N-1,3} \end{bmatrix}, \qquad \boldsymbol{\delta} = \begin{bmatrix} \delta_x \\ \delta_y \\ \delta_z \end{bmatrix}$$

$$\mathbf{D} = \begin{bmatrix} \hat{d}_{2,1} - f_{1,v} \\ \hat{d}_{3,1} - f_{2,v} \\ \vdots \\ \hat{d}_{N,1} - f_{N-1,v} \end{bmatrix}, \qquad \mathbf{e} = \begin{bmatrix} \epsilon_{2,1} \\ \epsilon_{3,1} \\ \vdots \\ \epsilon_{N,1} \end{bmatrix}.$$

The weighted least-square estimator for (30) produces

$$\boldsymbol{\delta} = \left[\mathbf{A}^T\mathbf{R}^{-1}\mathbf{A}\right]^{-1}\mathbf{A}^T\mathbf{R}^{-1}\mathbf{D}. \tag{31}$$

Given an initial position guess (x_v, y_v, z_v), compute $\boldsymbol{\delta}$ with (31). Then update the position estimate according to

$$x_v = x_v + \delta_x, \quad y_v = y_v + \delta_v, \quad z_v = z_v + \delta_z.$$

Continually refine the position estimate until δ is sufficiently small.

In optimization-based approaches, an objective/cost function is first defined. One such cost function may be chosen to be

$$\epsilon = \sum_{i=1}^{N}\left[\sqrt{(x_i - x)^2 + (y_i - y)^2 + (z_i - z)^2} - c(t_i - t_0)\right]^2. \tag{32}$$

Then, a minimization algorithm is applied to achieve the optimal solution to the position estimation. Both unconstrained and constrained minimization algorithms can be employed [46]. The Quasi-Newton DFP (Davidon–Fletcher–Powell) algorithm was exploited in the UWB precision assets location system [2]. A Gauss–Newton type Levenberg–Marquardt method [47] was studied in [48]. Convex optimization was considered in [49] and the simplex method [50] was employed in [51]. These algorithms can be found in the Matlab optimization toolbox. Since the DFP algorithm will be used in the simulations, a brief description of the algorithm is given here.

The solution to the position estimate is solved iteratively, and at time instant/iteration $k+1$, the position is updated by

$$\mathbf{p}_{k+1} = \mathbf{p}_k - \alpha \mathbf{B}_k \mathbf{g}_k \tag{33}$$

where $\mathbf{p}_k = [\hat{x}^{(k)}, \hat{y}^{(k)}, \hat{z}^{(k)}]^T$ is the vector of the estimated position coordinates at the kth iteration, α is the step size, and $\mathbf{g}_k$ is the gradient of the objective function defined as

$$\mathbf{g}_k = \nabla \epsilon(x, y, z) = \left[\frac{\partial \epsilon}{\partial x}, \frac{\partial \epsilon}{\partial y}, \frac{\partial \epsilon}{\partial z} \right]^T_{\mathbf{p}=\mathbf{p}_k}. \tag{34}$$

Also, $\mathbf{B}_k$ is the so-called Hessian that is updated according to

$$\mathbf{B}_{k+1} = \mathbf{B}_k + \frac{\mathbf{h}_k \mathbf{h}_k^T}{\mathbf{h}_k^T \mathbf{q}_k} - \frac{\mathbf{B}_k \mathbf{q}_k \mathbf{q}_k^T \mathbf{B}_k}{\mathbf{q}_k^T \mathbf{B}_k \mathbf{q}_k}, \tag{35}$$

where

$$\mathbf{h}_k = \mathbf{p}_{k+1} - \mathbf{p}_k, \qquad \mathbf{q}_k = \mathbf{g}_{k+1} - \mathbf{g}_k. \tag{36}$$

12.3.2 Tracking

We consider a scenario where the sensor nodes are mobile. To make use of accurate node position information for efficient network operation and management, node position information needs to be updated regularly. The updating frequency should be based on the mobility/speed of the nodes of interest. Suppose that the maximum speed of the nodes in the desired sensor network is 18 km/h (that is, 5 m/s). If the position error is required to be below 1 m, we need to update the position information at least 5 times/s. A higher position accuracy requirement will require a higher updating frequency. Different updating frequency may be applied to different nodes if the maximum velocities of the nodes are different and they are known *a priori*. This strategy would improve the network efficiency.

Positioning performance can be further improved by smoothing the tracks of the moving nodes. Smoothing techniques including Kalman filtering [4, 52, 53] and linear least-squares (LLS) techniques [4, 54] can be employed.

12.3.3 Simulation Results

In this section, we examine the performance of location and tracking. We use one of the realistic field structures, a snow covered slope of dimensions about 400 m × 100 m × 100 m [4]. The anchor nodes will be deployed along both sides of the slope and mounted on poles of varying height. Figure 12.9 shows the track for examination. The skier moves from A to B (120 m) at a speed of 8 meters per second (m/s). The skier moves from B to C (160 m) at a speed of 10 m/s and finally from C to D (120 m) at a speed of 8 m/s.

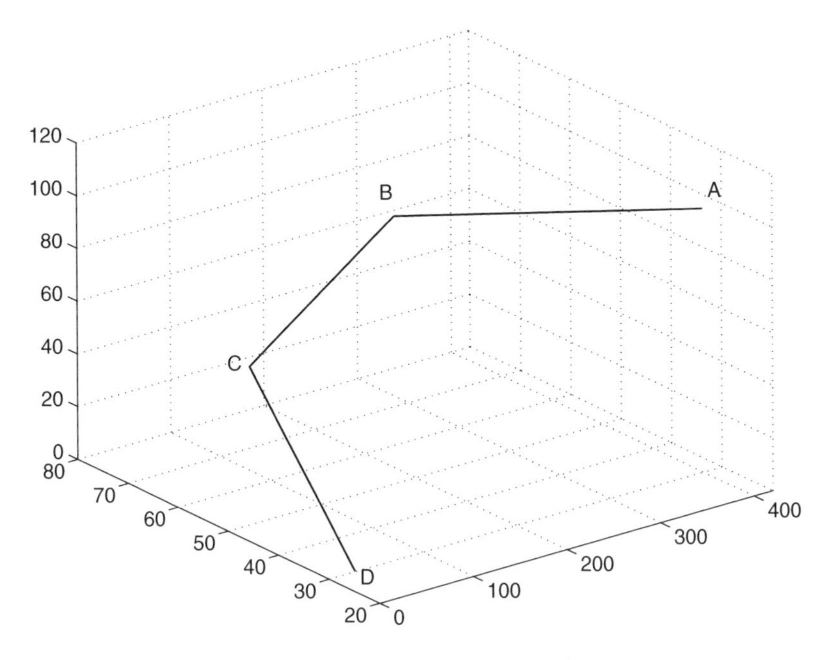

Figure 12.9 Track for examination

Table 12.1 Averaged root-mean-square error (RMSE) of four algorithms at SNR of 16 dB. FNs: fixed nodes. TS: Taylor series method. SI: spherical-interpolation method. DM: direct method

No of FNs	4	5	6	8
DFP	4.36 (m)	1.65 (m)	0.63 (m)	0.15 (m)
TS	5.43	2.1	0.88	0.15
SI		1.55	0.33	0.09
DM	7.54			

First we examine the performance of the different position estimation algorithms under realistic circumstances. Two hundred different combinations of fixed node positions were tested and then the results were averaged. Tables 12.1 and 12.2 compare the averaged results of the four algorithms at SNR of 16 dB. The distance errors were produced using the noncoherent TOA estimation approach that was described in Section 12.2.2. Two performance indexes are examined: one is the root-mean-square error (RMSE) and the other is the failure rate. The failure rate includes the cases where there is no solution or the solution is unreasonable. With the iterative methods, the failure rate includes situations where the algorithm does not converge to a solution, the maximum number of function evaluations/iterations is exceeded, or the results are beyond the area examined. With the noniterative methods, 'failures' include cases when the solutions are beyond the monitored

Table 12.2 Averaged failure rates of four algorithms at SNR of 16 dB

No of FNs	4	5	6	8
DFP	10.8 (%)	6.6 (%)	5.2 (%)	4.2 (%)
TS	14.3	3.1	1.5	0.9
SI		3.4	1.3	0.9
DM	46.6			

Table 12.3 Averaged RMSEs before and after smoothing. KF: Kalman filtering

Before smoothing	LS	KF
4.13 (m)	2.40	2.30
2.70 (m)	1.58	1.30
1.40 (m)	1.02	0.99

area, both solutions are complex-valued, the two solutions are reasonable but not close to each other, or inversion of singular matrices is involved.

For the parameters examined, the SI method provides the best trade-off between performance and complexity when there are at least five fixed nodes. To achieve submeter accuracy, at least six fixed nodes are needed with SNR up to 16 dB.

Let us consider position smoothing by making use of Kalman filtering and the LLS smoothing. Table 12.3 shows the averaged RMSEs before and after smoothing. The estimated tracks (before smoothing) are produced by using the SI algorithm with five fixed nodes under three sets of fixed node configurations. Although Kalman filtering performs better, the performance gain is marginal compared to the increase in complexity.

12.4 Location in Distributed Architectures

In this section, we consider localization in ad hoc and distributed networks. An ad hoc network is a collection of wireless nodes that self-configure to form a network without the aid of any infrastructure. This sort of network is characterized by large size, need for distributed coordination and ubiquitous connectivity, power constraints, and the ability to be ad hoc deployable. In the following subsections, an overview of related localization approaches is first presented. Then a positioning algorithm is proposed and finally some simulation results are provided.

12.4.1 Overview

In the recent years, there have been numerous algorithms (either centralized or distributed) to localize sensor nodes in wireless sensor networks. In [49, 55], a centralized scheme is proposed, which collects the entire topology in a server to minimize the errors using convex optimization. In [56–58], instead of directly solving the set of constraints of the whole wireless network, multi dimensional scaling (MDS) is exploited. This technique uses local connectivity or distance measures to generate relative maps that represent the relative positions of nodes. The main problem of the algorithms discussed earlier is the need to have some powerful node or server to perform the large computation. In [59], distributed algorithms are divided in two subfamilies; range-based and range-free algorithms. In range-free localization [34, 60], beacon nodes broadcast their positions to their neighbors that keep an account of all received beacons. Then, nodes calculate their positions based on the received beacon locations, the hop-count from the corresponding beacon and the average distance per hop. In [61], the distance per hop is averaged by taking into account the local density of nodes. In range-based algorithms, the distance between two neighboring sensors is first estimated, for example, by using TOA measurements. In [62] and more recently in [63], a distributed mechanism is proposed for GPS-free positioning in mobile ad hoc networks. A slightly modified version of the GPS-free algorithm is proposed in [64]. In [65], the distance between a sensor and a beacon is directly calculated using basic triangle rules and simple geometry. Collinearity is exploited in [66] and factor graphs are employed in [67].

12.4.2 Proposed Algorithm

It is assumed that each node (sensor and beacon) has the ability of measuring both TOA and AOA with its 1-hop neighbors. In the first phase, the beacons broadcast their coordinates. Then, all sensors establish the path (with the least number of hops) to the beacons. As a result, a sensor should have the coordinates of at least three beacons and the path to reach them. After the path is developed, all sensors and beacons belonging to the path calculate the TOA and AOA to the neighboring nodes in the path. Then, using the TOA and AOA measurements, a sensor is able to estimate the Euclidean distance to each of at least 3 beacons.

Figure 12.10 gives an illustration of the algorithm. In the figure, sensor S_1 established a path to each of the three beacons (B_1, B_2 and B_3). Then, it stores the relevant information

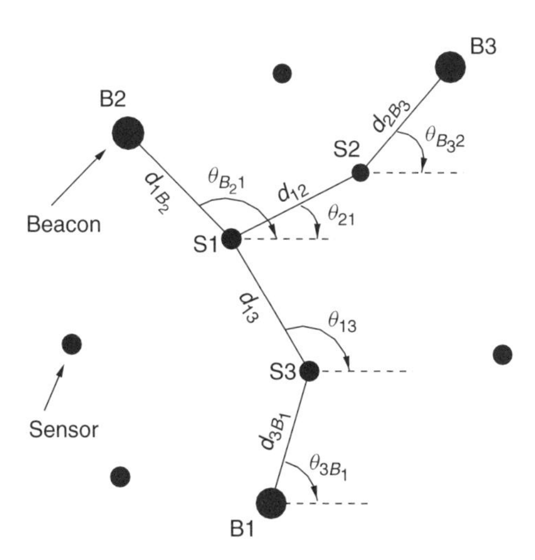

Figure 12.10 Illustration of the proposed location scheme

in its database in the form

$$S_1 \longrightarrow B_1: \qquad \Phi_1 = \{(S_1, S_3, B_1), (\hat{\theta}_{13}, \hat{\theta}_{3B_1}), (\hat{d}_{13}, \hat{d}_{3B_1})\}$$

$$S_1 \longrightarrow B_2: \qquad \Phi_2 = \{(S_1, B_2), (\hat{\theta}_{B_21}), (\hat{d}_{1B_2})\}$$

$$S_1 \longrightarrow B_3: \qquad \Phi_3 = \{(S_1, S_2, B_3), (\hat{\theta}_{21}, \hat{\theta}_{B_32}), (\hat{d}_{12}, \hat{d}_{2B_3})\}.$$

The distance from S_1 to the three beacons can be determined by

$$\hat{d}_{1B_1} = \sqrt{\left(\hat{d}_{13}\cos\hat{\theta}_{13} + \hat{d}_{3B_1}\cos\hat{\theta}_{3B_1}\right)^2 + \left(\hat{d}_{13}\sin\hat{\theta}_{13} + \hat{d}_{3B_1}\sin\hat{\theta}_{3B_1}\right)^2}$$

$$\hat{d}_{1B_2} = \hat{d}_{1B_2}$$

$$\hat{d}_{1B_3} = \sqrt{\left(\hat{d}_{12}\cos\hat{\theta}_{21} + \hat{d}_{2B_3}\cos\hat{\theta}_{B_32}\right)^2 + \left(\hat{d}_{12}\sin\hat{\theta}_{21} + \hat{d}_{2B_3}\sin\hat{\theta}_{B_32}\right)^2}.$$

After obtaining the distance to each of the beacons, the position coordinates of the node can be estimated using the algorithms described in Section 12.3. When considering the optimization-based approaches, the cost function is defined as

$$\epsilon(x, y) = \sum_{k=1}^{N_B}\left[\hat{d}_k - \sqrt{(x_k - x)^2 + (y_k - y)^2}\right]^2, \tag{37}$$

where $\hat{d}_k$ is the estimated distance between the desired sensor and the kth beacon, N_B is the number of beacons available in the network, and (x, y) and (x_k, y_k) are the unknown coordinates of the sensor of interest and the known coordinates of the kth beacon, respectively. Although we focused on 2-D sensor location, it is straightforward to extend the algorithm to 3-D positioning.

12.4.3 Simulation Results

The monitored area has dimensions of $100(w) \times 100(l)$ m. 100 nodes are randomly positioned in the area together with a number (5, 15, and 20) of beacon nodes depending on the simulation scenarios. The transmission radius of any node in this network is equal to 30 m and it is kept constant. The positioning algorithms considered are the DFP algorithm and the DM. The performance evaluation assumes that the TOA and AOA measurement errors are white Gaussian random variables of mean zero and variance σ_{TOA}^2 and σ_{AOA}^2, respectively. At each test point, 30 simulations are conducted with new random topology of the network and the performance is then averaged. In the case of the DFP algorithm, 50 iterations are used to update the estimated coordinate.

Figure 12.11 shows the root-mean-square error of the coordinates estimations using either the DM or DFP algorithm. In general, the root-mean-square errors of both algorithms are under 2 m and the DFP algorithm results in higher accuracy. We also notice that the curves are quite flat (except for the case of DM with $\sigma_{TOA} = 0$) as the AOA measurements error increases. This means that the algorithms, to localize the nodes, are not sensitive to the AOA measurement error.

Figure 12.12 shows the RMS error with respect to σ_{TOA} at a given σ_{AOA}. In this case, the slope of the curves becomes sharper. This shows that the algorithms are comparatively more sensitive to TOA measurement error.

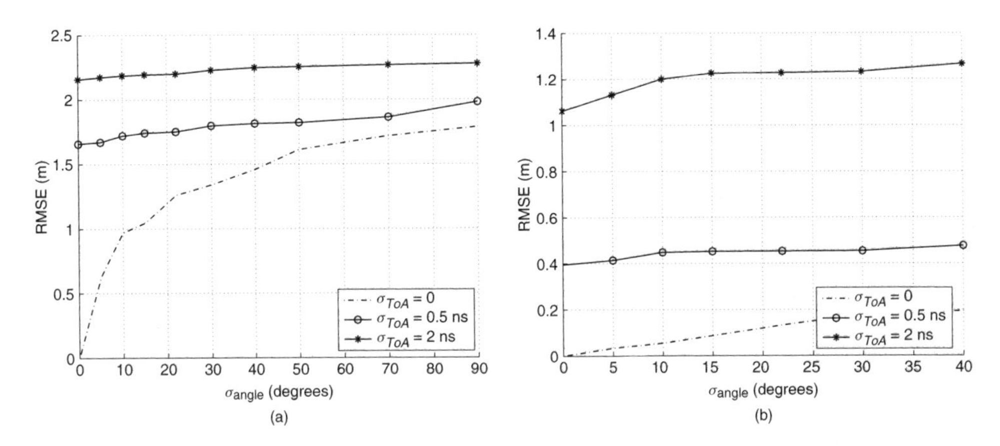

Figure 12.11 RMSE of direct method (a) and DFP algorithm (b) with respect to AOA errors. 5 beacons in the network and three different variances of the TOA error ($\sigma_{TOA} = \{0, 0.5, 2\}$ ns)

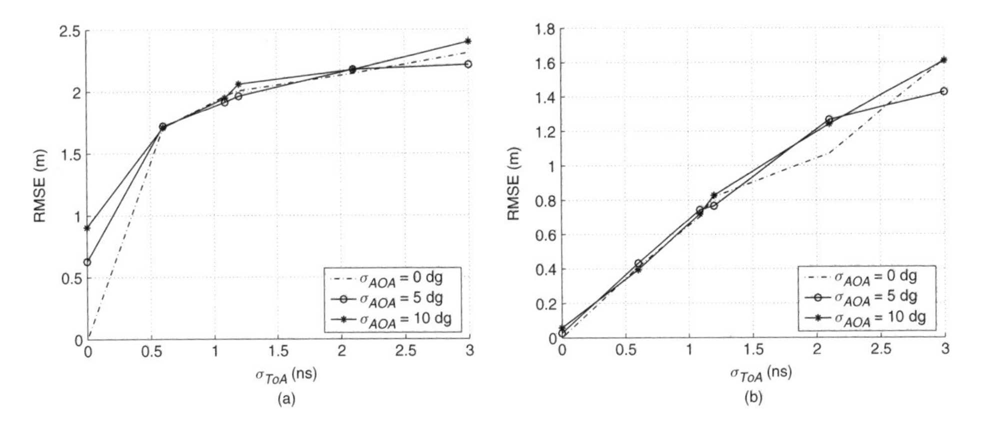

Figure 12.12 RMSE of direct method (a) and DFP algorithm (b) with respect to TOA errors. 5 beacons in the network and three different variances of the AOA error ($\sigma_{AOA} = \{0, 5, 10\}$ degrees)

12.5 Theoretical Positioning Accuracy

Positioning algorithms were discussed early in this chapter. However, the obtainable accuracy was not addressed. This is done in this section.

The accuracy of a positioning method is dependent on measurement and systematic errors. Systematic errors include such factors as errors in sensor locations and network synchronization. The location accuracy with respect to measurement errors has been widely studied [38, 39, 68]. However, the effects of systematic errors have not been widely considered. A tool to analyze the systematic errors is presented in [39], but analysis is not performed therein. This method is closely related to error propagation laws[2] [69]. In this

[2] See also physics.nist.cov/cuu/Uncertainty/, the web page of Physics laboratory of National Institute of Standards and Technology.

section, we briefly explain this tool and apply it to analysis of the location accuracy of a hyperbolic (time-delay–based) location system.

12.5.1 Analysis Tool

Let $\mathbf{x}$ be the estimator of the actual location $\mathbf{x}^0$ either in two or three dimensions. The bias is $\mathbf{b} = \mathrm{E}\{\mathbf{x} - \mathbf{x}^0\}$, the covariance $\mathbf{C} = \mathrm{E}\{(\mathbf{x} - \mathrm{E}\{\mathbf{x}\})(\mathbf{x} - \mathrm{E}\{\mathbf{x}\})^{\mathrm{T}}\}$, and the total mean-squared error $\mathbf{V} = \mathrm{E}\{(\mathbf{x} - \mathbf{x}^0)(\mathbf{x} - \mathbf{x}^0)^{\mathrm{T}}\} = \mathbf{b}\mathbf{b}^{\mathrm{T}} + \mathbf{C}$. Therefore, the analysis of the bias and covariance suffices to determine the accuracy of location systems.

Let the measurements without uncertainties be defined as $r_i^0 = f_i(\mathbf{x}^0, \mathbf{q}^0)$, or in a vector form

$$\mathbf{r}^0 = \mathbf{f}(\mathbf{x}^0, \mathbf{q}^0), \tag{38}$$

where $\mathbf{q}$ denotes the system parameters (such as sensor locations) and $\mathbf{q}^0$ denotes the actual system parameters. In other words, the parameters $\mathbf{q}$ are used when the position is calculated, but these may differ from the actual ones $\mathbf{q}^0$, for example, the position solution (shown early in this chapter) assumes that a node is at $\mathbf{x}_i$ although it, in reality, is at $\mathbf{x}_i^0$.

The model (38) is, in general, nonlinear. Therefore, it is linearized using the first two terms of the Taylor series of $\mathbf{r} = \mathbf{f}(\mathbf{x}, \mathbf{q})$ around $\mathbf{x}^0$ and $\mathbf{q}^0$. It follows that

$$\mathbf{r} \approx \mathbf{f}(\mathbf{x}^0, \mathbf{q}^0) + J_{\mathbf{x}}(\mathbf{x}^0, \mathbf{q}^0)(\mathbf{x} - \mathbf{x}^0) + J_{\mathbf{q}}(\mathbf{x}^0, \mathbf{q}^0)(\mathbf{q} - \mathbf{q}^0), \tag{39}$$

where $J_{\mathbf{x}}(\mathbf{x}, \mathbf{q})$ and $J_{\mathbf{q}}(\mathbf{x}, \mathbf{q})$ are the Jacobians of $\mathbf{f}(\mathbf{x}, \mathbf{q})$ with respect to $\mathbf{x}$ and $\mathbf{q}$. Solving this linear equation for $\mathbf{x} - \mathbf{x}^0$ gives

$$\mathbf{x} - \mathbf{x}^0 \approx \left(J_{\mathbf{x}}^{\mathrm{T}}(\mathbf{x}^0, \mathbf{q}^0) J_{\mathbf{x}}(\mathbf{x}^0, \mathbf{q}^0)\right)^{-1} J_{\mathbf{x}}^{\mathrm{T}}(\mathbf{x}^0, \mathbf{q}^0)\left(\mathbf{r} - \mathbf{f}(\mathbf{x}^0, \mathbf{q}^0) - J_{\mathbf{q}}(\mathbf{x}^0, \mathbf{q}^0)(\mathbf{q} - \mathbf{q}^0)\right). \tag{40}$$

In reality, the measurements $\mathbf{r}$ are disturbed by measurement errors $\mathbf{e}$ such that $\mathbf{r} = \mathbf{r}^0 + \mathbf{e}$. Substituting this into (40) yields

$$\mathbf{x} - \mathbf{x}^0 \approx \left(J_{\mathbf{x}}^{\mathrm{T}}(\mathbf{x}^0, \mathbf{q}^0) J_{\mathbf{x}}(\mathbf{x}^0, \mathbf{q}^0)\right)^{-1} J_{\mathbf{x}}^{\mathrm{T}}(\mathbf{x}^0, \mathbf{q}^0)\left(\mathbf{e} - J_{\mathbf{q}}(\mathbf{x}^0, \mathbf{q}^0)(\mathbf{q} - \mathbf{q}^0)\right). \tag{41}$$

This is the result that allows us to analyze the bias and covariance of the positioning error. The measurement errors $\mathbf{e}$ are typically modeled as zero-mean random variables with the covariance $\mathbf{C}_{\mathbf{e}}$. Alternatively, the measurement errors may contain bias $\mathbf{b}_{\mathbf{e}}$, which could be included into the systematic errors in the analysis that follows.

The term

$$\mathbf{b}_{\mathbf{r}} = J_{\mathbf{q}}(\mathbf{x}^0, \mathbf{q}^0)(\mathbf{q} - \mathbf{q}^0) \quad (+\mathbf{b}_{\mathbf{e}}) \tag{42}$$

presents the measurement bias. The location bias resulting from the measurement bias (42) is

$$\mathbf{b} = -\left(J_{\mathbf{x}}^{\mathrm{T}}(\mathbf{x}^0, \mathbf{q}^0) J_{\mathbf{x}}(\mathbf{x}^0, \mathbf{q}^0)\right)^{-1} J_{\mathbf{x}}^{\mathrm{T}}(\mathbf{x}^0, \mathbf{q}^0)\mathbf{b}_{\mathbf{r}}. \tag{43}$$

The positioning error covariance becomes

$$\mathbf{C} = \left(J_{\mathbf{x}}^{\mathrm{T}}(\mathbf{x}^0, \mathbf{q}^0) J_{\mathbf{x}}(\mathbf{x}^0, \mathbf{q}^0)\right)^{-1} J_{\mathbf{x}}^{\mathrm{T}}(\mathbf{x}^0, \mathbf{q}^0)\mathbf{C}_{\mathbf{e}} J_{\mathbf{x}}(\mathbf{x}^0, \mathbf{q}^0)\left(J_{\mathbf{x}}^{\mathrm{T}}(\mathbf{x}^0, \mathbf{q}^0) J_{\mathbf{x}}(\mathbf{x}^0, \mathbf{q}^0)\right)^{-\mathrm{T}}. \tag{44}$$

It can be shown[3] that $\mathbf{C}$ may be upper bounded by

$$\mathbf{C}_u = n\sigma_e^2\left(J_{\mathbf{x}}^{\mathrm{T}}(\mathbf{x}^0,\mathbf{q}^0)J_{\mathbf{x}}(\mathbf{x}^0,\mathbf{q}^0)\right)^{-1}, \tag{45}$$

where n is the number of measurements (the dimension of $\mathbf{r}$) and σ_e^2 is the maximum (diagonal) element of $\mathbf{C_e}$. The upper bound is in the sense that the matrix difference $\mathbf{C}_u - \mathbf{C}$ is positive semidefinite. A tighter upper bound and a lower bound can be obtained by substituting $n\sigma_e^2$ by the maximum and minimum eigenvalues of $\mathbf{C_e}$, respectively.

It can be concluded that the mean-squared error is upper bounded by

$$\mathbf{V}_u = n(b^2+\sigma_e^2)\left(J_{\mathbf{x}}^{\mathrm{T}}(\mathbf{x}^0,\mathbf{q}^0)J_{\mathbf{x}}(\mathbf{x}^0,\mathbf{q}^0)\right)^{-1}, \tag{46}$$

where b is a maximum diagonal element of $\mathbf{b_r}\mathbf{b_r}^{\mathrm{T}}$. Equation (46) can be used to find an approximate performance of a location system after the maximum single measurement error bias b and maximum estimation error variance σ_e^2 have been evaluated. The term $\mathbf{Q} = \left(J_{\mathbf{x}}^{\mathrm{T}}(\mathbf{x}^0,\mathbf{q}^0)J_{\mathbf{x}}(\mathbf{x}^0,\mathbf{q}^0)\right)^{-1}$ presents effects of the sensor geometry to the location accuracy and may be called as the geometric dilution of precision (GDOP) [39]. Reference [39] gives results related to it for the hyperbolic and direction finding location systems.

12.5.2 Hyperbolic Location Accuracy

The calculation of the position using time differences (TD) was explained early in this chapter. Herein, the TD positioning concept is specified again for accuracy analysis. The hyperbolic location systems use differences of TOAs measured either directly [70] or by subtracting measured TOAs [39]. Therefore, they may be called *time difference* (TD) location systems. The methods require that the measuring node network is synchronized or nodes have known relations between their clocks. Let the propagation time be denoted t_0. The signal travels the distance $d_i^0 = \|\mathbf{x}^0 - \mathbf{x}_i^0\|$, where $\|\ \|$ denotes the Euclidean vector norm, between the emitter and the node i at $\mathbf{x}_i$ at time τ_i. The receiver therefore observes the signal at time $t_i = t_0 + \tau_i + \delta_i$, where δ_i denotes the network synchronization timing error. Let node j be a reference node. Subtracting t_j from the other measurements and multiplying the results by the propagation speed of the signal c, the measurements of the hyperbolic location systems are modeled as

$$r_i \equiv r_{i,j} = c(t_i - t_j) = \|\mathbf{x}-\mathbf{x}_i\| - \|\mathbf{x}-\mathbf{x}_j\| + c(\delta_i - \delta_j) + e_i. \tag{47}$$

If there are N nodes, then there are $N-1$ measurements, the minimum possible number being $N = D+1$, where D is the dimension of the location problem, that is, two or three.

Often, in UWB, the RTT is used since precise network synchronization is a challenging task in sensor networks. In that case, a message is sent to a target, which retransmits it (or a corresponding response) back. The time difference between transmission and reception is $\Delta = 2\tau + T_{\mathrm{process}}$, where the latter is the processing time spend by the target. If this is not

[3] The proof is omitted herein.

precisely known, an effect similar to network synchronization error occurs. Therefore, the following analysis based on network synchronization error is valid for RTT measurements as well.

As explained early in this chapter, the calculation of the position requires that the node locations $\mathbf{x}_i$ are known. However, there might be errors on those. In addition, the network timing errors, shown in (47), were ignored early. The effects of these errors to the accuracy are now evaluated. The system parameters $\mathbf{q} = [\mathbf{x}_1^{\mathrm{T}} \ldots \mathbf{x}_N^{\mathrm{T}}\ \delta_1 \ldots \delta_N]^{\mathrm{T}}$ include the possibly erroneous node locations $\mathbf{x}_i$ and the clock errors δ_i. The actual node locations are denoted $\mathbf{x}_i^0$, that is, $\mathbf{x}_i = \mathbf{x}_i^0 + \delta_{\mathbf{x}_i}$, where $\delta_{\mathbf{x}_i}$ is the difference between the presumed and actual node locations. For brevity, let $J_{\mathbf{x}} = J_{\mathbf{x}}(\mathbf{x}^0, \mathbf{q}^0)$. Then, the Jacobian with respect to the unknown location is

$$J_{\mathbf{x}} = \begin{bmatrix} \dfrac{(\mathbf{x}^0 - \mathbf{x}_1^0)^{\mathrm{T}}}{d_1^0} - \dfrac{(\mathbf{x}^0 - \mathbf{x}_j^0)^{\mathrm{T}}}{d_j^0} \\ \vdots \\ \dfrac{(\mathbf{x}^0 - \mathbf{x}_{j-1}^0)^{\mathrm{T}}}{d_{j-1}^0} - \dfrac{(\mathbf{x}^0 - \mathbf{x}_j^0)^{\mathrm{T}}}{d_j^0} \\ \dfrac{(\mathbf{x}^0 - \mathbf{x}_{j+1}^0)^{\mathrm{T}}}{d_{j+1}^0} - \dfrac{(\mathbf{x}^0 - \mathbf{x}_j^0)^{\mathrm{T}}}{d_j^0} \\ \vdots \\ \dfrac{(\mathbf{x}^0 - \mathbf{x}_N^0)^{\mathrm{T}}}{d_N^0} - \dfrac{(\mathbf{x}^0 - \mathbf{x}_j^0)^{\mathrm{T}}}{d_j^0} \end{bmatrix}. \tag{48}$$

The ith element of the bias $\mathbf{b_r}$ is

$$[\mathbf{b_r}]_i = \frac{-(\mathbf{x}^0 - \mathbf{x}_i^0)^{\mathrm{T}}\delta_{\mathbf{x}_i}}{d_i^0} + \frac{(\mathbf{x}^0 - \mathbf{x}_j^0)^{\mathrm{T}}\delta_{\mathbf{x}_j}}{d_j^0} + c(\delta_i - \delta_j). \tag{49}$$

The terms may be contrastive, resulting in a small bias. The terms may, as well, be restorative resulting in a large bias. The clock error term, the last term in (49), attains its maximum if the clock errors are to opposite directions and minimum if the clock errors are in the same direction. If the clock error $\delta_i - \delta_j$ is 3 ns, 1 μs, or 1 ms and the signal is a radio signal, then the corresponding biases are 1 m, 300 m, and 300 km, respectively. This clearly shows why very accurate network synchronization is required in these systems.

Because of the Schwartz inequality, the first two terms in (49) are bounded as

$$\|(\mathbf{x}^0 - \mathbf{x}_i^0)^{\mathrm{T}}\delta_{\mathbf{x}_i}\|/d_i^0 \leq \frac{1}{d_i^0}\|(\mathbf{x}^0 - \mathbf{x}_i^0)\|\|\delta_{\mathbf{x}_i}\| = \|\delta_{\mathbf{x}_i}\|,$$

with equality if and only if $\delta_{\mathbf{x}_i} = \alpha(\mathbf{x}^0 - \mathbf{x}_i^0)$, where α is a constant. Consequently, the errors through the node locations are at maximum if $\delta_{\mathbf{x}_i}$ are parallel to $\mathbf{x}^0 - \mathbf{x}_i^0$. These errors vanish if $\delta_{\mathbf{x}_i}$ and $\mathbf{x}^0 - \mathbf{x}_i^0$ are orthogonal. The bias due to node location errors are not necessarily very large since the sensors (or base stations) can typically be set within few meters. However, the sensor location accuracy cannot be worse than the accuracy required from the system. Otherwise, the bias in the sensor location accuracy may dominate the error budget.

The accuracy may be approximated, as already explained, by calculating the GDOP term for different positioning geometry using the Jacobian $\mathbf{J}_{\mathbf{x}}$, the bias b, and measurement error variance.

12.6 Conclusions

This chapter comprehensively studied localization in cellular/sensor networks by making use of UWB technology. It covered several topics including TOA estimation, positioning approaches, and positioning accuracy. Also included are the authors' recent contributions.

To exploit the fine time resolution of UWB signals, TOA/TDOA measurements should be employed to locate UWB devices. For low cost and low-complexity applications, the noncoherent TOA estimation scheme that was described may be employed. On the other hand, the proposed two-stage approach could be considered to achieve rapid acquisition and high accuracy, especially for low data rate scenarios. A trade-off between complexity and accuracy needs to be worked out in choosing a positioning algorithm for the practice that is of interest. In evaluating positioning performance, systematic errors should be taken into account to provide a more realistic performance reference.

Acknowledgment

This work is partly funded by the EU 6th framework project PULSERS.

References

[1] R. J. Fontana, "Recent System Applications of Short-Pulse Ultra-Wideband (UWB) Technology," *IEEE Trans. Microw. Theory Technol.*, vol. 52, no. 9, pp. 2087–2104, 2004.

[2] R. J. Fontana, E. Richley, and J. Barney, "Commercialization of an Ultra Wideband Precision Asset Location System," *Proc. IEEE Conference UWB systems and Technologies*, Reston, VA, 2003.

[3] I. Oppermann, L. Stoica, A. Rabbachin, Z. Shelby, and J. Haapola, "UWB Wireless Sensor Networks: UWEN–a Practical Example," *IEEE Commun. Magn.*, vol. 42, pp. 527–532, Dec. 2004.

[4] K. Yu, J. P. Montillet, A. Rabbachin, P. Cheong, and I. Oppermann, "UWB Location and Tracking for Wireless Embedded Networks," *EURASIP J. Signal Process. Special Issue Signal Process. UWB Commun.*, to appear.

[5] M. Mauve, J. Widmer, and H. Hartenstein, "A Survey on Position Based Routing in Mobile Ad-Hoc Networks," *IEEE netw. Magn.*, vol. 15, pp. 30–39, Nov.–Dec. 2001.

[6] X. Li, *Super Resolution TOA Estimation with Diversity Techniques for Indoor Geolocation Applications*. PhD thesis, Worcester Polytechnic Institute, 2003.

[7] E. A. Homier and R. A. Scholtz, "Rapid Acquisition of Ultra-Wideband Signals in the Dense Multipath Channel," *Proc. IEEE 2nd Ultra Wideband Systems and Technologies (UWBST'02)*, Baltimore, MD, pp. 245–249, May 2002.

[8] R. Fleming, C. Kushner, G. Roberts, and U. Nandiwada, "Rapid Acquisition for Ultra-Wideband Localizers," *Proc. IEEE 2nd Ultra Wideband Systems and Technologies (UWBST'02)*, Baltimore, MD, pp. 105–109, May 2002.

[9] I. Maravic, M. Vetterli, and K. Ramchandran, "Channel Estimation and Synchronization with Sub-Nyquist Sampling and Application to Ultra-Wideband Systems," *Proc. IEEE Symposium on Circuits and Systems (ISCAS)*, Vancouver, Canada, pp. 381–384, May 2004.

[10] J.-Y. Lee and R. A. Scholtz, "Ranging in a Dense Multipath Environment Using an Uwb Radio Link," *IEEE J. Select. Areas Commun.*, vol. 20, pp. 1677–1683, Dec. 2002.

[11] M. Weisenhorn and W. Hirt, "Robust Noncoherent Receiver Exploiting UWB Channel Properties," *Proc. Joint UWBST&IWUWBS*, Kyoto, Japan, vol. 2, pp. 156–160, May 2004.

[12] Y. Nakache, P. Orlik, W. Gifford, A. Molisch, I. Ramachandran, G,. Fang, and J. Zhang, "Low-Complexity Ultrawideband Transceiver with Compatibility to Multiband-OFDM," *Proc. Joint UWBST & IWUWBS*, Kyoto, Japan, pp. 151–155, May 2004.

[13] A. Rabbachin, R. Tesi, and I. Oppermann, "Bit Error Rate Analysis for UWB Systems with a Low Complexity, Non-Coherent Energy Collection Receiver," *Proc. IST Mobile & Wireless Communications Summit*, Lyon, France, vol. 1, pp. 223–227, Jun. 2004.

[14] T. Q. S. Quek and M. Z. Win, "Ultrawide Bandwidth Transmitted-Reference Signaling," *Proc. IEEE International Conference on Communications (ICC)*, Paris, France, vol. 6, pp. 3409–3413, Jun. 2004.

[15] R. Hoctor and H. Tomlinson, "Delay Hopped Transmitted Reference Experimental Results," *Proc. IEEE 2nd Ultra Wideband Systems and Technologies (UWBST'02)*, Baltimore, MD, pp. 93–98, May 2002.

[16] R. Hoctor and H. Tomlinson, "Delay-Hopped Transmitted-Reference RF Communications," *Proc. IEEE 2nd Ultra Wideband Systems and Technologies (UWBST'02)*, Baltimore, MD, pp. 265–269, May 2002.

[17] J. D. Choi and W. E. Stark, "Performance of Ultra-Wideband Communications with Suboptimal Receivers in Multipath Channels," *IEEE Journal on Selected Areas in Communications*, vol. 20, pp. 1754–1766, Dec. 2002.

[18] R. Moddemeijer, "On the Determination of the Position of Extrema of Sampled Correlators," *IEEE Trans. Signal Process.*, vol. 39, pp. 216–219, Jan. 1991.

[19] G. Jacovitti and G. Scarano, "Discrete time Techniques for time Delay Estimation," *IEEE Trans. Signal Process.*, vol. 41, pp. 525–533, Feb. 1993.

[20] G. Giunta, "Fast Estimation of Time Delay and Doppler Stretch Based on Discrete-Time Methods," *IEEE Trans. Signal Process.*, vol. 46, pp. 1785–1797, Jul. 1998.

[21] H. Saarnisaari, "Some Design Aspects of Mobile Local Positioning Systems," *Proc. IEEE Position Location and Navigation Symposium*, Monterey, CA, pp. 1688–1692, 2004.

[22] S. M. Kay, *Fundamentals of Statistical Signal Processing: Detection Theory*. Prentice Hall, Upper Saddle River, NJ, 1998.

[23] J. G. Proakis, *Digital Communications*. 3rd edition, McGraw-Hill, 1995.

[24] A. Polydoros and C. L. Weber, "A Unified Approach to Serial Search Spread-Spectrum Code Acquisition-Part II: A Matched-Filter Receiver," *IEEE Trans. Commun.*, vol. 32, pp. 550–560, May 1984.

[25] E. Sourour and S. C. Gupta, "Direct-Sequence Spread-Spectrum Parallel Acquisition in Nonselective and Frequency-Selective Rician Fading Channels," *IEEE J. Select. Areas Commun.*, vol. 10, pp. 535–544, Apr. 1992.

[26] A. F. Molisch, K. Balakrishnan, D. Cassioli, C.-C. Chong, S. Emami, A. Fort, J. Karedal, J. Kunisch, H. Schantz, U. Schuster, and K. Siwiak, "IEEE 802.15.4a channel model – final Report." http://www.ieee802.org/15/pub/TG4a.html, 2004.

[27] B. T. Fang, "Simple Solutions for Hyperbolic and Related Position Fixes," *IEEE Trans. Aerosp. Electron. Syst.*, vol. 26, pp. 748–753, Sept. 1990.

[28] K. Yu and I. Oppermann, "Performance of UWB Position Estimation Based on TOA Measurements," *Proc. Joint UWBST & IWUWBS*, Kyoto, Japan, pp. 400–404, 2004.

[29] J. O. Smith and J. S. Abel, "Closed-form Least Squares Source Location Estimation from Range Difference Measurements," *IEEE Trans. Acoust. Speech Signal Process.*, vol. 35, pp. 1661–1669, Dec. 1987.

[30] B. Friedlander, "A Passive Localization Algorithm and its Accuracy Analysis," *IEEE J. Oceanic Eng.*, vol. 12, pp. 234–245, Jan. 1987.

[31] H. C. Schau and A. Z. Robinson, "Passive Source Location Employing Intersecting Spherical Surfaces from Time-of-Arrival Differences," *IEEE Trans. Acoust. Speech Signal Process.*, vol. 35, pp. 1223–1225, Aug. 1987.

[32] Y. T. Chan and K. C. Ho, "A Simple and Efficient Estimator for Hyperbolic Location," *IEEE Trans. Signal Process.*, vol. 42, pp. 1905–1915, Aug. 1994.

[33] Y. Huang, J. Benesty, G. W. Elko, and R. M. Mersereau, "Real-Time Passive Source Localization: A Practical Linear-Correction Least-Squares Approach," *IEEE Trans. Speech Audio Process.*, vol. 9, pp. 943–956, Nov. 2001.

[34] D. Niculescu and B. Nath, "Ad Hoc Positioning System (APS)," *Proc. IEEE GLOBECOM*, San Antonio, TX, pp. 2926–2931, 2001.

[35] A. Savvides, C.-C. Han, and M. B. Strivastava, "Dynamic Fine-Grained Localization in Ad-Hoc Networks of Sensors," *Proc. ACM SIGMOBILE*, Rome, Italy, pp. 166–179, 2001.

[36] K. Langendoen and N. Reijers, "Distributed Localization in Wireless Sensor Networks: A Quantitative Comparison," *Comput. Netw.*, vol. 43, pp. 499–518, 2003.

[37] S. M. Kay, *Fundamentals of Statistical Signal Processing: Estimation Theory*. Prentice Hall, Upper Saddle River, NJ, 1993.

[38] W. H. Foy, "Position-Location Solutions by Taylor-Series Estimation," *IEEE Trans. Aerosp. Electron. Syst.*, vol. 12, pp. 187–194, Mar. 1976.

[39] D. J. Torieri, "Statistical Theory of Passive Location Systems," *IEEE Trans. Aerosp. Electron. Syst.*, vol. 20, pp. 183–198, Mar. 1984.

[40] D. E. Manolakis, "Efficient Solution and Performance Analysis of 3-D Position Estimation by Trilateration," *IEEE Trans. Aerosp. Electron. Syst.*, vol. 32, pp. 1239–1248, Oct. 1996.

[41] M. A. Spirito, "On the Accuracy of Cellular Mobile Station Location Estimation," *IEEE Trans. Veh. Technol.*, vol. 50, pp. 674–685, May 2001.

[42] K. W. Cheung, H. C. So, W. K. Ma, and Y. T. Chan, "Least Squares Algorithms for Time-of-Arrival-Based Mobile Location," *IEEE Trans. Signal Process.*, vol. 52, pp. 1121–1128, Apr. 2004.

[43] P. E. Gill, W. Murray, and M. H. Wright, *Practical Optimization*. Academic Press, London, 1981.

[44] R. Fletcher, *Practical Methods of Optimization*. John Wiley & Sons, Chichester, England, 1987.

[45] S. Boyd and L. Vandenberghe, *Convex Optimization*. Cambridge University Press, 2004.

[46] J. J. Caffery and G. L. Stuber, "Overview of Radiolocation in CDMA Cellular Systems," *IEEE Commun. Magn.*, vol. 36, no. 5, pp. 38–45, May 1998.

[47] D. Marquardt, "Algorithm for Least-Squares Estimation of Nonlinear Parameters," *SIAM J. Appl. Math.*, vol. 11, pp. 431–441, 1963.

[48] K. Yu and I. Oppermann, "UWB Positioning for Wireless Embedded Networks," *Proc. IEEE RAWCON*, Atlanta, Georgia, 2004.

[49] L. Doherty, K. S. J. Pister, and L. E. Ghaoui, "Convex Position Estimation in Wireless Sensor Networks," *IEEE INFOCOM*, Anchorage, AK, pp. 1655–1663, 2001.

[50] J. Nelder and R. Mead, "A Simplex Method for Function Minimization," *Comput. J.*, vol. 7, pp. 308–313, 1965.

[51] H. Wu, C. Wang, and N.-F. Tzeng, "Novel Self-Configurable Positioning Technique for Multi-Hop Wireless Network," *IEEE Trans. Netw.*, submitted.

[52] M. Hellebrandt and R. Mathar, "Location Tracking of Mobiles in Cellular Radio Networks," *IEEE Trans. Veh. Technol.*, vol. 48, pp. 1558–1562, Sept. 1999.

[53] M. McGuire and K. Plataniotis, "Dynamic Model-Based Filtering for Mobile Terminal Location Estimation," *IEEE Trans. Veh. Technol.*, vol. 52, pp. 1012–1031, Jul. 2003.

[54] M. Hellebrandt, R. Mathar, and M. Scheibenbogen, "Estimating Position and Velocity in a Cellular Radio Network," *IEEE Trans. Veh. Technol.*, vol. 46, pp. 65–71, Feb. 1997.

[55] P. Biswas and Y. Ye, "Semidefinite Programming for Ad Hoc Wireless Sensor Network Localization," *Proc. IEEE Information Processing in Sensor Networks (IPSN)*, Berkeley, CA, pp. 46–54, Apr. 2004.

[56] X. Li, H. Shi, and Y. Shang, "A Map-Growing Localization Algorithm for Ad-Hoc Wireless Sensor Networks," *Proc. IEEE Parallel and Distributed Systems,(ICPADS)*, Newport Beach, CA, pp. 395–402, Jul. 2004.

[57] X.-G. Ji and H. Zha, "Sensor Positioning in Wireless Ad-Hoc Sensor Networks Using Multidimensional Scaling," *Proc. IEEE INFOCOM 2004, Twenty-Third Annualjoint Conference of the IEEE Computer and Communications Societies*, Hong Kong, China, vol. 4, pp. 2652–2661, Mar. 2004.

[58] Y. Shang, J. Meng, and H. Shi, "A New Algorithm for Relative Localization in Wireless Sensor Networks," *Proc. IEEE Parallel and Distributed Processing Symposium*, Nice, France, pp. 26–30, Apr. 2004.

[59] T. He, C. Huang, B. M. Blum, J. Stankovic, and T. Abdelzaher, "Range-Free Localization Schemes for Large Scale Sensor Networks," *Proc. IEEE Mobicom 2003*, San Diego, CA, pp. 81–95, Sept. 2003.

[60] D. Niculescu and B. Nath, "Ad Hoc Positioning System (Aps) Using Aoa," *Proc. IEEE Twenty-Second Annual Joint Conference of the IEEE Computer and Communications Societies (INFOCOM)*, San Francisco, CA, vol. 3, pp. 1734–1743, Apr. 2003.

[61] R. Nagpal, H. Shrobe, and J. Bachrach, "Organizing a Global Coordinate System from Local Information on an Ad-Hoc Sensor Network," *Proc. International Workshop on Information Processing in Sensor Networks*, Palo Alto, CA, Apr. 2003.

[62] S. Capkun, M. Hamidi, and J. P. Hubaux, "Gps-Free Positioning in Mobile Ad-Hoc Networks," *Proc. IEEE Proceedings of the Hawaii International Conference on System Sciences*, Maui, Hawaii, pp. 3481–3490, Jan. 2001.

[63] D. Moore, J. Leonard, D. Rus, and S. Teller, "Robust Distributed Network Localization with Noisy Range Measurements," *Proc. ACM SenSys'04*, Baltimore, MD, pp. 50–61, Nov. 2004.

[64] R. Iyengar and B. Sikdar, "Scalable and Distributed Gps Free Positioning for Sensor Networks," *Proc. IEEE ICC'03*, Anchorage, AK, vol. 1, pp. 338–342, May 11-15, 2003.

[65] Y. Zhang and L. Cheng, "Place: Protocol for Location and Coordinate Estimation–a Wireless Sensor Network Approach," *Comput. Netw.*, vol. 46, pp. 679 – 693, Jul. 2004.

[66] C. Poggi and G. Mazzini, "Collinearity for Sensor Network Localization," *Proc. IEEE Vehicular Technology Conference, VTC 2003-Fall*, Orlando, FL, vol. 5, pp. 3040–3044, Oct. 2003.

[67] J.-C. Chen, C.-S. Maa, Y.-C. Wang, and J.-T. Chen, "Mobile Position Location Using Factor Graphs," *IEEE Electron. Lett.*, vol. 7, no. 9, pp. 431–433, 2003.

[68] D. Kaplan, *Understanding GPS Principles and Applications*. Artech House, Boston, MA, 1996.

[69] J. R. Taylor, *An Introduction to Error Analysis: the Study of Uncertainties in Physical Measurements*, 2nd edition, University Science Books, 1997.

[70] G. C. Carter, "Time Delay Estimation for Passive Sonar Signal Processing," *IEEE Trans. Acoust. Speech Signal Process.*, vol. 29, pp. 463–470, June 1981.

Index

Ultra-wideband Wireless Communications and Networks
Edited by Xuemin (Sherman) Shen, Mohsen Guizani, Robert Caiming Qiu and Tho Le-Ngoc